Mechanical Engineering for Students and Engineers

Mechanical Engineering for Students and Engineers

Edited by **Rene Sava**

NY RESEARCH PRESS
New York

Published by NY Research Press,
23 West, 55th Street, Suite 816,
New York, NY 10019, USA
www.nyresearchpress.com

Mechanical Engineering for Students and Engineers
Edited by Rene Sava

International Standard Book Number: 978-1-63238-485-0 (Hardback)

Printed in China.

Contents

Preface

Mechanical Engineering is perhaps the oldest existing branch of engineering. It encompasses designing, manufacturing and maintenance of machines. Technological advances have widened the scope of this discipline. This book brings forth the existing as well as modern approaches of this field. It will also provide interesting topics for research. In this book, using case studies and examples, constant effort has been made to make the understanding of the difficult concepts of mechanical engineering as informative as possible for readers. It aims to serve as a resource guide for students and experts alike and contribute to the growth of this discipline.

This book is a result of research of several months to collate the most relevant data in the field.

When I was approached with the idea of this book and the proposal to edit it, I was overwhelmed. It gave me an opportunity to reach out to all those who share a common interest with me in this field. I had 3 main parameters for editing this text:

1. Accuracy - The data and information provided in this book should be up-to-date and valuable to the readers.
2. Structure - The data must be presented in a structured format for easy understanding and better grasping of the readers.
3. Universal Approach - This book not only targets students but also experts and innovators in the field, thus my aim was to present topics which are of use to all.

Thus, it took me a couple of months to finish the editing of this book.

I would like to make a special mention of my publisher who considered me worthy of this opportunity and also supported me throughout the editing process. I would also like to thank the editing team at the back-end who extended their help whenever required.

Editor

The Effect of Adiabatic Obstacles on Natural Convection inside a Horizontal Elliptical Porous Annulus

Faysal Ayad Khaleel[1] & Amir Sultan Dawood[1]

[1] Mechanical Engineering Department, Collage of Engineering, Mosul University, Mosul, Iraq

Correspondence: Faysal AyadKhaleel, Mechanical Engineering Department, Mosul University, Mosul, Iraq.
E-mail: f_eng23@yahoo.com

Abstract

This study numerically examines the influence of installing two identical radial adiabatic obstacles on natural convection within a porous layer of constant thickness. This porous layer surrounds an isothermal hot elliptical cylinder and is surrounded by an isothermal cold one.The obstacles' length, angular position and radial position were changed in various amounts. A non-orthogonal curvilinear coordinate system is adopted, in order to treat the presence of the obstacles inside a two-dimensional elliptical annulus. Consequently, Finite Differences Techniques and ADI with Relaxation Method were used to approximate the computational regime of the non-Darcian-Boussinesq equation, governing the flow inside the computational porous domain. An efficient code of high accuracy was constructed to solve the approximated governing equations. The results showed that the installation of long obstacles in the meeting place of two convection cells causes an increase in heat transfer rates by up to 16%. Whereas the installation of short obstacles in a place lying close to the center of the major convection cell causes a reduction in heat transfer by up to 19%.

Keywords: porous media, elliptical annulus, non-darcian model, natural convection, elliptic grid generation, obstructions

Nomenclature

a	Half of the major axis, m
b	Half of the minor axis, m
Da	Darcy number
g	Gravitational acceleration, $\mathrm{m\ s^{-2}}$
J	The jacobian $(= \hat{x}_\xi \hat{y}_\eta - \hat{x}_\eta \hat{y}_\xi)$
K	Permeability of porous medium, $\mathrm{m^2}$
L	Length of the studied model (characteristic length), m
Nu	Local Nusselt number
P	Pressure, Pa
Pr	Prandtl number
Ra	Rayleigh number
Ra*	Modified Rayleigh number
T	Temperature, K
u	Horizontal velocity component, $\mathrm{m\ s^{-1}}$
v	Vertical velocity component, $\mathrm{m\ s^{-1}}$
$\mathbf{v}$	Velocity vector, $\mathrm{m\ s^{-1}}$

Accents

$\hat{T}$	Non-dimensional
$\overline{Nu}$	Average

Greek symbols

α	Thermal diffusivity, $\mathrm{m^2 s^{-1}}$
β	Thermal expansion coefficient, $\mathrm{K^{-1}}$
γ	$= \hat{x}_\xi{}^2 + \hat{y}_\xi{}^2$
δ	$= \hat{x}_\xi \hat{x}_\eta + \hat{y}_\xi \hat{y}_\eta$
ε	$= \hat{x}_\eta{}^2 + \hat{y}_\eta{}^2$
μ	Dynamic viscosity, Pa s
ν	Kinematic viscosity, $\mathrm{m^2 s^{-1}}$
ξ, η	Computational region coordinates
ρ	Density, $\mathrm{kg\ m^3}$
ϕ	Porosity of porous medium
ψ	Stream function, $\mathrm{m^2 s^{-1}}$
ω	Vorticity, $\mathrm{s^{-1}}$

Subscripts

c	Cold surface
f	fluid
h	Hot surface
i	Inner cylinder surface
o	outer cylinder surface

1. Introduction

Natural convection heat and mass transfer in porous media has been studied extensively. This is due to the increasing need to understand the complicated transport process for application of diverse insulation fields which include insulation of hot or cold pipes, aircraft cabin, nuclear and power plants and solar collector receivers. Moreover, the porous media are used or naturally exist in fields rather than those of insulation. For example, geothermal engineering, solid matrix heat exchanger (compact heat exchanger), oil extraction, underground disposal of nuclear waste materials, and many more.

A special case study of natural convection inside porous media is that medium bounded by two concentric (or eccentric) cylinders. The previous published studies of such case have demonstrated, both experimentally and theoretically, some details related to the nature of fluid flow and heat transfer inside porous media. Bishop and Charley (1966) investigated the phenomenon of natural convection in a porous layer put between two concentric horizontal cylinders. The results photographically showed the fluid motion inside the enclosure. Another study on the same geometrical shape and boundary conditions had been both experimentally and theoretically done by Caltagirone (1976). He investigated the axial (3rd) direction heat transfer and concluded that there is a critical value for the modified Rayleigh number, across which the 3rd direction flow is undergoing taking place.Bau et al. (1983) carried out the experiment of Caltagirone again with two eccentric cylinders. They observed the growth of the multicellular in the upper part of the model. This multicellular had gradually transformed to a unicellular as time passes. Charrier and Mojtabi (1991) had also studied the free convection (2D) flow in a horizontal saturated porous annulus. They joined their numerical study with an experimental work. Their work led to conclude that the transition from unicellular to multicellular flow depends strongly on the initial conditions. Moreover, they concluded that the heat transferred in the axial direction is an unsteady state phenomenon. Sakr et. al., (2008) experimentally and numerically studied the natural convection inside an annulus bounded by inner elliptical tube and outer circular one. Their results showed that the horizontal major axis position gives more efficient insulation than the vertical position. Mota and Saatdjian (1995, 1999) numerically studied the hysteresis loop inside a circular and elliptical annuli respectively. They concluded that the behavior of the flow and (Ra-Nu) curve depend on the initial thermal condition. Another study for Mota et al. (2000) concluded that the judicious horizontal stretching of one of the annulus's boundaries reduces heat losses with respect to concentric cylindrical annulus with the same amount of insulating material. Zhu et al. (2003) also numerically investigated the effect of the major axis position of the elliptic boundaries. They concluded that more vortices in streamlines is generated in the case of horizontal position. Moreover, they did mention that further investigations may be needed for the phenomenon to be clear and understood.

All the aforementioned theoretical studies used the Darcian model to describe the momentum equation of the flow. However, Kaviany (1986) assumed a circular annulus and studied the effect of the inertia, boundary and velocity-square terms. His results showed that all of these effects reduces heat losses with the boundary term being the most significant. Moreover, an acceptable agreement between the Darcian and non-Darcian regimes holds for low Rayleigh and Prandtl numbers. Braga and de Lemos (2006) had a circular porous annulus and numerically simulated the turbulent natural convection by using a macroscopic non-Darcian two-equation model. They studied the dependency of the laminarization phenomenon on Darcy number and porosity.

The effect of inserting an obstacle inside a rectangular, heated-cooled and porous cavity was studied by Bejan (1983). The results indicated an increase in heat losses in some cases of the horizontal obstacle's length. On the other hand, a vertical obstacle of $l/H = 1$ decreases the heat lost. Facas (1994) numerically tried to reduce the heat transfer from a hot pipe buried in a semi-infinite, saturated, porous medium, by installing two obstacles on the pipe's wall. The results indicated a maximal energy saving in the case of horizontal radial position.

In the present study the effect of inserting two radial adiabatic obstacles on natural convection within a porous layer of a constant thickness was numerically examined. These two obstacles were symmetrically set about the vertical axis of the study's model with various amounts of length and angular position. The porous layer surrounds an isothermal hot elliptical cylinder and surrounded by an isothermal cold one. This study also assumed the 2-D non-Darcian model with Boussinesq approximation to represent the momentum equation governing the flow within the porous medium.

2. Mathematical Formulation

The physical problem of the present study is considered by a two-dimensional elliptical cross section, as illustrated in Figure 1. This 2D annulus is completely defined by the following equations and parameters. Firstly, the following equation generates the inner boundary of the annulus:

$$\left(\frac{x_i}{a_i}\right)^2 + \left(\frac{y_i}{b_i}\right)^2 = 1 \tag{1}$$

The hydraulic radius ratio (R) is given as follows:

$$R = \sqrt{a_o b_o}/\sqrt{a_i b_i} \tag{2}$$

The thickness (t) of the porous layer is kept constant in each angular position and considered as a function of (R). So, it is given by:

$$t = \sqrt{\left(\frac{a_i + b_i}{2}\right)^2 + a_i b_i (R^2 - 1)} - \left(\frac{a_i + b_i}{2}\right) \tag{3}$$

For the thickness (t) being constant, the annulus's outer boundary is given by:

$$x_o = x_i + t.\sin(180 - \varphi) \tag{4}$$

$$y_o = y_i + t.\cos(180 - \varphi) \tag{5}$$

Where (φ) is the local slope angle of the inner boundary curve.

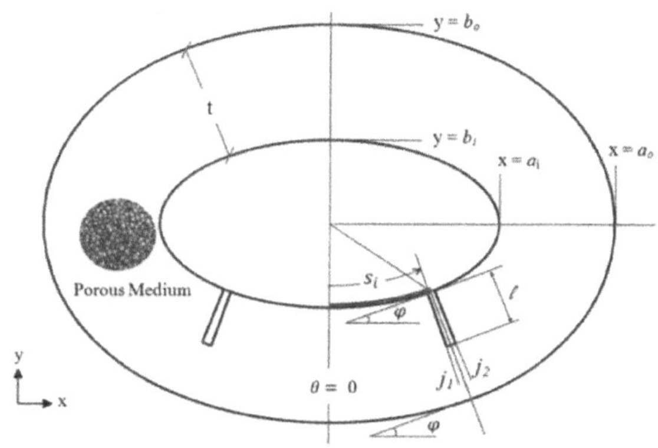

Figure 1. Geometrical configuration of the present study

The obstacles are set normal to the tangent of the boundary curvature as illustrated in Figure 1. Their length (l) are given with respect to the porous layer thickness (t). The angular position (θ) is described by the ratio of the circumferential length (s_i) to the whole length of the circumference (c_i), as the following:

$$\theta = \frac{s_i}{C_i} \times 360^o \tag{6}$$

Where:

$$s_i = \int_0^{x_i} \sqrt{\left(\frac{dy_i}{dx_i}\right)^2 + 1}\; dx_i \tag{7}$$

$$c_i = 4\int_0^{a_i} \sqrt{\left(\frac{dy_i}{dx_i}\right)^2 + 1}\; dx_i \tag{8}$$

The present study assumed steady state, laminar and incompressible flow, with homogenous saturated porous medium of non-conductive solid matrix. The annulus is of isothermal boundaries, and the inner is hot (T_h), whereas the outer is cold (T_c). The solid matrix is in thermal equilibrium with the fluid flowing inside it, and no viscous dissipation is assumed. The thermo-physical properties are kept constant except for density variation in the buoyancy force term (Boussinesq approximation).

According to the aforementioned assumptions and considerations, the equations govern the steady-state two-dimensional problem in Cartesian coordinates will be as follows (Bejan, 1995; Vafai, 2005):

Continuity

$$\frac{\partial u}{\partial x}+\frac{\partial v}{\partial y}=0 \tag{9}$$

Momentum

$$\frac{\partial P}{\partial x}=-\frac{\mu}{K}u-\frac{\rho_f F}{\sqrt{K}}u|\mathbf{v}|+\mu\nabla^2 u-\frac{\rho_f}{\emptyset^2}(\mathbf{v}.\nabla)u \tag{10}$$

and

$$\frac{\partial P}{\partial y}=-\frac{\mu}{K}v-\frac{\rho_f F}{\sqrt{K}}v|\mathbf{v}|+\mu\nabla^2 v-\frac{\rho_f}{\emptyset^2}(\mathbf{v}.\nabla)v+\rho_f g \tag{11}$$

Energy

$$\nabla^2 T=\frac{1}{\alpha_f}(\mathbf{v}.\nabla)T \tag{12}$$

Where F is the non-dimensional form of the drag constant defined by the Formula (13) used by Hadim and North, (2005).

$$F=\frac{1.75}{\sqrt{150}}\times\emptyset^{-3/2} \tag{13}$$

The non-dimensional forms of the physical quantities and variables, included in the above equations or needed in transforming them to their dimensionless forms, are defined as the following:

$$\hat{x}=\frac{x}{L},\hat{y}=\frac{y}{L},\hat{a}=\frac{a}{L},\hat{b}=\frac{b}{L},\hat{P}=\frac{PL^2}{\rho_f{\alpha_f}^2},\hat{\omega}=\frac{\omega L^2}{\alpha_f},\hat{u}=\frac{uL}{\alpha_f},\hat{v}=\frac{vL}{\alpha_f},\hat{\psi}=\frac{\psi}{\alpha_f}, \tag{14}$$
$$Ra=\frac{g\beta(T_h-T_c)L^3}{\nu\,\alpha_f},Da=\frac{K}{L^2},Ra^*=Ra.Da,Pr=\frac{\nu}{\alpha_f},\hat{T}=\frac{T-T_c}{T_h-T_c}$$

The pressure terms in Equations (10) and (11) can be eliminated by differentiating these equations with respect to (y) and (x) respectively, and then subtracting one from other. Hereafter, by using forms (14), the dimensionless governing equations in terms of vorticity and stream function, will respectively be as follows (Kaviany, 1986):

$$\hat{\omega}=-\hat{\nabla}^2\hat{\psi} \tag{15}$$

$$\hat{\omega}-Da.\hat{\nabla}^2\hat{\omega}=Ra^*\frac{\partial\hat{T}}{\partial\hat{x}}+\frac{F.\sqrt{Da}}{Pr}\left\{\hat{u}\frac{\partial|\hat{v}|}{\partial\hat{y}}-\hat{v}\frac{\partial|\hat{v}|}{\partial\hat{x}}\right\}-\frac{F.\sqrt{Da}.|\hat{v}|}{Pr}\hat{\omega}-\frac{Da}{Pr.\emptyset^2}(\hat{\mathbf{v}}.\hat{\nabla})\hat{\omega} \tag{16}$$

$$\hat{\nabla}^2\hat{T}=(\hat{v}.\hat{\nabla})\hat{T} \tag{17}$$

Where $(\hat{u}=\partial\hat{\psi}/\partial\hat{y})$ and $(\hat{v}=-\partial\hat{\psi}/\partial\hat{x})$.

The appropriate boundary conditions, in their non-dimensional forms, are listed below:

1) *Inner cylinder :*

$$\hat{T}=1,\quad \hat{\psi}=0,\quad \hat{u}=0,\quad \hat{v}=0,\quad \hat{\omega}=-\frac{\partial^2\hat{\psi}}{\partial\hat{n}^2} \tag{18}$$

2) *Outer cylinder :*

$$\hat{T}=0\ ,\hat{\psi}=0\ ,\ \hat{u}=0,\ \hat{v}=0,\quad \hat{\omega}=-\frac{\partial^2\hat{\psi}}{\partial\hat{n}^2} \tag{19}$$

3) *Obstacles' surfaces :*

$$\frac{\partial\hat{T}}{\partial\hat{n}}=0\ ,\hat{\psi}=0\ ,\ \hat{u}=0,\ \hat{v}=0,\ \hat{\omega}=-\frac{\partial^2\hat{\psi}}{\partial\hat{n}^2} \tag{20}$$

Where $\hat{n}$ is the non-dimensional local normal vector, and:

$$\frac{\partial}{\partial\hat{n}}=\sqrt{\left(\frac{\partial}{\partial\hat{x}}\right)^2+\left(\frac{\partial}{\partial\hat{y}}\right)^2} \tag{21}$$

A non-orthogonal curvilinear coordinate system is adopted, in order to treat the irregular boundaries of the studied elliptical annulus. Consequently, transformations from the physical regime, which is Cartesian, to a computational regime of (ξ,η) coordinates are performed by assuming $\xi = \xi(\hat{x},\hat{y})$ and $\eta = \eta(\hat{x},\hat{y})$.

The grid of this study, was generated by assuming elliptic Poisson's equations of a control method adopted by Hoffmann and Chiang (2000).The following partial differential equations were solved to obtain the desired grid:

$$\varepsilon(\hat{x}_{\xi\xi} + \Phi\hat{x}_{\xi}) - 2\delta\hat{x}_{\xi\eta} + \gamma(\hat{x}_{\eta\eta} + \Psi\hat{x}_{\eta}) = 0 \tag{22}$$

$$\varepsilon(\hat{y}_{\xi\xi} + \Phi\hat{y}_{\xi}) - 2\delta\hat{y}_{\xi\eta} + \gamma(\hat{y}_{\eta\eta} + \Psi\hat{y}_{\eta}) = 0 \tag{23}$$

Where Ψ and Φ are respectively defined as:

$$\Psi = -\left.\frac{\hat{x}_{\eta\eta}\hat{x}_{\eta} + \hat{y}_{\eta\eta}\hat{y}_{\eta}}{\hat{x}_{\eta}^{\,2} + \hat{y}_{\eta}^{\,2}}\right|_{\xi=\text{constant}} \tag{24}$$

$$\Phi = -\left.\frac{\hat{x}_{\xi\xi}\hat{x}_{\xi} + \hat{y}_{\xi\xi}\hat{y}_{\xi}}{\hat{x}_{\xi}^{\,2} + \hat{y}_{\xi}^{\,2}}\right|_{\eta=\text{constant}} \tag{25}$$

Suitable clustering functions were used for raising the grid resolution in some places inside the studied region, where the physical variables are of highly stepper gradient (Petrovic & Stupar, 1996). Figure 2 illustrates the resulted grid.

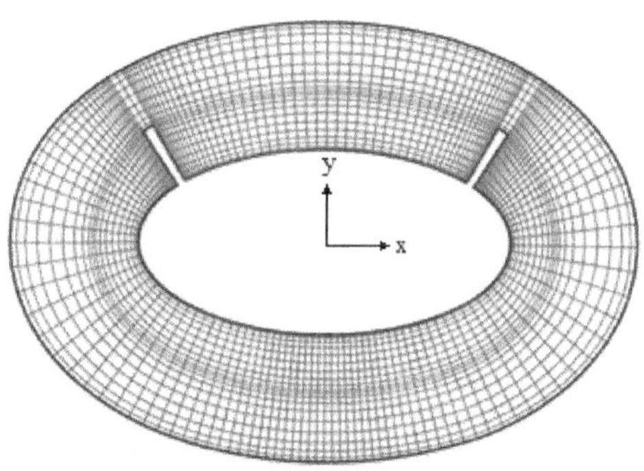

Figure 2. Schematic sketch of the computational grid generated by Poisson's elliptic equation

The local heat transfer at the inner and outer surfaces are represented by the local Nusselt number. And the latter is expressed in computational regime as the following (Zhu et al., 2004):

$$\text{Nu} = -\frac{\sqrt{\gamma}}{J}\left(\frac{\partial \hat{T}}{\partial \eta}\right) \tag{26}$$

The average Nuselt number is then determined by integrating the local Nusselt number along the surface of the inner or outer cylinder, taking in consideration the elimination of the adiabatic obstacles' bases from calculations.

3. Numerical Procedure

After the governing differential equations, with the appropriate boundary conditions, had been transformed to the computational regime(ξ,η), they were approximated utilizing finite differences techniques of second order truncation error. The approximated equations were solved numerically by using ADI with under relaxation method. The relaxation factor depended on the geometric variables. So, it had different values, each is determined by trial and error. However, the optimal under relaxation factor in all cases was less than unity.

The independence of the numerical results from the grid size was also verified by trial and error. (49×200) was the chosen size, with which, the energy balance was, at worst, preserved to less than 1.6% in all calculations.

The convergence criterion, adopted in this study, was defined as follows:

$$\left|\frac{\Theta^{\tau+1} - \Theta^{\tau}}{\Theta^{\tau+1}}\right| \leq 10^{-6}$$

Where Θ is any local dependent variable and τ is the number of the last iteration. On the other hand, the maximum number of iterations depended on the values of the relaxation factor. The deeper the latter, the more the former needed to verify the error set. It is worth to mention that Matlab programming language was utilized to construct the computer code which generated the numerical results.

4. Code Validation

In order to validate the code, several previous works should be accommodated by resetting the input variables of the geometry and porous medium. Figure 3 shows the streamlines and isotherms obtained by the present study in compare with those of Braga and de Lemos (2006), for $(\mathrm{Da} = 3 \times 10^{-7}, \emptyset = 0.2,\ \mathrm{Ra}^* = 500\ \&\ \mathrm{R} = 2)$.

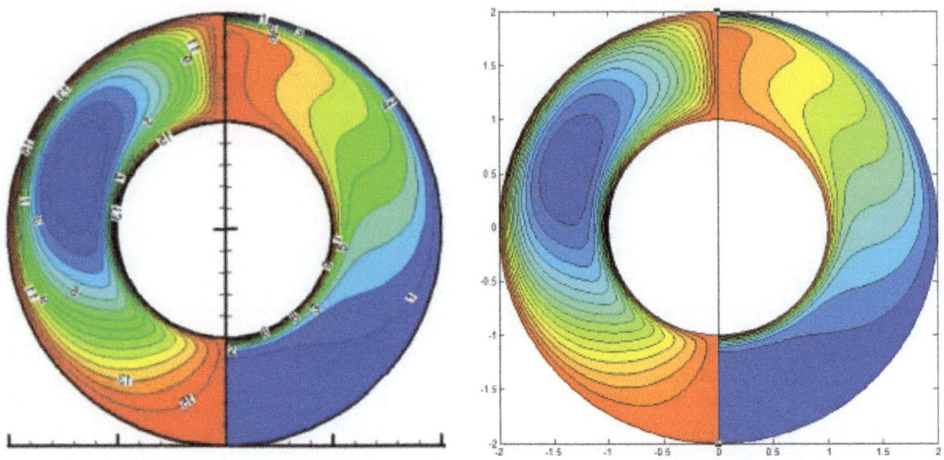

Figure 3. Streamlines and isotherms' comparison of the present study (to the right) and Braga-do Lemos, (2006) (to the left) for $(\mathrm{Da} = 3 \times 10^{-7})$, $(\emptyset = 0.2)$, $(\mathrm{Ra}^* = 500)$ and $(\mathrm{R} = 2)$

The results of the average Nusselt number, obtained by the code of the present study, were also validated for a range of modified Rayleigh number. So, Figure 4 shows a comparison between the results of the present study and those experimentally and theoretically obtained by some previous works. Regarding the presence of the obstacles inside porous annuli, no comparable study is available up to the authors' knowledge.

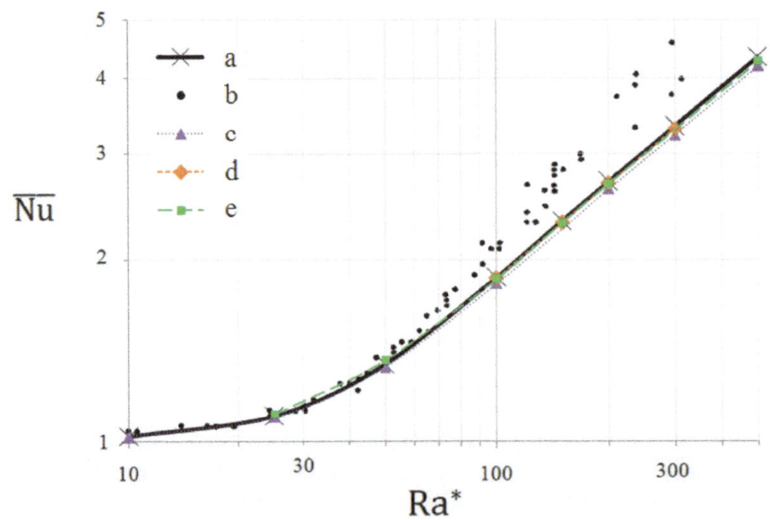

Figure 4. Average Nusselt number versus modified Rayleigh number for$(\mathrm{Da} = 3 \times 10^{-7})$, $(\emptyset = 0.2)$ and $(\mathrm{R} = 2)$. (a): Present work, (b): Experimental (Caltagirone, 1976), (c): Numerical (Caltagirone, 1976), (d): Unicellular flow (Charrier-Mojtabi et. al., 1991) and (e): Laminar model (Braga-de Lemos, 2006)

Performing the above comparisons indicated a good agreement with the previous works and ensured that the constructed code is reliable enoughto solve the problem of the present work.

5. Results and Discussion

For the purpose of simplifying the presentation of geometric cases, a special symbolic system is adopted in this study. For example, (0.67-PI-90) means that the twin obstacles are of length (l = 0.66 t), and they are fixed on the surface of the inner cylinder at (θ =90°). Within this code, (PI) would be replaced by (PO), if the twin obstacles were fixed on the outer cylinder. However, this study examines the influence of the obstacles' existence by assuming three angular positions (60°, 90° and 120°), two radial positions (PI and PO) and three lengths (0.33, 0.5 and 0.66).

The influence of modified Rayleigh number in porous annuli has already been investigated in details by many authors (e.g.: Caltagirone, 1976, Charrier-Mojtabi, 1991, Mota & Saatdjian, 1995, 1999, 2000; Kaviany, 1986). So, in order to focus on the obstacles effect, the results discussed here are obtained at a single modified Rayleigh number of 500 and a single hydraulic radius ratio of 2:1. The combined effect of Rayleigh number, radius ratio and the obstacles would be all examined in other future studies. Darcy number, Prandtl number and porosity are also fixed for the same reason. They respectively are: (10^{-4}, 0.71 and 0.9).

The modified Rayleigh number 500 is high enough for the convective effects and the cellular flow pattern to be clear. Moreover, the installation of the obstacles is more needed and applicable at such a high Rayleigh number. Figure 5 shows a tri-cellular flow pattern occurring at (Ra^* = 500) and (R = 2). From this figure one can notice that the angular position 90° is close to the center of the major convective cell where the maximum amount of stream function is. As so as, the direction of the flow, is almost normal to the obstacles, whether it is applied there. On another hand, the position 120° is occupied by the meeting boundaries of two convection cells. Besides that, the direction of the flow there, is tangential to the obstacles. These two different positions were chosen by the authors to be occupied by their twin obstacles for the recently aforementioned reasons. In addition to the positions 90° and 120°, the position 60° was also studied in order to examine how the existence of the obstacles in the lower part of the model does affect the flow pattern and hence the heat transfer rate. Based on what was foregoing, the following sections covers all the mentioned cases.

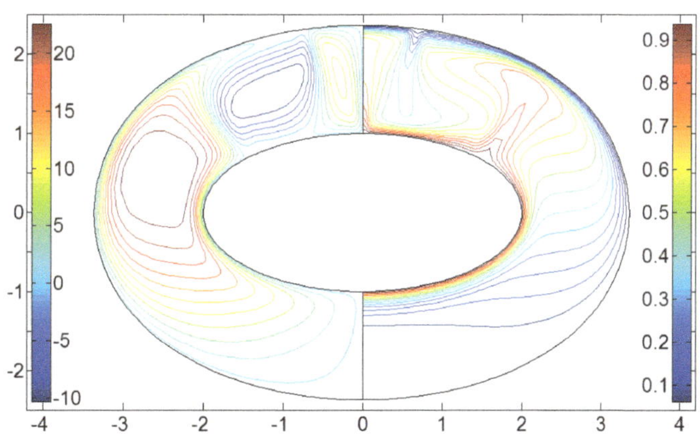

Figure 5. Streamlines (left) and isotherms (right) for $Ra^* = 500$ and R = 2, tri-cellular flow pattern without obstacles. It should be noticed that the clockwise flow direction has been negatively signed and vice versa

a. The angular position 60°

The aim of applying the obstacles in such a position is to confine and stagnate some of fluid under the inner cylinder and eliminate the convection currents there. The three length cases of PI-60 had quite performed that aim. The isotherms in Figures 6 (a, b and c) ensures that the heat almost conductively transfers in the lower part of the model. The other radial position's cases (PO-60), respectively shown in Figures 6 (d, e and f), hadn't do the same, but they just distorted the streamlines and slightly the cases of lengths and radial positions had already changed the flow pattern from tri-cellular to bi-cellular obstructed the major convection cells. On another hand, all.

Table 1 shows that the average Nusselt number had decreased by considerable and approximately equal rates. Whereas the decrease in the maximum stream function was clearly different between the two radial positions (PI and PO). All these observations lead to the conclusion that the presence of the obstacles in the lower part of the model reduces the heat transfer significantly by reducing the number of convection cells, not necessarily by obstructing the fluid or by confining it there.

Table 1. Percentage of variation in the average Nusselt number and the maximum stream function

		$l = 0.33$		$l = 0.50$		$l = 0.66$	
		$\%(\Delta\overline{Nu})$	$\%(\Delta\psi_{max})$	$\%(\Delta\overline{Nu})$	$\%(\Delta\psi_{max})$	$\%(\Delta\overline{Nu})$	$\%(\Delta\psi_{max})$
45°	PI	-08.20	-5.73	-08.12	-5.38	-08.07	-5.15
	PO	-08.71	-2.8	-11.04	-2.27	-07.94	-1.85
60°	PI	-08.90	-8.62	-09.18	-7.93	-09.16	-6.96
	PO	-09.93	-3.03	-08.43	-2.41	-08.50	-1.55
90°	PI	-18.26	-13.51	-16.37	-9.27	-14.44	-4.59
	PO	-14.42	-4.36	-12.34	-2.78	-11.33	-1.84
120°	PI	+15.66	-11.53	+12.37	-11.44	+07.98	-12.14
	PO	-01.29	-8.89	+05.76	-8.46	+03.55	-11.54
145°	PI	-01.00	-7.85	-01.47	-8.11	-01.58	-7.37
	PO	-06.06	-1.7	-06.27	-1.65	-02.81	-6.14

$\overline{Nu}|_{without\ obstacles} = 3.23$

$\psi_{max}|_{without\ obstacles} = 24.8$

b. The angular position 90°

The presence of the obstacles according to case (0.33-PI-90) causes obstructing to the flow in a spot where the velocity was relatively high. Moreover, presence of the obstacles in that position, with any length, causes lengthening to the flow path by diverting it from the alignment of the hot cylinder, in that spot, to the alignment of the adiabatic obstacle's surface (see Figures 6 (g, h and i)). As a result, the decrease in the average Nusselt number is larger than the decrease in the maximum stream function, as it is clearly shown by Table 1. On another hand, the long (or moderate) obstacles, applied in the position (PI-90), partially separates the annulus into upper active part and lower inactive one. This separation causes a short circuit to the convection currents in the upper part. Consequently, such separation badly affects the insulation efficiency. This conclusion was supported by the results of Bejan (1983) exactly which concerning with full separation. Furthermore, Table 1 in the present study shows that the shorter the obstacles, the less the average Nusselt number in case of (PI-90). Summarily, the above conclusions may indicate an equilibrium, in which the effect of the obstructing and the effect of separation are on contrast.

Regarding the position (PO-90), the short obstacles, installed on the outer cylinder, obstructs the flow, but slightly less than what caused by installation in position (PI-90). The reason would be explained as follows: the side (major) convection cell expands along a curved enclosure. Although the installation of a short obstacle in position (PO-90) slightly chokes the flow, it reduces the severity of enclosure curvature. On contrast, installation in position (PI-90) increases the severity of enclosure curvature where the side convection cell expands along. On the other hand, installing long (or moderate) obstacles leads to a partial separation for the studied model into upper and lower parts. Figures 6 (j, k and l) show the posed cases in terms of isotherms and streamlines.

c. **The angular position 120°**

The position (PI-120) is considered the worst for the goal of reducing heat transfer. Table 1 shows an increase in Nusselt number up to 15.66% for some cases, despite of the drop in the maximum stream function. In order to understand the reason, the few following lines should be explicated firstly. The flow, within the small convection cells in the upper part, is subject to two opposite effects: the first is the contact with the major cell in the lower part, which directs the rotation of the adjacent upper cell to be negative. And this is actually what is taking place in case of no obstacle installed. The second is the effect of the geometry of the space including the upper convection cells. By looking at Figure 5, one can conclude that, if the moderate convection cell was left to spin under only the effect of geometry, it would have already taken the positive direction. This contrary in effect significantly reduces the amount of fluid flow in the upper part. Therefore, the installation of the obstacles, in the position (PI-120), increases the amount of flow in the upper part by creating small secondary cell at the obstacle head and eliminating the influence of contact with the major cell in thelower part. Consequently, this installation allows the upper main cell to spin freely in the positive direction under only the influence of the geometry of the space.

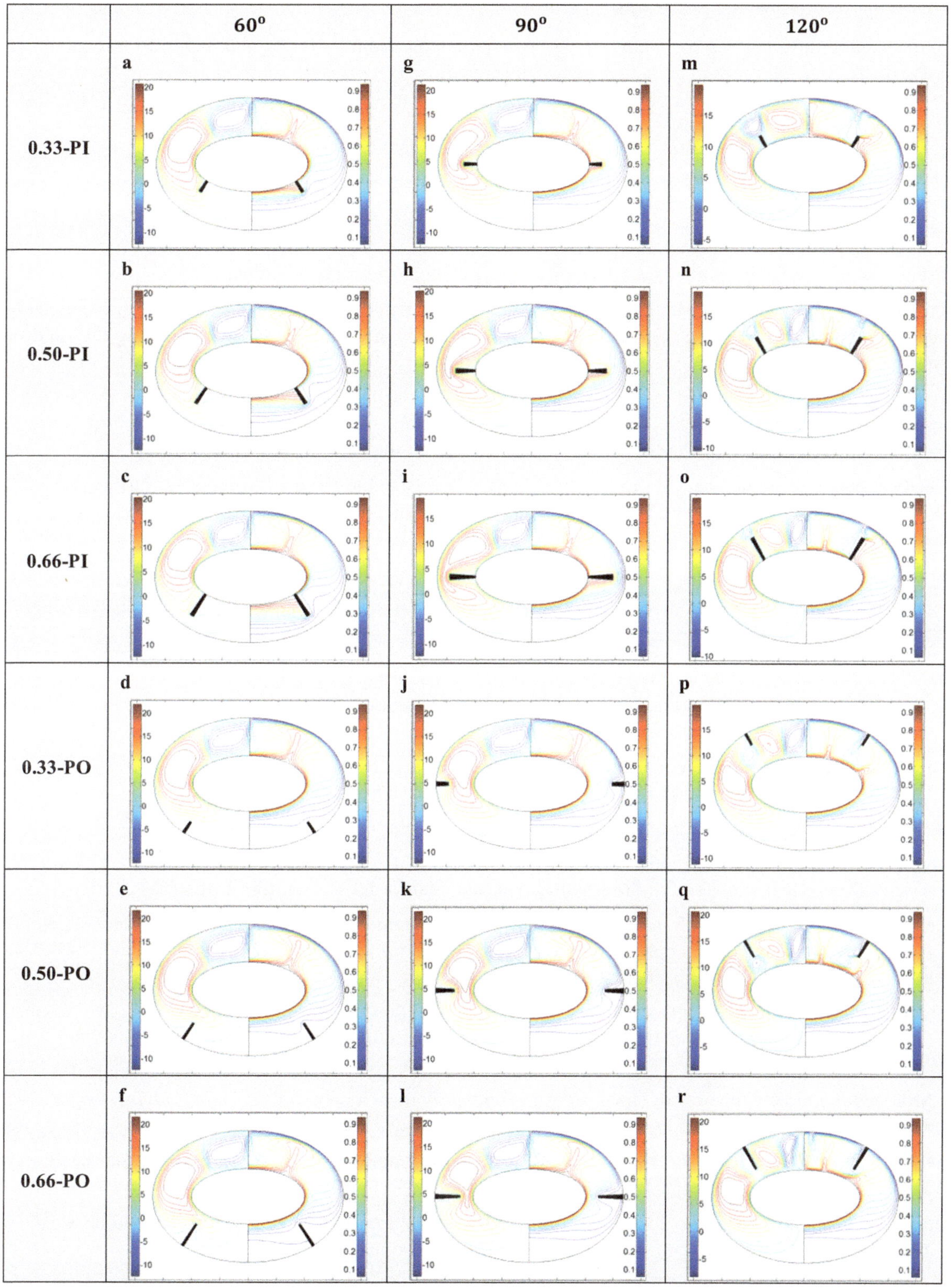

Figure 6. Streamlines (to the left) and isotherms (to the right) for various cases of angular position, radial position and length. $(\mathrm{Da} = 10^{-4})$, $(\emptyset = 0.9)$, $(\mathrm{Ra}^* = 500)$ and $(\mathrm{R} = 2)$

All the foregoing above, leads to understand that the reduction in the maximum stream function, due to the presence of the obstacles, indicates reduction in the amount of fluid flow within the major cell. And this is regardless of the increase in the amount of the upper part flow, which is the actual reason of the increase in the overall average Nusselt number (see Table 1).

No fundamental difference exists between the two radial positions (PI-120 and PO-120) except the case (0.33-PO-120). Since, this case didn't cause enough separation. But on contrary, it caused a creation of a small additional convection cell of a significant effect on the major convection cell in the lower part. Table 1 shows that this case is unpractical to carry out.

6. The Conclusions

A non-orthogonal curvilinear coordinate system is adopted, in order to treat the presence of the obstacles inside two-dimensional elliptical annulus. Consequently, an efficient code of high accuracy has been developed to solve the non-Darcian-Boussinesq equation, governing the flow inside porous media. The code had already been validated then the results were discussed. Hereafter, several conclusions are listed below:

1) The presence of the obstacles in the lower part of the model reduces the heat losses significantly by changing the flow pattern and reducing the number of convection cells, not necessarily by obstructing the fluid or by confining it there.
2) When the obstacles is installed in the angular position 90°, the shorter the obstacles, the less the overall heat losses. However, the installing at the inner cylinder, in such an angular position, reduces the heat losses more than the reduction caused in case of installing the obstacles at the outer cylinder.
3) The best position for the obstacles to be installed in, with respect to the goal of reducing heat losses, is 90°. Whereas the worst is 120°. So, in such a problem, one can preserve $18.26\,\%$ of the lost energy, while another may lose 15.66% of it.

Acknowledgements

The authors acknowledge reconcile of God, besides to the support of Mechanical Engineering Department in The University of Mosul.

References

Bau, H., Mcblane, G., & Saferstein, I. (1983). *Numerical Simulation Of Thermal Convection In An Eccentric Annulus Containing Porous Media.* ASME 83 WA/HT 34.

Bejan, A. (1983). Natural Convection Heat Transfer In A Porous Layer With Internal Flow Obstructions. *Int. J. Heat Mass Transfer, 26*(6), 815-822. http://dx.doi.org/10.1016/S0017-9310(83)80105-7

Bejan, A. (1995). *Convection Heat Transfer*. New York, USA: John Wiley & Sons, Inc.

Bishop, F. H., & Charley, C.T. (1966). *Photographic Of Natural Convection Between Concentric Cylinders* (pp. 63-78). Proc., Heat Transfer Fluid Mechanics Institute, Stanford, CA: Stanford Univ. Press.

Braga, E. J., & de Lemos, M. J. S. (2006). Simulation of Turbulent Natural Convection In A Porous Cylindrical Annulus Using A Macroscopic Two-Equation Model. *Int. J. Heat Mass Transfer, 49*, 4340-4351. http://dx.doi.org/10.1016/j.ijheatmasstransfer.2006.04.032

Caltagirone, J. P. (1976). Thermoconductive Instabilities In Porous Medium Bounded by Two Concentric Horizontal Cylinders. *J. Fluid Mech., 76*, 337- 362. http://dx.doi.org/10.1017/S0022112076000669

Charrier, M. C., Mojtabi, A., Azaiez, M., & Labrosse, G. (1991). Numerical And Experimental Study Of Multicellular Free Convection Flows In An Annular Porous Layer. *Int. J. Heat Mass Transfer, 34*(12), 3061-3074. http://dx.doi.org/10.1016/0017-9310(91)90076-Q

Facas, G. N. (1994). Reducing the Heat Transfer From a Hot Pipe Buried in a Semi-infinite, Saturated, Porous Medium. *J. Heat Transfer, 116*, 473-476. http://dx.doi.org/10.1115/1.2911420

Hadim, H. & North, M. (2005). Forced Convection in a Sintered Porous Channel with Inlet and Outlet Slots. *Int. J. Thermal Sciences, 44*, 33-42. http://dx.doi.org/10.1016/j.ijthermalsci.2004.04.016

Hoffmann, K. A., & Chiang, S. T. (2000). *Computational Fluid Dynamics I.* Wichita, Kansas, USA: A Publication of Engineering Education System.

Kaviany, M. (1986). Non-Darcian Effects On Natural Convection In Porous Media Confined Between Horizontal Cylinders. *Int. J. Heat Mass Transfer, 29*(10), 1513-1519. http://dx.doi.org/10.1016/0017-9310(86)90066-9

Mota, J. P. B., & Saatdjian, E. (1995). Natural Convection In Porous Cylindrical Annuli. *Int. J. Num. Meth. Heat Fluid Flow, 5*, 3-12. http://dx.doi.org/10.1108/EUM0000000004054

Mota, J. P. B., Esteves, I. A. A. C., Portugal, C. A. M., Esperança, J. M. S. S., & Saatdjian, E. (2000). Natural

Convection Heat Transfer In Horizontal Eccentric Elliptic Annuli Containing Saturated Porous Media. *Int. J. Heat Mass Transfer, 43*, 4367-4379. http://dx.doi.org/10.1016/S0017-9310(00)00068-5

Petrovic, Z., & Stupar, S. (1996). *Computational Fluid Dynamics I.* Mech. Eng. Faculty, University Of Belgrade.

Saatdjian, E., Lam, R., & Mota, J. P. B. (1999). Natural Convection Heat Transfer In The Annular Region Between Porous Confocal Ellipses. *Int. J. Numer. Meth Fluids, 31*, 513-522. http://dx.doi.org/10.1002/(SICI)1097-0363(19990930)31:2<513::AID-FLD889>3.0.CO;2-7

Sakr, R. Y., Berbish, N. S., Abd-Aziz, Ali A., & Hanafi, A. S. (2008). Experimental and Numerical Investigation of Natural Convection Heat Transfer in Horizontal Elliptic Annuli. *Journal of Applied Sciences Research, 4*(2), 138-155.

Vafai, K. (2005). *Handbook of Porous Media* (2nd ed.). Taylor & Francis Group, USA: CRC Press.

Zhu, Y. D., Shu, C., Qiu, J., & Tani, J. (2003), Numerical Simulation of Natural Convection between Two Elliptical Cylinders Using DQ Method. *Int. J. Heat Mass Transfer, 47*(2004) 797-808. http://dx.doi.org/10.1016/j.ijheatmasstransfer.2003.06.005

2

Pilot Study on Indoor Flows and Turbulence Statistics of a Vertical Axis Wind Turbine

YinMun H'ng[1], Yusri Yusup[1], H. P. S. Abdul Khalil[1] & TjoonTow Teng[1]

[1] School of Industrial Technology, Universiti Sains Malaysia, Penang, Malaysia

Correspondence: Yusri Yusup, School of Industrial Technology, Universiti Sains Malaysia, Penang 11800, Malaysia. E-mail: yusriy@usm.my

Abstract

Turbulence statistics were applied to the downwind flow of a Darrieus-type vertical axis wind turbine (VAWT) to provide insights about VAWT micro-flow characteristics. Micro-turbulence measurements of the flow in front of and behind a VAWT were made using a sonic anemometer. Turbulence data for downwind distances of 0.5 m and 2.0 m from the wind source were collected with and without the VAWT, respectively, using a blower (Type 1 flow) and industrial fans (Type 2 flow) as the wind source. The voltage V, generated by the VAWT was recorded every 15 min. Only the center line of the flow was considered. The Type 2 flow induced a constant spectral power at a high dimensionless frequency range of $0.007 < f < 0.04$, perhaps due to the fan configuration, which had a larger wind-swept area. The results show an increase in lateral turbulence, σ_v, downwind of the VAWT compared to its longitudinal, σ_u, and vertical, σ_w, counterparts. Using spectral analysis, the VAWT was found to reduce the horizontal (u and v) turbulent component energies in the dimensionless frequency range of 0.003–1. Interesting dip and peak features were observed for Type 1 flow, but only dip features were observed for Type 2 flow in the resulting spectra. Higher wind speeds increased the voltage generated at a rate of 0.5.

Keywords: VAWT, micro-turbulence, spectral power, dimensionless frequency, voltage

1. Introduction

1.1 Darrieus-Type VAWT

There are two types of vertical axis wind turbines (VAWTs): 1) Darrieus and 2) Savonious. The Darrieus VAWT is a lift-type turbine of which the aerodynamic characteristics are similar to a helicopter blade. The vertical airfoil-shaped blades move forward into oncoming wind, producing small but changing positive angles of attack on the blades. This creates a net lift force on the blade, inducing it toward the wind turbine mast and resulting in a positive torque on the shaft. When the wind flows across the VAWT blades, a positive torque is produced; a small or negative torque is produced when the blades spin parallel to the wind.

VAWTs are omnidirectional, which means they do not need to be positioned into the main wind direction and are able to capture winds from any direction (Paulides, Encica, Jansen, Lomonova, & Van Wijck, 2009; Stankovic, Campbell, & Harries, 2009; Stitt, 1999). Thus, they are well suited for areas where the winds are directionally unpredictable. VAWT are increasingly common in the wind energy market, especially for domestic or small scale installations, such as on rooftops and in urban and suburban locations. Moreover, VAWTs are smaller than the traditional horizontal axis wind turbines, so they can be easily integrated onto buildings. Placing a VAWT at a higher elevation such as on a rooftop, provides access to stronger winds with greater mechanical energy. The Darrieus VAWT exhibits an inherent "torque ripple" caused by the continuously changing angles of attack between the turbine blades and the wind (Eriksson, Bernhoff, & Leijon, 2008); torque ripple is defined as the amount of torque measured by subtracting the minimum torque from the maximum torque in one revolution.

Horizontal axis wind turbines are self-starting unlike some VAWTs, which depend on wind conditions. However, the lift-driven VAWT can self-start and operate in low wind speed conditions. When required, an induction motor, which must be connected to its own energy source, is employed to ensure the start-up of the VAWT.

Winds are generally turbulent in the built urban environment (Christen, Rotach, & Vogt, 2009; Rotach et al., 2005; Roth, 2000). Although studies of urban canopy flows have simulated wind in urban areas (Britter & Hanna, 2003; Lopez & Garcia, 2001; Sanz, 2003), knowledge of the turbulence generated by VAWTs may help to improve their performance and efficiency in such areas. One study reported an increased power-collection capability for these turbines in close proximity because they maximize power generation per surface area (Dabiri, 2011). The study of the turbines in close proximity is essential in urban installations of wind turbines due to urban real-estate limitations. In addition, instrumental measurements of the vortices, turbulence and eddies generated by VAWTs are rarely reported (Ferreira, van Bussel, & van Kuik, 2006), as most published works have focused on horizontal axis wind turbines (Ebert & Wood, 1997; Vermeer, Sorensen, & Crespo, 2003; Yang, Sarkar, & Hu, 2012).

In this experiment, turbulence statistics were applied to the upwind and downwind flows of a VAWT. Turbulent statistics are commonly used in the micro-meteorological field. The turbulence generated by a VAWT can be quantified by the fluctuation of the three-directional wind components of the mean wind speed, which are, u', v', and w'. Although most VAWTs are positioned at a low hub height, where the atmospheric conditions may not have any influence, taller and larger VAWTs are progressively being developed. Furthermore, a study has shown that the power output of wind turbines can be affected by atmospheric stability, which in turn governs atmospheric turbulence, thus linking the two fields (Wharton & Lundquist, 2012). This paper reports results the turbulence statistics, including variances and spectra, generated by a generic Darrieus-type VAWT.

1.2 Theoretical Framework: Turbulence Statistics

Wind speed fluctuations create turbulence in the mean wind flow. These fluctuations are composed of random wind vectors in three-dimensional space. Steady-state wind flow has a low turbulence intensity, while unsteady-state wind has a high turbulence intensity. Turbulence intensity is a common term used to express the turbulence level.

Coordinates can be superimposed onto these vectors by adding reference axes, namely the orthogonal x-y-z axes. Thus, the wind vectors can be mapped onto the three perpendicular directions of the x-y-z axes, with notations of u, v, and w, respectively, to denote the longitudinal, lateral, and vertical directions.

If flow conditions are statistically stationary with time and Taylor's frozen hypothesis applies for turbulence intensities $I_{u,v,w} = \sigma_{u,v,w}/\bar{u} < 0.5$, where $\bar{u}$ is mean wind speed, turbulence can be quantified by the deviation or standard deviation of the instantaneous wind velocity u', v', and w' from the mean wind direction $\bar{u}$, $\bar{v}$, and $\bar{w}$ at a single measurement point. The instantaneous wind velocities are u, v, and w.

The standard deviations of the instantaneous velocities u, v, and w are a quantification of the turbulence in the three directions and are usually denoted as σ_u, σ_v, and σ_w. This method of turbulence quantification has been used in the micro-meteorological field since the early 1900s in homogeneous and heterogeneous landscapes (Roth, 2000).

The friction velocity, or u_*, indicates the magnitude of momentum transferred and is also called the shear velocity because wu is (shear stress/density). It can be calculated using Equation (1) where the bars over the letters indicate ensemble averages. This equation can be used in the experimental setup because of the stable nature of the flow generated; the values obtained would be local or unique to a certain measurement position within the flow.

$$u_* = \sqrt[4]{\overline{w'u'}^2 + \overline{w'v'}^2} \quad (1)$$

Friction velocity, or u_*, is a usual "scaling" parameter used when creating turbulence vertical profiles of the surface layer of the atmosphere because it is known to be constant with height in the latter layer. In the urban roughness sublayer, the atmospheric layer above the urban or rough surface, its value is lower than the former (Roth, 2000). Reducing the mean wind flow energy is the general method employed to quantify VAWT efficiency. In the context of this experiment, u_* can be used to judge the combined horizontal (u and v) and vertical (w) reductions in kinetic energy downwind of the VAWT.

Turbulence spectral analysis, which is another method used to analyze the instantaneous wind velocity components, reveals important information such as the peak frequency and the rate of decay or generation of spectral power with dimensionless frequency. The method used by Kaimal et al. (1972) has been the standard among micro-meteorologists and was used in the current study for which the ordinate was the non-dimensional spectral power, $nS_{u,v,w}(n)/\sigma_{u,v,w}^2$ or S_P, where $S(n)$ is the spectral power as a function of frequency (Hz), n, and the abscissa is the dimensionless frequency, $f = nL/\bar{u}$, where L is the distance from the wind source. The blade passing frequency (blade rotation frequency) can also be ascertained from the resulting spectra. When flow conditions are statistically stationary, the eddy size can be estimated from S_P and its corresponding turbulence statistics, σ_i, L, and $\bar{u}$ through dimensional analysis. The eddy size in can be written as Equation (2).

$$l_e = \sqrt{S_P \times \sigma_i} \times \frac{L}{\bar{u}} \quad (2)$$

where
$l_{e,i}$ = eddy length, m
$i = u$, v or w

2. Methodology

2.1 Experimental Set-Up

This experiment was conducted in a laboratory to avoid the influence of external winds and other uncontrollable factors that might affect the readings taken by the sonic anemometer (Young Model 81000). The anemometer is able to measure high frequency wind velocity components of up to 20 Hz. Data logging was conducted by connecting the sonic anemometer to a personal computer (PC). An industrial blower and three industrial fans were used as the wind source, with the height of the middle fan being altered. The VAWT was a Darrieus-type turbine, as shown in Figure 1, and was acquired from iWind Energy (M) Sdn. Bhd. (model no.: iW301). The specifications of the turbine are listed in Table 1. The power, P, versus revolutions per minute, rpm, and the rpm versus $\bar{u}$ of the tested VAWT were provided by manufacturer and are shown in Figures 2 (a) and (b), respectively. As detailed later in this paper, the rpm values ranged from 50–100 rpm, while the apparent wind speed was between 2–3 m/s for Type 1 flow (using blower). For Type 2 flow (using industrial fans), the rpm values ranged from 100–150 rpm, while the range of the apparent wind speed was between 5–6 m/s.

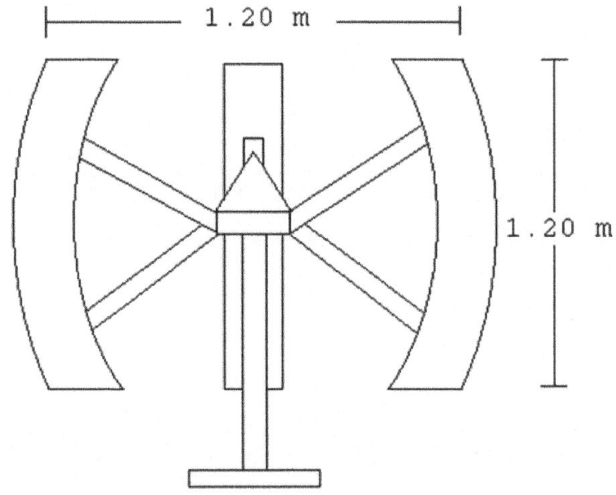

Figure 1. Schematic of the three-bladed Darrieus-type vertical axis wind turbine (VAWT) employed in this experiment

Table 1. Vertical axis wind turbine (VAWT) specifications

Specification	
Rated power output	300 W
Rated wind speed	12 m/s
Cut-in wind speed	2.5 m/s
Survival wind speed	30 m/s
Total swept area	1.44 m^2

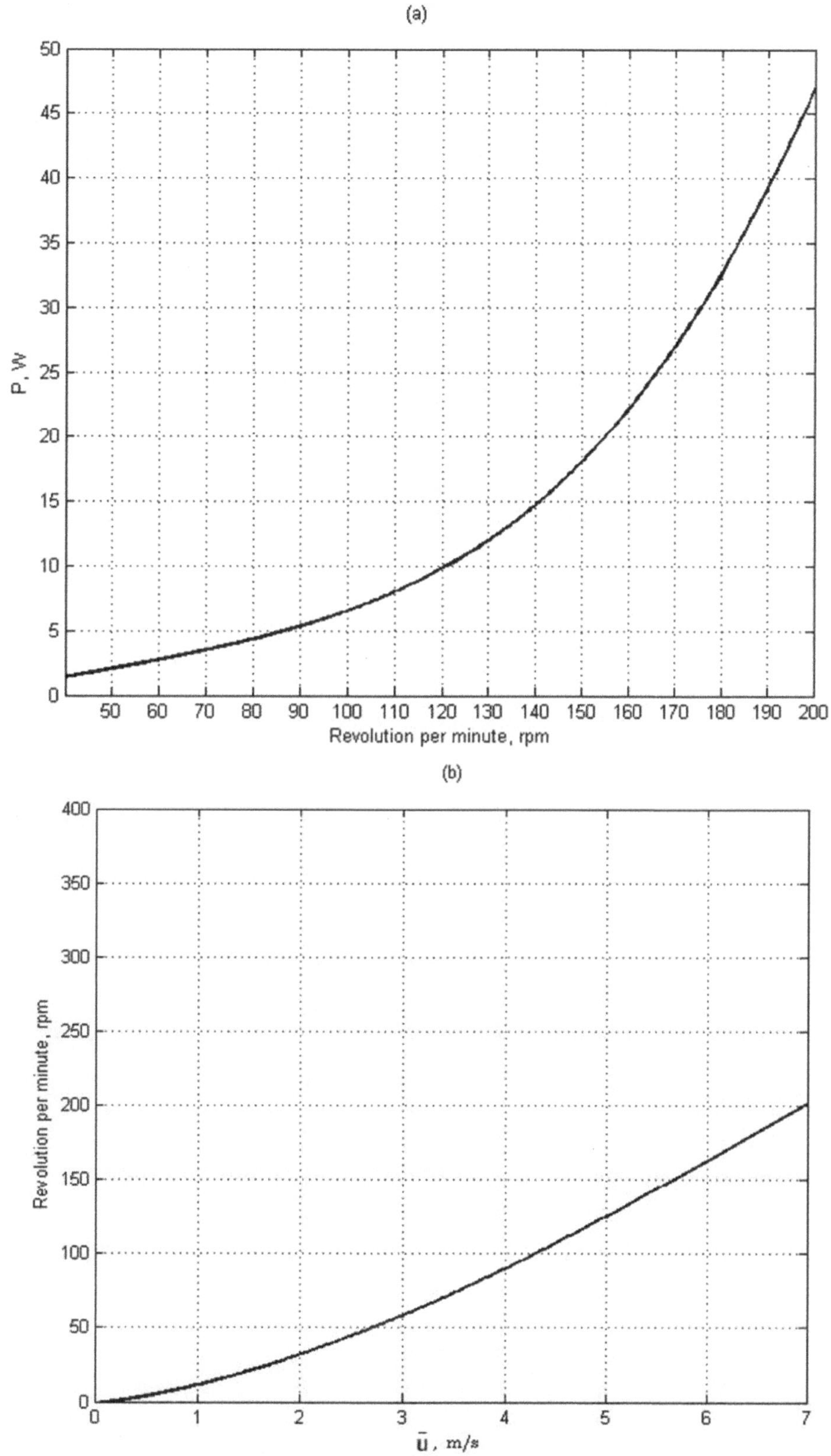

Figure 2. (a) Power, *P*, versus rpm and (b) rpm versus mean wind speed, $\bar{u}$, of the tested vertical axis wind turbine (VAWT) as stated by the manufacturer

Throughout the experiment, two types of wind flows were simulated using an industrial blower and industrial fans at three different speeds obtained by adjusting the speed control buttons: low ≡ 1, medium ≡ 2, and high ≡ 3. For Type 1 flow, the blower was aligned with the center of the sonic anemometer (shown in Figure 3). Despite being very high at the outlet, the speeds were slower downwind. The flow in this case was not uniform, and because this was a pilot study, only the center line of the flow point was considered. For Type 2 flow, three industrial fans were arranged in a row, and the height of the middle fan was lowered (shown in Figure 4). Type 2 flow covered a larger wind source area compared to Type 1 flow. We used three industrial fans in Type 2 flow because we want to maintain a uniform flow condition as the flow disperses downwind. A similar fan configuration, implemented at the indoor testing facility of Universiti Teknologi Malaysia, was used as a reference to this research work (Wahab & Chong, 2003). The turbulence statistics for the flows for different positions at a wind speed of 3, the high setting, are summarized in Table 2. The accuracy of the velocity measurements of the sonic anemometer used is ±1% root mean square, and the mean velocities in the y and z directions were negligible. Different blower speed options from low to high were later discovered to produce the same mean wind speeds. Only the highest speed option was chosen for further analysis. Measurements were taken three times to obtain averages.

The difference between the three ergodic experiments was small, as indicated by the small standard error of the mean wind speed, $\bar{u}$ (refer to Table 2). Although the flow may not have been uniform, this was a preliminary study to explore the turbulence statistics and spectra profiles of a VAWT, which have never been measured before. Furthermore, averaging the flow for 30 min gave stable results (Table 2), suggesting that non-simultaneous measurements before and after the VAWT were not necessary for either type of flow. This experimental setup is similar to that used in a wind tunnel study by Cao et al. (2011) in which the wind velocity ranged from 3–20 m/s with 0.5% turbulence. One benefit of using a blower is that the turbulent component frequency and energy at the nozzle should be similar, unlike natural wind. Meanwhile, the positioning of the three industrial fans for Type 2 flow covered a larger wind source area.

Table 2. Summary of the turbulence standard deviations and friction velocities (at 30-min averages) for positions A, B, C, and D at speed 3 (high setting) for Type 1 and Type 2 flow; note that all parameters of A and C have similar values even though measurements were non-simultaneous for Type 1 flow; the results are reported as the mean ± standard error, where standard error = $\sigma/n^{1/2}$ where σ = standard deviation and n = number of data points

Position (Type of flow)	L (m)	$\bar{u}$ (m/s)	σ_u (m/s)	σ_v (m/s)	σ_w (m/s)	u_* (m/s)
A (T1)	0.5	12.6±0.2	0.613±0.004	0.432±0.002	0.378±0.007	0.281±0.014
B (T1)	2.0	5.04±0.04	1.24±0.01	0.724±0.004	0.763±0.001	0.305±0.008
C (T1)	0.5	13.0±0.1	0.602±0.004	0.422±0.005	0.376±0.000	0.259±0.003
D (T1)	2.0	3.84±0.04	1.05±0.02	1.77±0.02	0.672±0.024	0.656±0.016
A (T2)	0.5	6.41±0.018	2.387±0.048	0.722±0.005	0.573±0.004	0.387±0.006
B (T2)	2.0	3.72±0.019	1.237±0.003	0.620±0.001	0.611±0.003	0.359±0.001
C (T2)	0.5	6.21±0.043	1.210±0.040	0.486±0.001	0.484±0.007	0.555±0.012
D (T2)	2.0	0.46±0.002	0.363±0.003	0.202±0.002	0.230±0.003	0.059±0.004

The results are reported as the mean ± standard error, where standard error = $\sigma/n^{1/2}$ where σ = standard deviation and n = number of data points. T1 indicates Type 1 flow (using a blower), and T2 indicates Type 2 flow (using industrial fans).

Two measurement positions from the blower (Type 1 flow) and industrial fans (Type 2 flow) and two configurations with and without the VAWT were chosen: position A (L = 0.5 m), B (L = 2.0 m), C (L = 0.5 m) and D (L = 2.0 m downwind) (refer to Figure 3 and Figure 4). For positions A and B, the turbulence data were collected without the VAWT to serve as experimental controls, while for positions C and D, the VAWT was installed. The voltage, V, was recorded every 15 min with the VAWT.

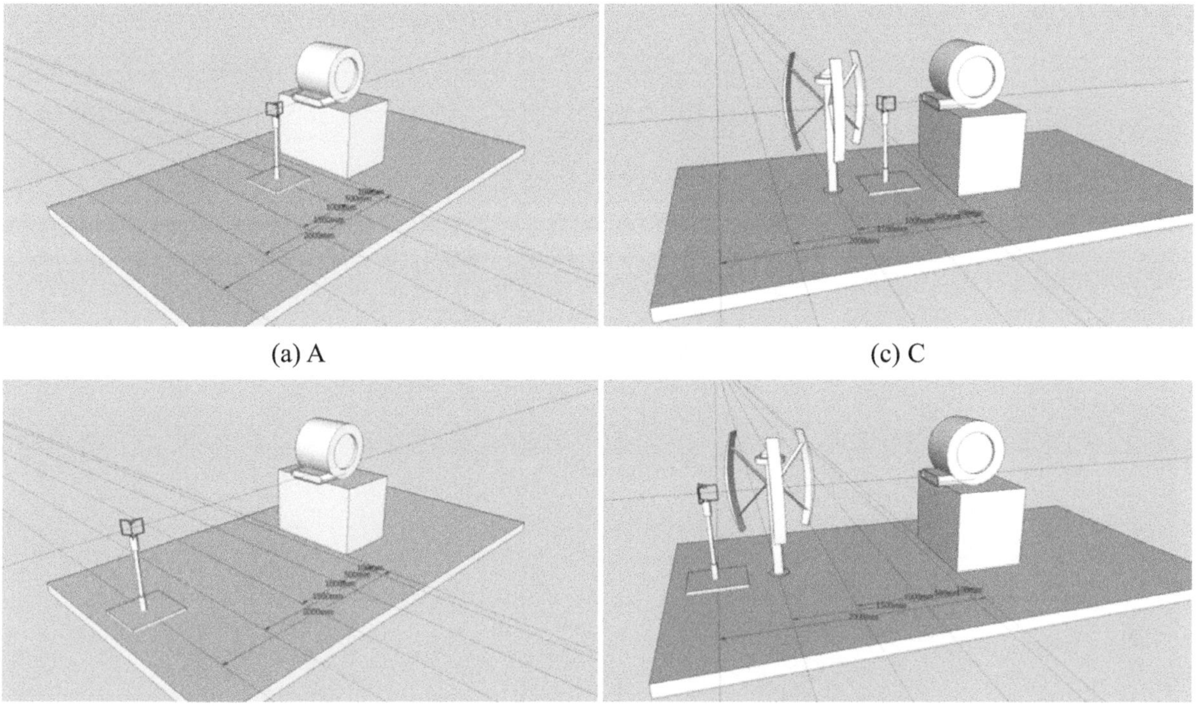

(a) A (c) C

(b) B (d) D

Figure 3. Experimental setup depicting the vertical axis wind turbine (VAWT), blower, and sonic anemometer for Type 1 flow at positions A, B, C and D; runs for positions A and B were without the VAWT, while positions C and D included the VAWT

The sonic anemometer was connected directly to a PC using a RS232 cable and the "Hyperterm" computer program as an interface. Wind component velocities were recorded at a sample rate of 10 Hz (0.1 s). During the experiment, a radius of 0.5 m around the blower, industrial fans, sonic anemometer, and VAWT was cleared to avoid any wind flow obstructions or wakes. An averaging time of 30 min was used to compute the turbulence statistics. As discussed later, this is a reasonable averaging time to use in this situation because it creates a statistically stationary condition for the momentum fluxes. This averaging time has also been used in micro-meteorological studies of urban areas (Yusup, 2012).

For the entire experiment, the angular velocity of the turbine was 5.2–6.3 rad/s or 50–60 rpm for Type 1 flow and 7.3–15.7 rad/s or 70–150 rpm for Type 2 flow, as determined from the display panel of the VAWT. The angular velocity, ω, can be calculated from Equation (3).

$$\omega = 2\pi f \tag{3}$$

where

ω = angular velocity (rad/s)

f = rotational frequency (Hz or s^{-1})

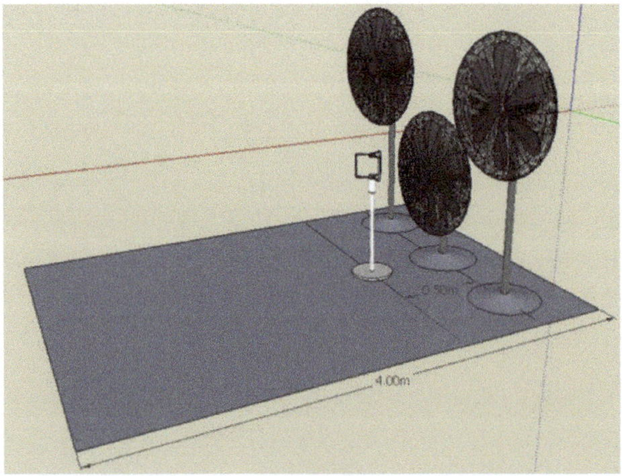

(a) A

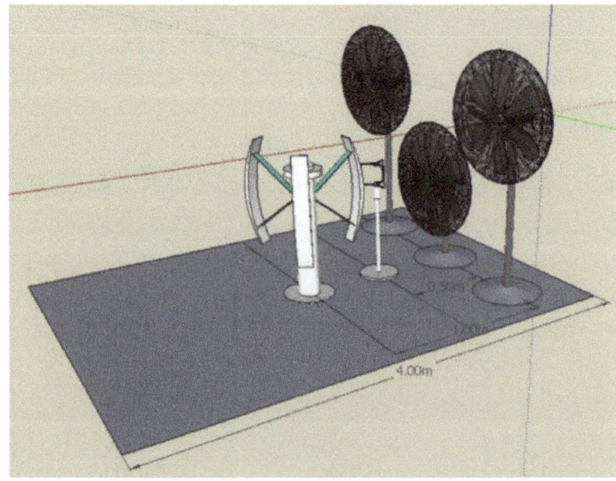

(c) C

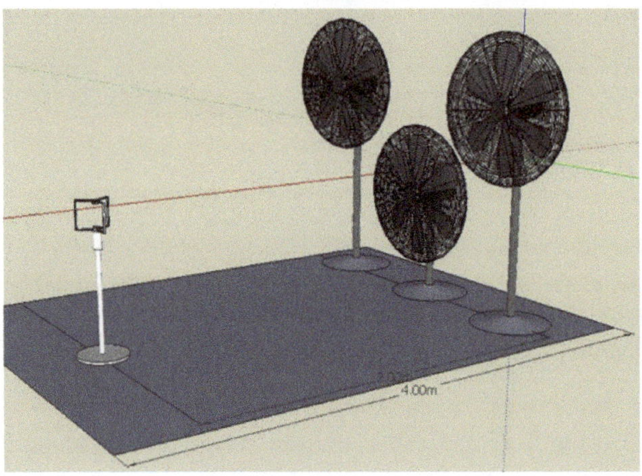

(b) B

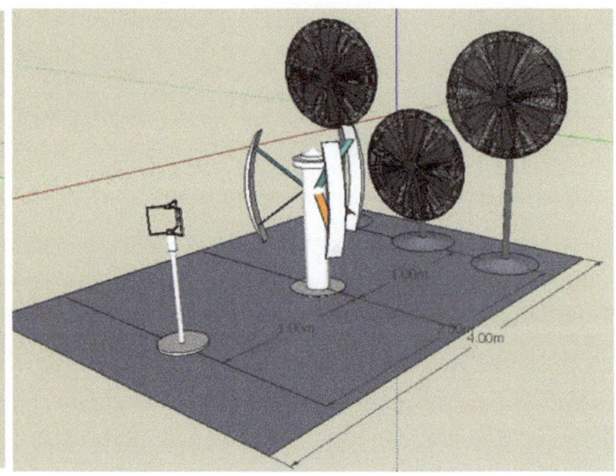

(d) D

Figure 4. Experimental setup depicting the vertical axis wind turbine (VAWT), industrial fans, and sonic anemometer for Type 2 flow at positions A, B, C and D; runs for positions A and B were without the VAWT, while positions C and D included the VAWT

From the power, rpm, and wind velocity curves given by the manufacturer, the apparent wind velocity was determined to be 3–4 m/s (from Figure 2 (b)), which was different from the range of 11–12 m/s measured by the sonic anemometer before the VAWT (positions A and C), indicating that only a fourth of the wind speeds were harnessed by the VAWT. As detailed in this paper, the rpm range for Type 1 flow was between 50–100 rpm, and the apparent wind speed was between 2–3 m/s. For Type 2 flow, the rpm range was between 100–150 rpm, and the apparent wind speed was between 5–6 m/s. These results were further corroborated by the amount of P generated, which was within the range of 1–3 W for Type 1 flow and 7–15 W for Type 2 flow (see Figure 2 (a)). From these values, the estimated tip speed ratio (TSR), λ, for mean wind speeds of 3 m/s for Type 1 flow and 5 m/s for Type 2 flow was calculated using Equation (4) to be 1.0–1.2 and 0.9–1.9.

$$\lambda = \frac{\omega r}{\bar{u}} \tag{4}$$

where

r = rotor radius = 0.6 m

$\bar{u}$ = mean wind speed

The TSR of a wind turbine depends on the number of blades; the optimal TSR values of two-bladed, three-bladed, and four bladed rotors are 6.28, 4.19, and 3.14, respectively (M. Ragheb & A. M. Ragheb, 2011). The optimal λ for a three-bladed wind turbine is well above the λ_{op} found in this experiment.

The power coefficient, C_p, of a VAWT can be calculated using Equation (5). The C_p calculated for both flows was approximately 9% based on mean wind speeds of 3 m/s for Type 1 flow and 5 m/s for Type 2 flow. The extractable power was 2 W for Type 1 flow and 10 W for Type 2 flow.

$$C_p = \frac{P}{W} = \frac{P}{\frac{1}{2} . \rho . S . \bar{u}^3} \quad (5)$$

where

P = power, W

W = kinetic power

ρ = density of air ≈ 1.2 kg/m^3

S = swept area for turbine = 1.44 m^2

2.2 Data Analysis

In this work, high-frequency three-dimensional wind speed measurements (u, v, and w) were collected. Data were processed using Matlab 2008b (MathWorks, USA). Data processing includes data filtration, turbulence statistics calculations, and spectral analysis. Data quality checks and filtration (such as data spikes removal) were also performed. The standard deviations of the fluctuations of u, v, and w from the mean, i.e., σ_u, σ_v, and σ_w, were then calculated. The Fast Fourier Transform or FFT function in Matlab was used for spectral analysis. FFT converts time series signals (or data) by changing the time domain to the frequency domain and the magnitude domain to the spectral power domain. The plot of the spectral power versus frequency would reveal the signal's periodic pattern, i.e., distribution of energy at different frequencies.

2.3 Scope and Limitations of Study

Wind speeds produced by the industrial blower were not uniform. Since this is a pilot study, only the center line of the flow is considered. The amount of turbulence generated at the entrance of the experimentally observed section was not controlled, i.e., the upwind area before the VAWT.

3. Results and Discussion

3.1 Standard Deviations of the Wind Component Fluctuations for the Darrieus VAWT

The turbulence statistics are summarized in Table 2. The values shown are averaged. The turbulence statistics σ_u, σ_v, and σ_w were similar for repeated experiments, with low standard deviations in the range of 0.001–0.010 for the three wind speeds. These statistics did have a relatively higher variation on the order of 0.02 for the standard errors for position D (Type 1 flow with the VAWT) compared to the other positions. This variation was expected because these turbulence statistics should include the turbulence generated by the VAWT. The turbulence intensities, I, were also found to be < 0.5 for this dataset, validating Taylor's frozen hypothesis and the use of a 30 min averaging time. If fans are used as a wind source, an averaging time of greater than 90 min is required to stabilize the turbulence spectra at lower dimensionless frequencies, $f < 0.01$.

For brevity, only the summarized results for the speed ≡ value of 3 (high) are listed. These values show consistent trends with the two lower speed settings. Type 1 flow had a higher mean wind speed ($\bar{u}$) and lower turbulence standard deviations (σ_u, σ_v, σ_w) than Type 2 flow for Position A; while for Position B, the $\bar{u}$ was lower for Type 2 flow compared to Type 1 flow, although the σ_u, σ_v and σ_w were mostly the same. The friction velocities had the same magnitude for both types of flow. For Type 1 flow, the mean wind speed, $\bar{u}$, decreased by 60% from 12.6 m/s to 5.04 m/s from position A to position B downwind of the blower for runs without the VAWT. Conversely, for runs with the VAWT (positions C and D), there was a 70% reduction in wind speed from 13.0 m/s to 3.84 m/s. The turbulence parameters increased with distance, L and $\bar{u}$. There was a 51% increase in σ_u from 0.613 m/s to 1.24 m/s without the VAWT, but only 43% increase from 0.602 m/s to 1.05 m/s when the VAWT was used. As for σ_v, a 41% increase from 0.432 m/s to 0.724 m/s was observed without the VAWT, while a 76% increase from 0.422 m/s to 1.77 m/s was observed with the VAWT. For σ_w, a 50% increase from 0.378 m/s to 0.763 m/s was observed without the VAWT, while a 44% increase from 0.376 m/s to 0.672 m/s was observed with the VAWT. For Type 2 flow using industrial fans and without the VAWT, $\bar{u}$ decreased from 6.41 m/s at position A to 3.72 m/s at position B, for a total decrease of 42%. With the VAWT installed, $\bar{u}$ fell by 90% from 6.21 m/s at position C to 0.46 m/s at position D. Turbulence decreased with distance by 48% from 2.387 m/s at position A to 1.237 m/s at position B without the VAWT, while it decreased by 70% from 1.21 m/s at position C to 0.363 m/s at position D with the VAWT. The parameter u_*, which is proportional to wind speed (Yusup, 2012), increased by 40 to 55% with the VAWT and by < 10% without the VAWT, indicating the transfer of a significant amount of vertical momentum when wind flows through the turbines. This result agrees with the findings of Dabiri (2011) that show VAWTs' ability to extract planform (from the top) kinetic energy.

The highest increase of the three turbulence statistics was found in σ_v for Type 1 flow when the VAWT was used.

A relative decrease in σ_u and σ_w was observed for Type 1 flow; and σ_u, σ_v and σ_w decreased for Type 2 flow when the VAWT was within the flow. The 40% increase in u_* was caused by heightened lateral wind fluctuations, v', of the spinning VAWT.

3.2 Spectral Characteristics of the Wind Components

Using FFT, we obtained spectral power and frequency. These plots showed the distribution of spectral power at different frequencies also revealing energy peaks and dips at discrete frequencies. A plot of the magnitude of a time-series input signal versus dimensionless frequency could be used to reveal the signal's periodic pattern. Figure 5 shows the spectral power for both upwind and downwind flows before normalization. Dissipation still occurred from upwind positions such as A and C, to downwind positions, such as B and D, at higher frequencies. The spectral power decreased with distance due to the lower wind component fluctuations (turbulence intensity) associated with the lower mean wind speeds downwind. We could not compare the upwind and downwind u, v, and w spectra without normalizing them with σ_i^2.

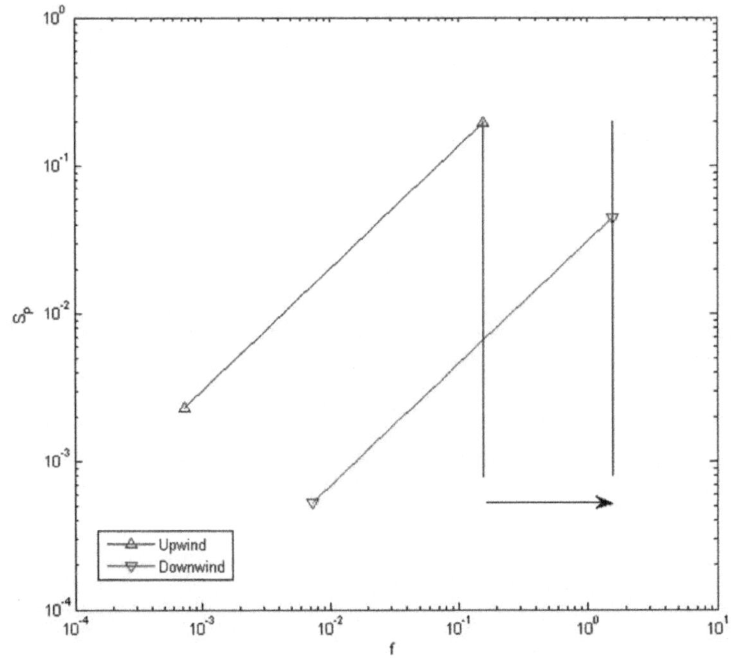

Figure 5. Schematic of the spectral power for both upwind and downwind flows before normalization; energy dissipation occurred from upwind to downwind or was displaced to higher frequencies (to the right)

The three speeds for Type 1 flow (low, 1: $\bar{u}$ = 11.7 m/s; medium, 2: $\bar{u}$ = 12.1 m/s; and high, 3: $\bar{u}$ = 12.6 m/s) did not exhibit much difference in terms of normalized spectral power, S_P, and f thus only speed 3 is displayed and discussed because the other speeds showed similar trends and magnitudes. The normalized dimensionless frequency, f, ranged from 0.00001 to 10. The plot in Figure 6 shows the upwind and downwind (without VAWT) spectra for Type 1 and Type 2 flows, which are separated by a factor of ten (one period) for the dimensionless frequency, f, parameter. The resulting spectra trend of Type 1 flow is similar to that of Type 2 flow, although the wind mean speeds were higher and the three wind vector fluctuation standard deviations (σ_u, σ_v, and σ_w) were lower for the Type 1 flow. Figure 6 exhibits higher dimensionless frequency ranges for the downwind flow compared to the upwind flow, although both show a similar increasing trend with dimensionless frequency. The dissipation occurred at higher frequencies with distance from wind source, similar to other flow types such as atmospheric wind, and at to lower frequencies for the same distance category (downwind or upwind) from the wind sources (blower or industrial fans). For Type 2 flow, the spectral power was constant, or the eddy size was constant, at a high dimensionless frequency range from 0.007 to 0.04. This might be due to the fan configuration, which covered a larger wind-swept area.

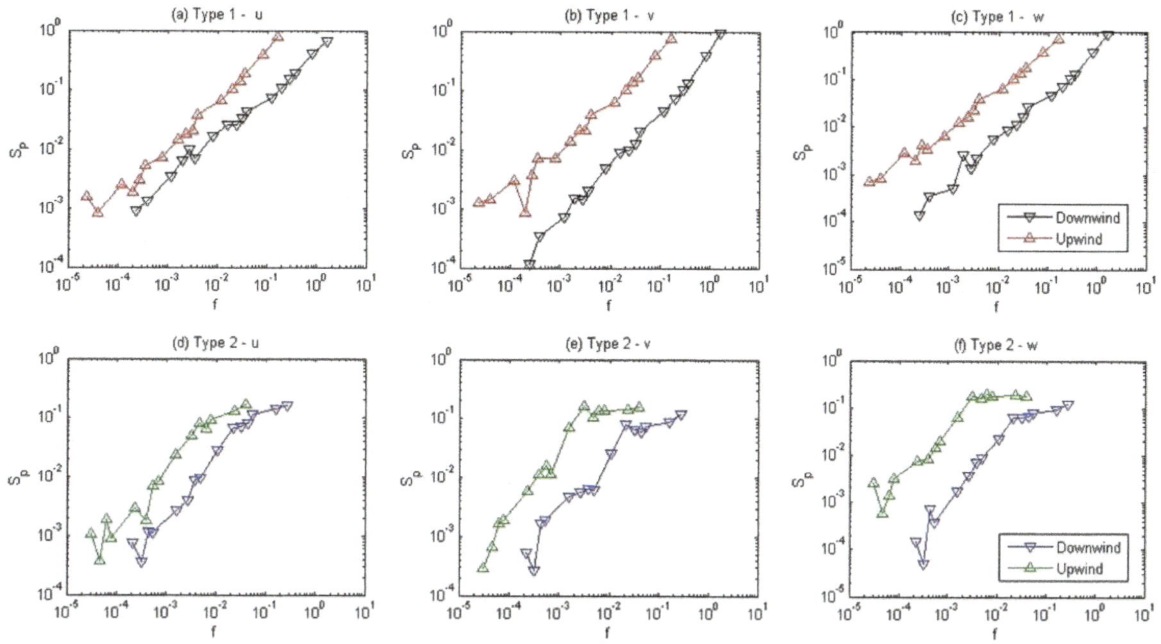

Figure 6. Upwind and downwind spectral power (normalized) of components *u*, *v*, and *w*, for Type 1 and Type 2 flows without the VAWT for positions A and B; only "high" or speed 3 shown here; a similar trend can be seen for both upwind and downwind flows, although because the flow were offset by a factor of ten, the axis of *f* (dimensionless frequency, where $f = nL/\bar{u}$. Frequency (Hz), *n*, was normalized by distance/mean wind speed, $L/\bar{u}$, to facilitate comparison

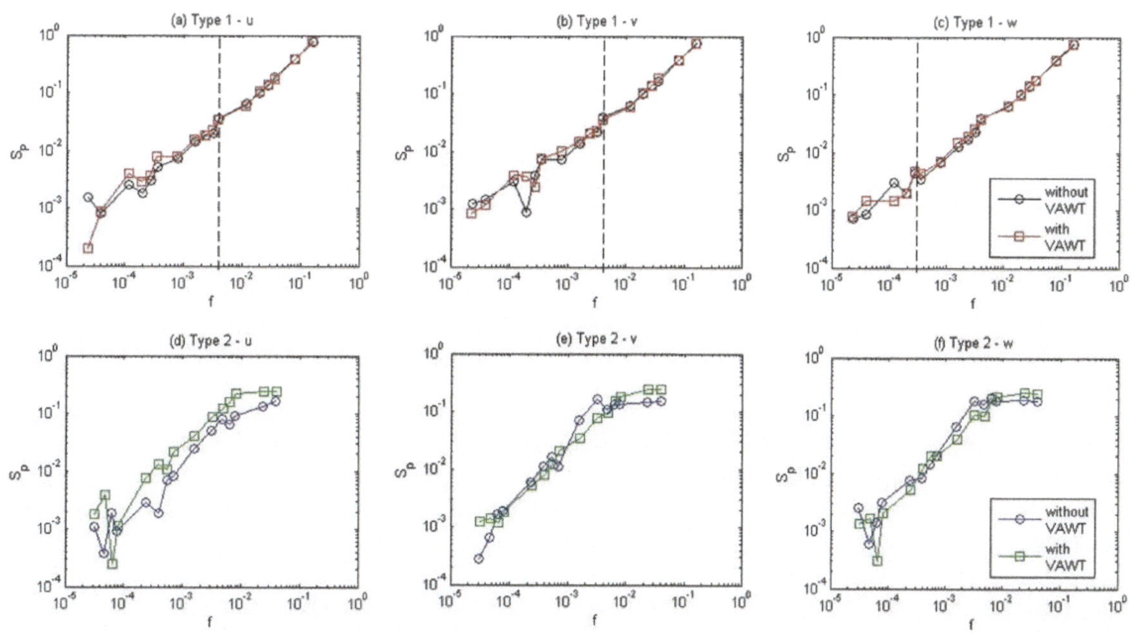

Figure 7. Upwind spectral power (normalized) of components *u*, *v*, and *w*, for Type 1 and Type 2 flows, with and without the VAWT, for positions A and C; only "high" or speed 3 shown here, which exhibited a spectral trend not affected by the presence of the VAWT upwind; the vertical dash lines for (a) and (b) are for $f = 0.004$ and for (c) $f = 0.0003$ (where $f = nL/\bar{u}$), indicating the influence of surrounding obstacles for Type 1 flow; the frequency (Hz), *n*, was normalized by distance/mean wind speed, $L/\bar{u}$ to facilitate comparison

The spectra of *u*, *v*, and *w*, with and without the VAWT, are shown in Figure 7 for positions A and C at 0.5 m downwind from the blower and industrial fans. Low dimensionless frequency values of S_P were scattered; $f < 0.004$ for *u* and *v*, while $f < 0.0003$ for *w* for Type 1 flow. These ranges indicate external influences on S_P in the latter low dimensionless frequency ranges originating from the surrounding area. The influences could have been wakes formed by obstacles or the blower itself. The components u and v should be more affected by these influences than w, just as they would be for atmospheric wind flow on rough surfaces such as urban morphologies (Yusup, Daud, Zaharim, & Talib, 2008). Figure 7 shows a similar pattern when the VAWT was not within the flow (position A) and when it was within the flow (position C), implying that the spectral characteristics of the upwind flows were unaffected by the VAWT. As mentioned earlier, because Type 2 flow had a larger wind-swept area, the spectral power was approximately constant at high dimensionless frequency values of $0.007 < f < 0.04$.

The observations for position B were controls for the observations for position D (flow after the VAWT). The spectra of position D shown in Figure 8 exhibit interesting features, such as the dip and peak of S_P for Type 1 flow at the dimensionless frequencies and power listed in Table 3. For Type 1 flow, the dip and peak of S_P decreases with the decreasing dimensionless frequency in distinct dimensionless frequency ranges of $0.003 < f < 0.02$ for u and v. As for w, the dimensionless frequency range is in a narrower band, $0.003 < f < 0.005$. The maximum magnitude of S_P for u and v after the dip is on the order of 5 times and 17 times w, respectively. In Figure 8, with the VAWT, and at a normalized dimensionless frequency, $f = 1$, there is a small peak, which is obvious for v, corresponding to the blade passing frequency of 45 rpm. For Type 2 flow, dips only appear at dimensionless frequencies of 0.0025 for u and 0.0035 for v and w. Every 10.7 min and at 0.00156 Hz, Type 1 flow with the VAWT produced a peak in spectral power, while every 1.5 min and at 0.0109 Hz, Type 2 flow with the VAWT produced a dip in spectral power. Hence, the VAWT with Type 2 flow absorbed more kinetic energy than Type 1 flow due to the larger swept area of the Type 2 flow.

Table 3. Maximum and minimum normalized spectral power, S_P, and normalized frequency, f, of u, v, and w of Type 1 and Type 2 flow at position D for speed 3 (high setting)

Type of flow	$\bar{u}$ (m/s)	Wind component	Maximum (peak)		Minimum (dip)	
			f	S_P	f	S_P
		u	0.003	0.1	0.02	0.007
1	3.84	v	0.003	0.02	0.02	0.002
		w	0.003	0.006	0.005	0.003
		u			0.0025	0.0006
2	0.46	v	-	-	0.0035	0.00006
		w			0.0035	0.00025

- data not available.

Based on the S_P values of Figure 6 for positions A and B, the turbulent eddies of u, v, and w dissipated in the low dimensionless frequency ranges at the same distance from the wind source but shifted to the right from upwind position A to downwind position B, indicating dissipation in the high dimensionless frequency range with distance from the wind source. The rates of dissipation at positions A and C with the VAWT were similar for u, v, and w (refer to Figure 7). Referring to Figure 8, the rates of dissipation at position D for the u and v components were in the high dimensionless frequency range (dip) compared to w for Type 1 flow. Meanwhile, the dips observed in Type 2 flow were almost at the same dimensionless frequency range for u, v and w.

The high spectral power at the high dimensionless frequencies observed in the spectra can be explained by the nature of the source: an industrial blower. This blower consists of small spinning rotors that generate high energy eddies at high dimensionless frequencies, as reflected in the spectra given here. As the wind moves downwind, the eddies dissipate and concurrently decrease in size and frequency. The spectral trend and magnitude scatter should be at a minimum because they originate from the same source. Any scatter observed in this experiment could have been caused by external influences that primarily affected the spectral power at lower frequencies, $f < 0.004$ for u and v and $f < 0.0003$ for w (refer to Figure 7), when the industrial blower was used in Type 1 flow. We observed more scatter in the spectral trend for Type 2 flow at lower dimensionless frequencies ($f < 0.0006$) compared to that of Type 1 flow. Considering the low dimensionless frequency "noise", the results show that the VAWT was able to reduce the horizontal turbulent component energies (u and v) in a dimensionless frequency range of 0.003–1. The VAWT introduced general dominant eddy-normalized frequencies of dips and peaks independent of the mean wind speed. These center line eddies of flow had low dimensionless frequencies ($f \approx 0.002$) and were large in size at $L = 2$ m and $\bar{u} = 3.84$ m/s for Type 1 flow at position D: 17 cm for u, 10 cm for v, and 3 cm for w. When the air left the blower's nozzle, the dimensions of u, v, and w were 0.4 cm for Type 1 flow at position C. The eddies in Type 2 flow were smaller compared to those in Type 1 flow; u was 6.4 cm at $f = 0.0025$, and v and w were 1.5 cm and 3.3 cm, respectively at $f = 0.0035$.

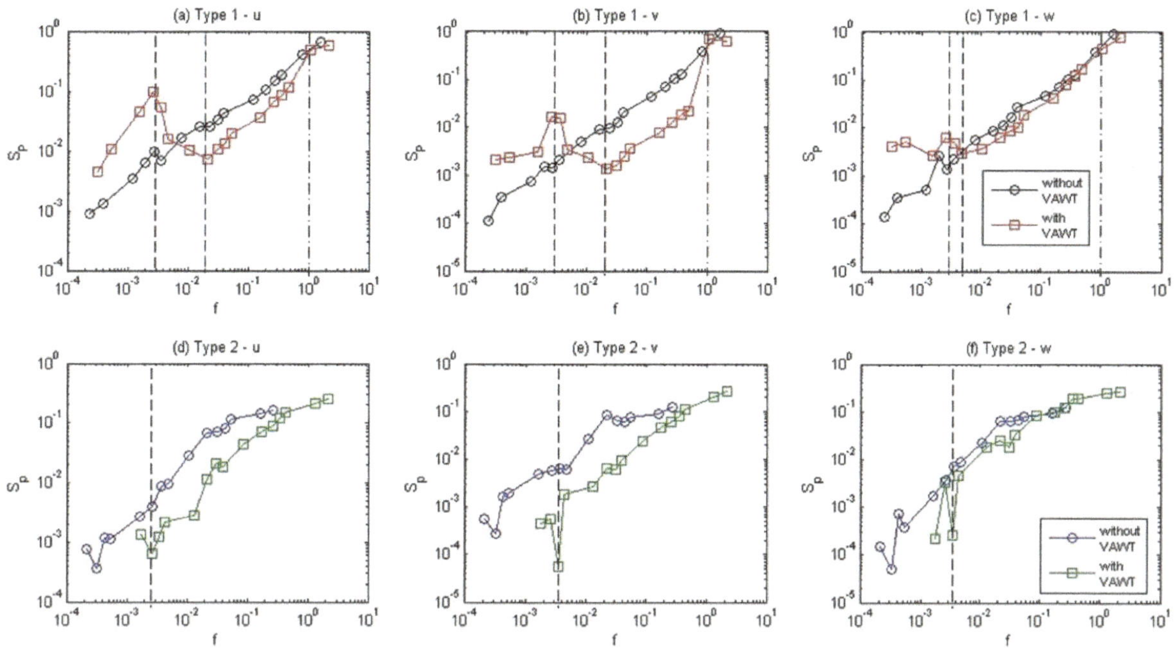

Figure 8. Downwind spectral power (normalized) of components u, v and w for Type 1 and Type 2 flows with and without the VAWT for positions B and D; only "high" or speed 3 shown here; these plots show the dip and peak frequencies induced by the VAWT at dimensionless frequencies marked by the vertical dashed lines; the dashed dot bolded lines show the blade passing frequency; frequency (Hz), n, was normalized by distance/mean wind speed, $L/\bar{u}$, to facilitate comparison

Various eddy sizes at different f values are listed in Table 4, which lists the original size of the eddy at the blower's nozzle, the size upwind of the industrial fans (position C), and the size at the dip and peak frequencies after the VAWT (position D). For the normalized spectra, the high values of σ_v presented in Table 2 cannot be seen here because the values were normalized to compare the distinct features of the spectral characteristics. Figure 9 shows that component v had the highest turbulence intensity, followed by u and w, after removing normalizing factors. Components u and v exhibited similar spectral trends with a distinct set of large eddies at $f = 0.0015$ Hz and small eddies at $f = 0.01$ Hz for u and v. These results reveal a long period pattern within the flow after the VAWT that included large and small eddies by a factor of 10 on the frequency scale.

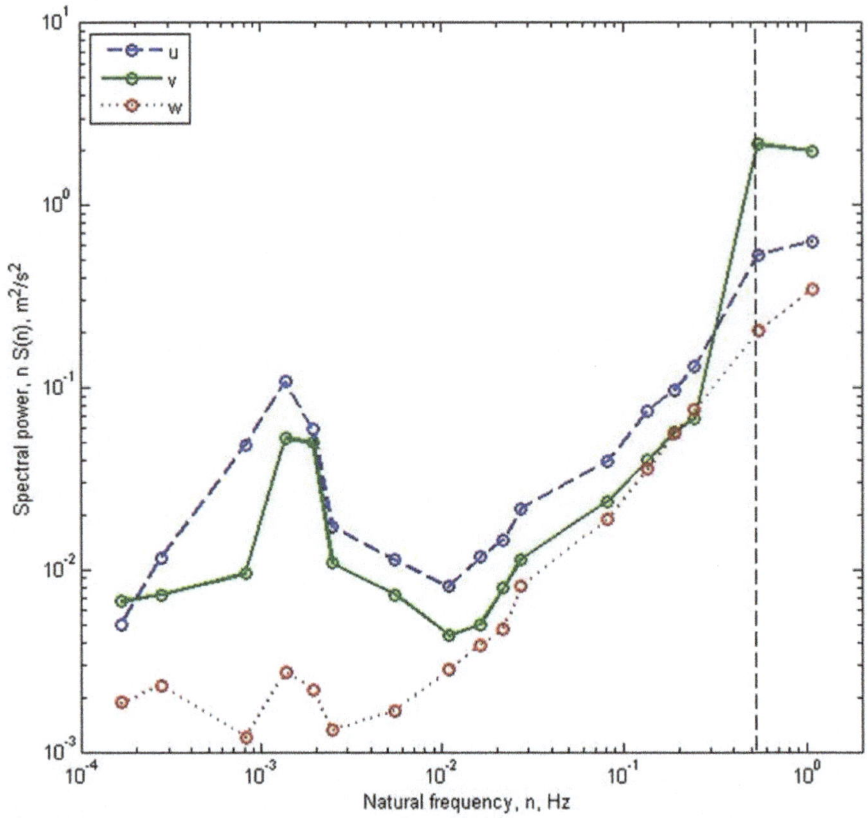

Figure 9. Downwind spectral power of u, v, and w for Type 1 flow, with the VAWT (position D); only "high" or speed 3 shown here; these plots show the dip and peak frequencies induced by the VAWT; the vertical dashed line shows the blade passing a frequency of 0.55 Hz; component v has more turbulence in the high frequency range ($n > 0.4$ Hz) followed by u and w

Table 4. Eddy sizes of wind components u, v, and w for positions C and D with the VAWT and dimensionless frequencies of dips and peaks for Type 1 and Type 2 flows at speed 3 (high setting); at position D, there is an increase in eddy sizes for u, v, and w from dips to peaks for Type 1; position C is listed to show a comparison between the eddy sizes for upwind and downwind flows; note that there are only dips and no peaks at Position D for Type 2 flow

Position (m)	Type of flow	f	Features	$l_{e,u}$ (cm)	$l_{e,v}$ (cm)	$l_{e,w}$ (cm)
C (0.5)	1	0.02	none	1	0.8	0.7
		0.005	none	0.4	0.5	0.5
		0.003	none	0.4	0.4	0.4
D (2.0)	1	0.02	dip (for u and v)	5	3	-
		0.005	dip (for w)	-	-	2
		0.003	peak (for u,v,w)	17	10	3
C (0.5)	2	0.0025	none	2.3	1.3	1.5
		0.0035	none	2.6	1.6	1.8
D (2.0)	2	0.0025	dip (for u)	6.4	-	-
		0.0035	dip (for v,w)	-	1.5	3.3

- data not available.

3.3 Voltage Produced and Turbulence Statistics

In this section, data on the voltage, V, generated at position C is presented. The voltage was proportional only to the rpm of the turbine and not to power production. We observed a positive correlation between $\bar{u}$ and V at position C (refer to Table 5 and Figure 10). A higher wind speed resulted in a higher V; with every increment in $\bar{u}$, the voltage increased by 0.5, in agreement with the function of $V = f(\bar{u})$. The relationship between σ_u and V was independent in this study because no significant trend was observed. σ_u was not correlated with V, and the type of flow (Type 1 and Type 2) did not affect σ_u.

Table 5. Voltage (V), rpm (rotation per minute), and turbulence statistics, $\sigma_{u,v,w}$, for Type 1 and Type 2 flows at positions C (upwind) and D (downwind) with the VAWT for all speed settings, 1–3 (low to high)

Position	Speed	V (V)	rpm range	$\bar{u}$ (m/s)	σ_u (m/s)	σ_v (m/s)	σ_w (m/s)
C (T1)	1	3.4	-	11.8	0.575	0.399	0.361
	2	4.7		12.6	0.598	0.418	0.373
	3	5.9		13.0	0.602	0.422	0.376
D (T1)	1	3.4	-	3.69	1.08	1.70	0.67
	2	4.7		3.55	1.07	1.66	0.74
	3	5.9		3.77	0.90	1.74	0.72
C (T2)	1	5	74–80	2.58	1.45	0.525	0.519
	1	5	74–80	2.63	1.54	0.516	0.522
	1	5	74–80	2.59	1.51	0.527	0.515
	2	7	100–105	3.00	1.59	0.616	0.625
	2	7	100–105	3.02	1.62	0.608	0.627
	3	12	148–154	6.13	1.26	0.486	0.492
	3	12	148–154	6.28	1.13	0.484	0.470
	3	12	148–154	6.21	1.24	0.488	0.491
D (T2)	1	5	74–80	0.428	0.312	0.200	0.196
	1	5	74–80	0.421	0.316	0.196	0.197
	2	7	100–105	0.467	0.388	0.218	0.247
	2	7	100–105	0.474	0.388	0.220	0.246
	3	12	160–163	0.46	0.357	0.204	0.224
	3	12	160–163	0.46	0.359	0.206	0.230
	3	12	160–163	0.46	0.366	0.200	0.228

T1 indicates Type 1 flow (using blower) while T2 indicates Type 2 flow (using industrial fans).

- data not available.

4. Future Work

Because this was a pilot assessment of the turbulence generated by a VAWT using turbulence statistics, further experiments should be conducted in a more controlled environment, such as a wind tunnel, so that the wind speeds are uniform and the turbulence manageable. Furthermore, the apparent speeds should be tested at a higher magnitude than what was done in this study, although this will require a larger wind-swept area. Natural wind should be simulated in the wind tunnel; a peak normalized dimensionless frequency of approximately 1 with turbulence cascading into the higher and lower dimensionless frequencies would give a more realistic scenario of a VAWT operating in the field. In future work, the integral length scale should be studied instead of the eddy size. Multiple sonic anemometers should be installed in three dimensions downwind and upwind of the VAWT at uniform distances to generate a turbulence profile of the wind and its subsequent exhaust following the VAWT.

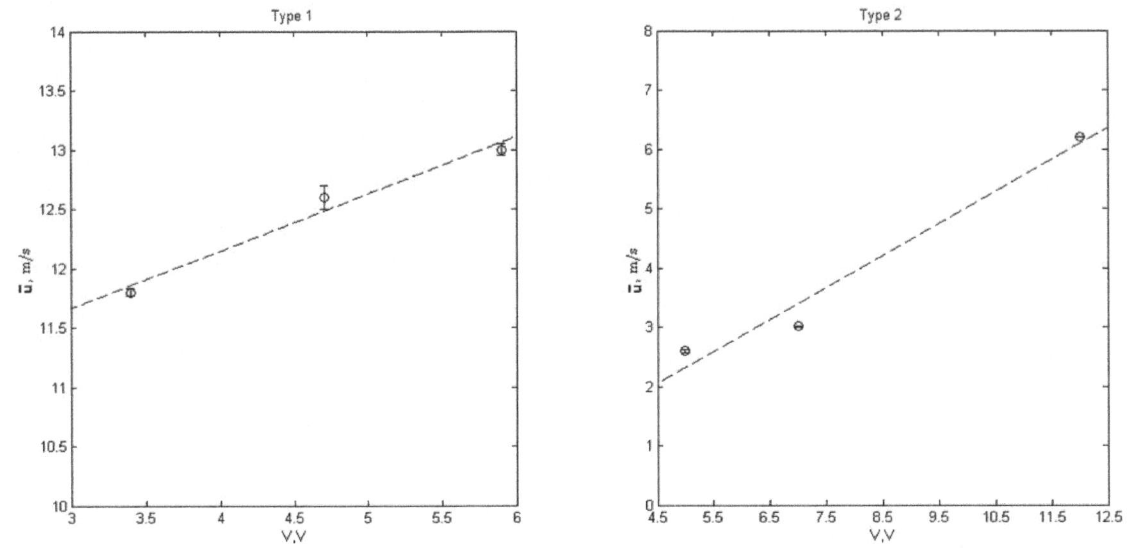

Figure 10. $\bar{u}$ versus Voltage, V for Type 1 and Type 2 flow at position C; dashed lines are best-fit linear regression lines; the slope for both Type 1 and Type 2 flows are approximately the same and equal to 0.5

5. Conclusions

The lateral turbulence, σ_v, in the flow increased downwind of the Darrieus-type VAWT compared to its longitudinal, σ_u, and vertical, σ_w, counterparts. The VAWT was able to reduce the horizontal turbulence component energies (u and v) in a dimensionless frequency range of 0.003–1 and produce two distinct sets of eddies sizes for Type 1 flow: large and small at $f = 0.003$ and $f = 0.02$, respectively. For Type 1 flow, the eddy sizes were large at $L = 2$ m and $\bar{u} = 3.84$ m/s: 17 cm for u, 10 cm for v, and 3 cm for w compared to an initial size at the source of 0.4 cm for all three components. For Type 2 flow, the eddy sizes for u, v and w were smaller at 6.4 cm (at $f = 0.0025$), 2.5 cm and 3.3 cm (at $f = 0.0035$). No peak at position D was observed for Type 2 flow. The VAWT introduced general dominant eddy-normalized dimensionless frequencies of dips and peaks independent of the mean wind speed. A long period pattern was also discovered within the flow after the VAWT that featured large and small eddies differing by a factor of 10 in the frequency scale. Finally, higher wind speeds resulted in higher V generated at a rate of 0.5. No significant trend was observed for the relationship between σ_u and V, but the type of flow did affect σ_u.

Acknowledgements

The research described in this paper was financially supported by the Malaysian Higher Education Institute's Fundamental Research Grant Scheme (FRGS) (grant no.: 203/PTEKIND/6711329) and the Universiti Sains Malaysia (USM) via a short-term grant (grant no.: 304/PTEKIND/6312099).

References

Britter, R. E., & Hanna, S. R. (2003). Flow and dispersion in urban areas. *Annual Review of Fluid Mechanics, 35*, 469-496. http://dx.doi.org/10.1146/annurev.fluid.35.101101.161147

Cao, N., Ting, D., & Carriveau, R. (2011). The performance of a high-lift airfoil in turbulent wind. *Wind Engineering, 35*(2), 179-196.

Christen, A., Rotach, M. W., & Vogt, R. (2009). The budget of turbulent kinetic energy in the urban roughness sublayer. *Boundary-Layer Meteorology, 131*(2), 193-222. http://dx.doi.org/10.1007/s10546-009-9359-5

Dabiri, J. O. (2011). Potential order-of-magnitude enhancement of wind farm power density via counter-rotating vertical-axis wind turbine arrays. *Journal of Renewable and Sustainable Energy, 3*(4), 043104. http://dx.doi.org/10.1063/1.3608170

Ebert, P. R., & Wood, D. H. (1997). The near wake of a model horizontal-axis wind turbine—I. Experimental arrangements and initial results. *Renewable Energy, 12*(3), 225-243. http://dx.doi.org/10.1016/S0960-1481(97)00046-3

Eriksson, S., Bernhoff, H., & Leijon, M. (2008). Evaluation of different turbine concepts for wind power. *Renewable and Sustainable Energy Reviews, 12*(5), 1419-1434. http://dx.doi.org/10.1016/j.rser.2006.05.017

Ferreira, C. J. S., van Bussel, G. J. W., & van Kuik, G. A. M. (2006). Wind tunnel hotwire measurements, flow visualization and thrust measurement of a VAWT in skew. *Journal of Solar Energy Engineering-Transactions of the Asme, 128*(4), 487-497. http://dx.doi.org/10.1115/1.2349550

Kaimal, J. C., Wyngaard, J. C., Izumi, Y., & Coté, O. R. (1972). Spectral characteristics of surface-layer turbulence. *Quarterly Journal of the Royal Meteorological Society, 98*(417), 563-589. http://dx.doi.org/10.1002/qj.49709841707

Lopez, F., & Garcia, M. H. (2001). Mean flow and turbulence structure of open-channel flow through non-emergent vegetation. *Journal of Hydraulic Engineering-Asce, 127*(5), 392-402.

Paulides, J. J. H., Encica, L., Jansen, J. W., Lomonova, E. A., & van Wijck, D. (2009). Small-scale urban venturi wind turbine: Direct-drive generator. *Proceedings IEEE International Electric Machines and Drives Conference, 2009. IEMDC '09, Miami, Florida.* http://dx.doi.org/10.1002/10.1109/IEMDC.2009.5075381

Ragheb, M., & Ragheb, A. M. (2011). Wind Turbines Theory—The Betz Equation and Optimal Rotor Tip Speed Ratio. In R. Carriveau (Ed.), *Fundamental and Advanced Topics in Wind Power* (Chapter 2). USA: InTech. http://dx.doi.org/10.5772/21398

Rotach, M. W. L., Vogt, R., Bernhofer, C., Batchvarova, E., Christen, A., Clappier, A., … Voogt, J. A. (2005). BUBBLE—An urban boundary layer meteorology project. *Theoretical and Applied Climatology, 81*(3-4), 231-261. http://dx.doi.org/10.1007/s00704-004-0117-9

Roth, M. (2000). Review of atmospheric turbulence over cities. *Quarterly Journal of the Royal Meteorological Society, 126*(564), 941-990. http://dx.doi.org/10.1002/qj.49712656409

Sanz, C. (2003). A note on k-epsilon modelling of vegetation canopy air-flows. *Boundary-Layer Meteorology, 108*(1), 191-197. http://dx.doi.org/10.1023/A:1023066012766

Stankovic, S., Campbell, N., & Harries, A. (2009). *Urban Wind Energy*. London: Sterling, VA: Earthscan.

Stitt, F. A. (1999). *Ecological Design Handbook: Sustainable Strategies for Architecture, Landscape Architecture, Interior Design, and Planning*. New York: McGraw-Hill.

Vermeer, L. J., Sorensen, J. N., & Crespo, A. (2003). Wind turbine wake aerodynamics. *Progress in Aerospace Sciences, 39*(6-7), 467-510. http://dx.doi.org/10.1016/S0376-0421(03)00078-2

Wahab, A. A., & Chong, W. T. (12-14 February 2003). Development of an indoor testing facility for low wind speed wind turbine research activities. *Proceedings of the 2nd Regional Conference on Energy Technology Towards a Clean Environment, Phuket, Thailand.*

Wharton, S., & Lundquist, J. K. (2012). Atmospheric stability affects wind turbine power collection. *Environmental Research Letters, 7*(1), 014005.

Yang, Z. F., Sarkar, P., & Hu, H. (2012). Visualization of the tip vortices in a wind turbine wake. *Journal of Visualization, 15*(1), 39-44. http://dx.doi.org/10.1007/s12650-011-0112-z

Yusup, Y. (2012). Aerodynamic drag coefficient over equatorial coastal industrialized and urban areas. *Journal of Wind Engineering and Industrial Aerodynamics, 110*, 25-39. http://dx.doi.org/10.1016/j.jweia.2012.07.006

Yusup, Y. B., Daud, W. R. W., Zaharim, A., & Talib, M. Z. M. (2008). Structure of the atmospheric surface layer over an industrialized equatorial area. *Atmospheric Research, 90*(1), 70-77. http://dx.doi.org/10.1016/j.atmosres.2008.04.003

3

The Improvement of the Wall Temperature Calculation Procedure for the Post-Crisis Region of the NPP Heated Channels

N. D. Agafonova[1] & I. L. Paramonova[1]

[1] Saint-Petersburg State Polytechnic University, St.Petersburg, Russia

Correspondence: N. D. Agafonova, St.Petersburg State Polytechnic University, St.Petersburg, 195251, Russia. E-mail: a-n-d-2011@mail.ru

Abstract

The reliable wall temperature prediction is important for the analysis of operation and safety of both nuclear reactors and direct-flow steam generators with high-temperature fluids. The onset of dangerous temperature excursion is possible in the post-crisis region of the heated channels when the inverted annular or the dispersed flow regime occurs. The wall temperature calculations for both flow regimes are represented. The data of Hammouda, Lapperiere, Chen, Miropolsky, Bennett and some other investigators were processed. The obtained average error and deviation are less than those which take place when special best estimate computer code KORSAR is used. Specified result was reached by the way of the engineering procedures correction and may be applied for the closure correlations improvement which is needed in two-fluid two-phase flow models. The conclusion is made that the correct choice of the integration step and the method of the critical quality determination produce effect on the adequate wall temperature prediction.

Keywords: dispersed flow regime, inverted annular flow regime, heated channel, nuclear reactor, post-crisis region, steam generator, wall temperature

1. Introduction

The thermal physical aspects are very important for the ground of the nuclear power systems under normal and emergency operation as well as the neutron-physical process study.

The reliable wall temperature prediction is urgent for the analysis of operation and safety of both nuclear reactors and direct-flow steam generators with high-temperature fluids. The onset of dangerous temperature excursion is possible in the post-crisis region of the heated channels when the inverted annular or the dispersed flow regime occurs.

The modern ideas about post-crisis heat transfer are based on the recognition of two fundamentally different mechanisms governing the onset of a situation in which a change occurs in the conditions of heat removal from a surface (the "poor" removal of heat by steam emerges instead of the "good" removal of heat by liquid, which leads to a growth of the wall temperature), namely, burnout and dryout. The burnout is connected with a transition from nucleate boiling of liquid to its less effective film boiling, and the dryout involves a disappearance of liquid film on a heating surface and transition from dispersed-annual to purely dispersed flow structure. In Russian scientific literature the post-dryout region is usually called a deteriorated heat transfer region and in foreign scientific literature – dispersed flow film boiling or DFFB.

It may be noticed that the level of the surface heat flux density at the burnout is much greater than at the dryout. The inverted annular flow regime is established at first in the post burnout region (also called inverted annual film boiling or IAFB) at the high wall heat flux density. It is characterized by a continuous liquid core separated from the wall by a vapor film. The wall temperature in the IAFB mode rapidly increases and becomes above Leidenfrost temperature. It is also known (Obot & Ishii, 1988; Nakla, Groeneveld, & Cheng, 2011) that the liquid core quickly collapses with the void fraction increase. The transition from the IAFB to DFFB mode takes place. This case differs from the post-dryout region in the level of wall temperature and, as effect, in the mechanism of the heat removal from the wall by drops: under direct contact of the drops with the surface for the relatively low temperatures or through the vapor layer for the high temperatures.

It is possible to indicate that the correct estimation of heat transfer at the inverted annular flow regime and at the dispersed flow regime, as well as determination of the IAFB zone existence are a key problems for the ground of nuclear power reactors working capacity and safety. The modern computer codes are used for thermal-hydraulic calculations of the nuclear equipment's work.

2. Heat Transfer and the Wall Temperature Calculations in the Inverted Annular Flow Regime

Since burnout is observed at high heat flux densities typical of the cores of water-cooled reactors, the increase of fuel rod cladding temperature may be very high, up to burn-through of the cladding. It should be noted that the bulk temperature of coolant flow may be well below the saturation temperature in this case.

The burnout conditions in the heat removal may occur in a nuclear reactor also at the surface heat loads lower than critical heat flux q_{cr} when the faulty core is flooded with cold water. In this case, with comparatively low heat flux, when the wall temperature exceeds the minimal temperature at which the liquid wets the surface, the occurrence of inverted annual film boiling is being also possible.

The Hammouda model (Hammouda, Groeneveld, & Cheng, 1997) is the only one being used when heat transfer under the conditions of inverted annular mode of two-phase flow can be described in a physically substantiated manner and at negative values of relative enthalpy x_{eq} as well. This model is used in the latest versions of modern computer best estimate codes for in-circuit hydrodynamics and heat transfer, such as the KORSAR and TRAC codes. Some problems connected with the wall temperature calculation when using the Hammouda model are considered by Agafonova & Paramonova (2013). It was shown that the temperature variation across the near-wall vapor layer and therefore the change in the physical properties of vapor may be considerable and produces an essential effect on the heat transfer and drag friction in vapor as compared with their values at constant physical properties of vapor (Petukhov, 1967; Kays & Leunge, 1963). According to our calculations (Agafonova & Paramonova, 2013) for the conditions of the experiments by Chen (2011) the wall temperature (t_w, °C) was nearly two or three times as large as compared with the interfacial surface (or the liquid core surface) temperature which is equal to the saturation temperature (t_s, °C). It is known (Petukhov, 1967; Kays & Leunge, 1963) that for given situation:

$$\frac{Nu}{Nu_{nc}} = \left(\frac{t_w}{t_s}\right)^m \qquad (1)$$

where Nu – the Nusselt number at the varying physical properties; Nu_{nc} – the Nusselt number at the constant physical properties.

In this connexion we offered the modification of the Hammouda's model by means of the temperature factor $\gamma_{IAFB} = t_w/t_s$ introduction (the temperature is expressed in Celsius degrees) and new approximations both for the influence coefficient θ in the range of Reynolds numbers Re_g from 0 to 3×10^5 and for the vapor Nusselt number, which were obtained in this work. These approximations are required since in the original model (Hammouda et al., 1997) $Re_g \leq 9\,900$ and in practice Re_g may be larger. All the necessary formulas in comparison with the original Hammouda model are provided in Table 1.

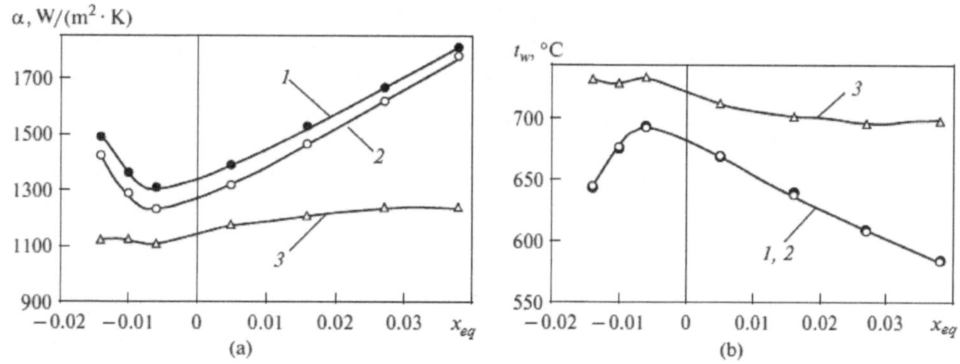

Figure 1. The dependences of the heat transfer coefficient (a) and the wall temperature (b) on the relative enthalpy (experimental data (Lapperiere & Groeneveld, 1984) at p = 10 MPa, ρw= 4 406 kg/(m^2·s), and q_w = 496 kW/m^2: 1 - experimental data, 2 - calculated results of the present study, and 3 - results from calculations with using Hammouda's original model (Hammouda et al., 1997)

Figure 1 shows a comparison between the calculated values of the heat transfer coefficient and the wall temperature with the experimental data obtained by (Lapperiere & Groeneveld, 1984). Introduced corrections in the original Hammouda's model guarantee the best agreement of the mentioned values with the experimental data.

In addition, the estimation of the channel length where the inverted annular flow regime with a smooth or wavy liquid core surface takes place was done according to the recommendations (Obot & Ishii, 1988). This estimation showed that the region with the smooth and wavy liquid core surface may be equal to or less than 0.2 m, but it is in this region that the heat transfer coefficient from the wall to vapor is minimum and the wall temperature is maximum The method under discussion of the wall temperature calculations is valid only in this relatively short region and the given fact has to be taken into account for the correct choice of the integration step to length when the computer codes for thermal-hydraulic prediction of the nuclear equipment's work are used.

Table 1. Calculation formulas for the heat transfer during film boiling determination (IAFB model)

Name of quantity	Hammouda's original model	Hammouda's modified model (the present study)
Heat flux from the wall to the vapor	$q_{wg} = \dfrac{Nu_g}{1-\dfrac{T_g-T_s}{T_w-T_g}\cdot\theta}\cdot\dfrac{\lambda_g}{2\delta}\cdot(T_w-T_g)$	$q_{wg} = \dfrac{Nu_g}{1+\theta\cdot\dfrac{q_{is}}{q_{wg}}}\cdot\dfrac{\lambda_g}{2\delta}\cdot\gamma_{IAFB}\cdot(T_w-T_g)$ or $q_{wg} = Nu_g\cdot\dfrac{\lambda_g}{2\delta}\cdot\gamma_{IAFB}\cdot(T_w-T_g)-\theta\cdot q_{is}$
Nusselt number	$Nu_g = \dfrac{5.071}{\mathrm{Pr}_g^{0.0439}}+0.0028\cdot\mathrm{Re}_g\cdot\mathrm{Pr}_g^{0.645}$	$Nu_g = 5.385+0.021\cdot\mathrm{Re}_g^{0.8}\cdot\mathrm{Pr}_g^{0.4}$
Influence coefficient θ	$\theta=0.347\cdot\mathrm{Pr}_g^{0.0026}-\dfrac{0.146\cdot\mathrm{Re}_g\cdot\mathrm{Pr}_g^{0.5418}}{9\ 900}$	$\theta=0.346-0.0163\cdot\ln\mathrm{Re}_g\cdot\mathrm{Pr}_g$
Auxiliary values	$\mathrm{Re}_g=\dfrac{W_g\cdot 2\delta}{\nu_g}$; $Pr_g=\dfrac{\nu_g}{a_g}$; $\gamma_{IAFB}=\dfrac{t_w}{t_s}$; $\delta=\dfrac{D}{2}\cdot(1-\sqrt{1-\varphi})$; $q_{is}=q_{if}+q_{ev}$; $q_{ev}=\dfrac{dG''\cdot r}{dL\cdot P}$; $q_{if}=0.023\cdot\dfrac{\lambda_f}{(D-2\cdot\delta)}\cdot\mathrm{Re}_f^{0.8}\cdot\mathrm{Pr}_f^{0.4}\cdot(T_s-T_f)$, where W_g – vapor velocity; ν_g – vapor kinematic viscosity coefficient; a_g – vapor thermal diffusivity coefficient; D – diameter of the channel; φ – void fraction	

3. Heat Transfer and the Wall Temperature Calculations in the Dispersed Flow Regime

There is a wide choice of different techniques for the wall temperature estimation in the region of dispersed two-phase flow. It is quite clear that the noticeable diversity takes place when the calculated values of the wall temperature are compared with the experimental data especially at high values of the wall heat flux density q_w. Analysis carried out in the present work confirms this conclusion. The experimental data (Bennett, Hewitt, Kearsey & Keeys, 1967) were processed with KORSAR code to obtain the wall temperature at DFFB mode (Figure 2).

In Figure 2 the length dependence of the wall temperature calculated by Mecholic (2011) with the modified program COBRA-TF is also represented. Both programs give the results fitted well into the experimental data in the dispersed-annular flow region and are not fit well for the dispersed flow. The difficulties of the dispersed flow mathematical modeling deal with the thermodynamic nonequilibrium and the deficiency of reliable recommendations for the droplets and wall interaction.

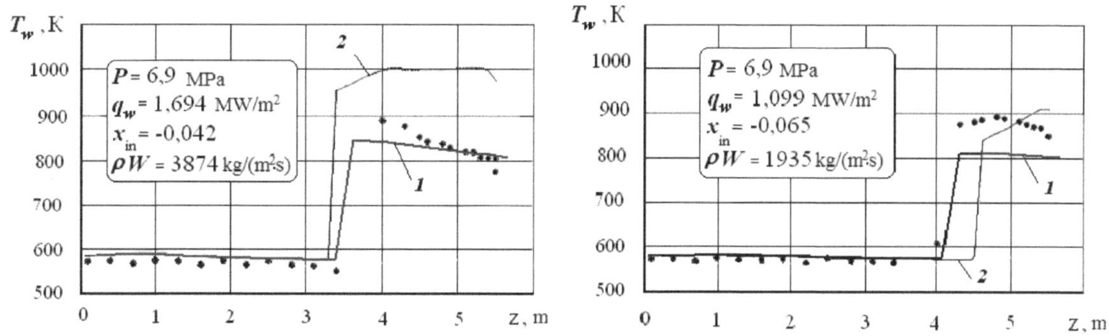

Figure 2. The comparison of the experimental wall temperatures (black points) (Bennett et al., 1967) and calculated values: 1 – with KORSAR code (the present work); 2 – with modified program COBRA-TF (Meholic, 2011)

The engineering (i.e. comparatively simple) model by Hein and Köhler (1986) seems to be the most correct among the nonequilibrium models for the dispersed flow film boiling. The main feature of this model is the determination of the vapor temperature (and then the wall temperature) in terms of the nonequilibrium factor $\omega = x_{act} / x_{eq}$. Hein and Köhler (1986) proposed the following equation for the determination of ω:

$$\omega = \frac{2}{1 + \sqrt{1 + 4 \frac{\overline{C}_{pg}}{r} \frac{q_w}{(\alpha_{gd} A_d / A_w)}}}, \qquad (2)$$

where term $\alpha_{gd} A_d / A_w$, i.e. the product of the heat transfer coefficient α_{gd} between the vapor and the liquid droplets and the area A_d of the evaporating droplets devided by area A_w of the heated surface, is calculated with relationships derived by Hein and Köhler from the experimental data. All formulas needed for use of this model in practice are represented in Table 2. It has to be mentioned that the model by Hein and Köhler gives a rather reliable prediction of the maximum wall temperature values and is recommended to use in VDI Heat Atlas (2010).

The model (Hein & Köhler, 1986) has some limitations to be directly used in the closing correlations system of the modern computer codes based on the two-liquid model of the two-phase flow. One of them is the determination of the effective heat transfer coefficient α_w^{HK} from the wall to the dispersed flow (without dividing in parts: to the vapor and to the liquid droplets). Another one – is the nonequilibrium factor constancy with the increase of the void fraction.

A modified version of Hein-Köhler's model for heat transfer in the dispersed flow mode region is proposed in the present study, central to which is the use of correction in calculating nonequilibrium factor of two-phase flow. The evaluation shows that in the dispersed flow (post-dryout region) the order of mean liquid droplet diameter magnitude is 10^{-6} m and the droplet velocity approximately is equal to the vapor velocity. Under these conditions the droplet breakup is absent and they evaporate without total amount change. This total amount is equal to the corresponding value at the point of dryout.

The correction γ_{DFFB} takes into account the decrease of total droplet surface area under evaporation and may be defined as:

$$\gamma_{DFFB} = 2 \cdot \left(\frac{1 - x_{act}}{1 - x_{act}^{cr}} \right)^{2/3} . \qquad (3)$$

The wall temperature calculations with proposed correction in the relationship (2) for the nonequilibrium factor ω determination are represented in Figure 3. Experimental regime parameters are shown in Table 3. All the necessary formulas are provided in Table 2.

The relative error under comparison of calculated and experimental wall temperatures (the temperature is expressed in Celsius degrees) is less than 12%.

Table 2. Calculation formulas for determining heat transfer during the dispersed flow (DFFB model)

Name of quantity	Original model by Hein-Köhler (1986)	Modified model by Hein-Köhler (the present study)
Heat flux fromthe wall to vapor	$q_{wg} = Nu_g \cdot \dfrac{\lambda_g}{D} \cdot (T_w - T_g)$	
The superheated vapor temperature	$T_g = T_s + \dfrac{r}{\overline{C}_{pg}} \left(\dfrac{1}{\omega} - 1 \right)$	
Nonequilibrium factor in dispersed flow	$\omega = \dfrac{2}{1 + \sqrt{1 + 4 \dfrac{\overline{C}_{pg}}{r} \dfrac{q_w}{(\alpha_{gd} A_d / A_w)}}}$	$\omega = \dfrac{2}{1 + \sqrt{1 + 4 \dfrac{\overline{C}_{pg}}{r} \dfrac{q_w}{(\alpha_{gd} A_d / A_w \cdot \gamma_{DFFB})}}}$, where $\gamma_{DFFB} = 2 \cdot \left(\dfrac{1 - x_{act}}{1 - x_{act}^{cr}} \right)^{2/3}$
The group of parameters $\alpha_{gd} A_d \ / \ A_w$	$\alpha_{gd} \dfrac{A_d}{A_w} = \begin{cases} 1.473 \cdot 10^{-7} \left(\dfrac{\rho W}{L} \right)^{1.33}; & \dfrac{\rho W}{L} < 1767 \cdot 10^3 ; \\ 3.078 \cdot 10^{-24} \left(\dfrac{\rho W}{L} \right)^{4}; & \dfrac{\rho W}{L} \geq 1767 \cdot 10^3 \end{cases}$	
Nusselt number	$Nu_g = \begin{cases} 4{,}36 \ ; & \mathrm{Re}_g \leq 2300 \ ; \\ \dfrac{(\xi/8) \cdot (\mathrm{Re}_g - 10^3) \cdot \mathrm{Pr}_g}{1 + 12.7 \sqrt{\xi/8} \cdot (\mathrm{Pr}_g^{2/3} - 1)} ; & \mathrm{Re}_g > 2300 \end{cases}$	
Friction coefficient	$\xi = (1.82 \cdot \lg \mathrm{Re}_g - 1.64)^{-2}$	
Auxiliary values	$L = \sqrt{\dfrac{\sigma}{g \cdot (\rho' - \rho'')}}$; $\mathrm{Re}_g = \dfrac{\rho w \cdot D}{\rho_g \cdot \nu_g} \cdot \left[x_{act} + (1 - x_{act}) \dfrac{\rho_g}{\rho'} \right]$; $\mathrm{Pr}_g = \dfrac{\nu_g}{a_g}$, where ρW – mass flow rate; ν_g – vapor kinematic viscosity coefficient; a_g – vapor thermal diffusivity coefficient	

4. The Droplet Heat Transfer Determination in the Dispersed Flow Region

The results received for the DFFB mode may be applied for the closure correlations improvement which is needed in two-fluid two-phase flow models.

Consider the total wall heat flux calculated with modified Hein-Köler model:

$$Q_w^{\text{H-K,m}} = \alpha_w^{\text{H-? m}} \cdot A_w \cdot \left(T_w - T_g \right). \tag{4}$$

On the other hand:

$$Q_w^{\text{H-K,m}} = Q_{wg} + Q_{gd} + Q_{wd} + Q_{d,rad}, \tag{5}$$

where Q_{wg} – heat flux from the wall to the vapor; Q_{gd} - heat flux from the vapor to the droplets; Q_{wd} - heat flux from the wall to the droplets; $Q_{d,rad}$ – radiation heat flux to the droplets.

The heat fluxes in the equation (5) are calculated as:

$$Q_{wg} = \alpha_{wg} \cdot (A_w - A_{wd}) \cdot (T_w - T_g) \approx \alpha_{wg} \cdot A_w \cdot (T_w - T_g), \tag{6}$$

where $A_{wd} << A_w$ – the area of the droplets and wall contact; α_{wg} - the heat transfer coefficient from the wall to the vapor; it may be defined according to Miropolsky (1975):

$$\alpha_{wg} = \frac{\lambda_g}{D} \cdot 0{,}028 \cdot \mathrm{Re}_g^{0,8} \cdot \mathrm{Pr}_g^{0,4} \cdot \left(\frac{\rho_{gw}}{\rho_g} \right)^{1,15}, \tag{7}$$

ρ_{gw} - the vapor density calculated at the wall temperature;

$$Q_{gd} = \alpha_{gd} \cdot A_d \cdot (T_g - T_s), \tag{8}$$

where α_{gd} - the heat transfer coefficient from the vapor to the droplets;

$$Q_{wd} = \alpha_{wd} \cdot A_{wd} \cdot (T_w - T_s), \tag{9}$$

where α_{wd} - the heat transfer coefficient from the wall to the droplets;

$$Q_{d,rad} = 0, \tag{10}$$

if the wall temperature is less than ~650°C and may be defined according to Sun, Gonzales-Santalo and Tien (1976) in the opposite case.

Table 3. Experimental regime parameters (Figure 3-4)

References	P, MPa	ρW, кг/(м^{2}с)	q_w, MW/m^2	x_{in}	Regime designation
(Nilsson, 1993)	3	1 503	0.77	-0.024	a
(Nilsson, 1993)		1 010	0.67	-0.026	b
(Bennett et al., 1967)	6.9	1 935	1.1	-0.065	c
		1 003	0.79	-0.08	d

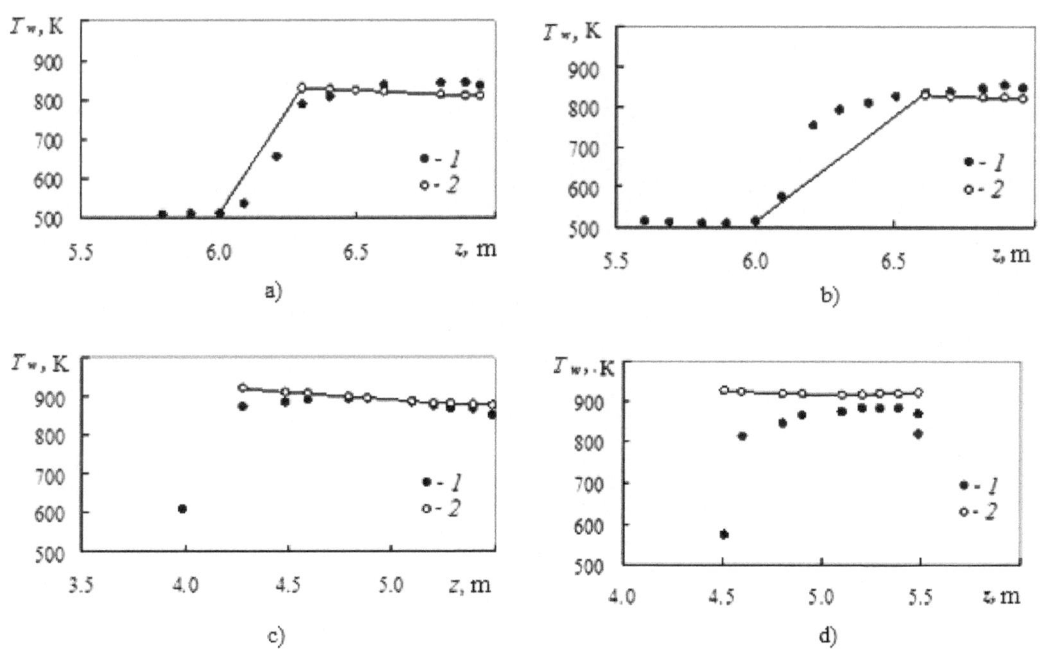

Figure 3. The comparison of the experimental (1) and calculated in the present study (2) wall temperature values: a) and b) - (Nilsson, 1993); c) and d) - (Bennett et al., 1967); (all regime parameters see in the Table 3)

After substitution all formulas for the heat fluxes in the equation (5) the effective heat transfer to the droplets α_d^{eff} may be defined as:

$$\alpha_d^{eff} = \alpha_{gd} \cdot \frac{A_d}{A_w} + \alpha_{wd} \cdot \frac{A_{wd}}{A_w} \cdot \frac{T_w - T_s}{T_g - T_s} + \frac{Q_{d,rad}}{(T_g - T_s) \cdot A_w} = (\alpha_w^{\text{H-K,m}} - \alpha_{wg}) \cdot \frac{T_w - T_s}{T_g - T_s}, \qquad (11)$$

where $\frac{A_d}{A_w} = 0.9 \cdot \frac{D}{d} \cdot (1 - \varphi)$ according to Ransom et al. (1985), d – droplet diameter.

This value is represented in Figure 4 as a function of the equilibrium relative enthalpy x_{eq} in the post-dryout region.

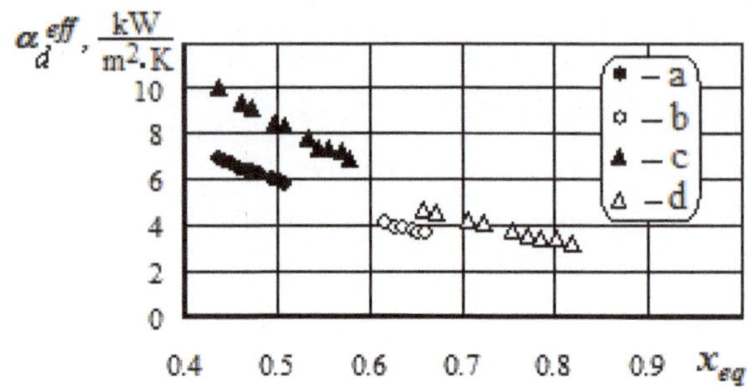

Figure 4. Effective heat transfer to the droplets α_d^{eff} as a function of the equilibrium relative enthalpy x_{eq} in the post-dryout region (the experimental regime parameters and the regime designation the same as in Figure 3 and Table 3)

It is seen that α_d^{eff} tends to zero as x_{eq} rises and void fraction tends to 1. Thus the transition to the heat removal from the wall by the superheated vapor takes place.

Later the value of $\alpha_d^{eff} - \frac{Q_{d,rad}}{(T_w - T_s) \cdot A_w}$ may be partitioned among the components of the relation (11), accounted for the droplet heat transfer with the vapor and the wall. Since today there are no reliable data observed experimentally for the droplets and wall interaction, values of the complex $\alpha_{wd} \cdot (A_{wd} / A_w)$ may be obtained if the proper recommendations for the droplet diameter in the dryout cross section and heat transfer between the droplets and the vapor will be assigned. Early decoding showed that the formula by Chen (2011) for the droplet diameter d

$$\frac{d}{d_{cr}} = \left(\frac{1 - x_{act}}{1 - x_{act}^{cr}}\right)^{1/3}; \quad d_{cr} = 798 \cdot \frac{\rho_g}{(\rho W) \cdot x_{act}^{cr}} \cdot \left(\frac{\sigma}{\rho'}\right)^{0.5} \cdot \left(\frac{\mu_g}{(\rho W) \cdot x_{act}^{cr} \cdot D}\right)^{a}; \qquad (12)$$

$a = 0.63 - 0.1 \cdot p$ for $p < 1.3$ MPa; $a = 0.5$ for $p \geq 1.3$ MPa and the correlation by Whitaker (1972) for the droplets and the vapor heat transfer

$$\alpha_{gd} = \frac{\lambda_g}{d_d} \cdot \left[2 + \left(0.4 \cdot \text{Re}_d^{1/2} + 0.06 \cdot \text{Re}_d^{2/3}\right) \cdot \text{Pr}_g^{2/5} \cdot \left(\frac{\mu_g}{\mu_{gw}}\right)^{1/4}\right]; \qquad (13)$$

$$\text{Re}_d = \frac{\rho_g \cdot W_R \cdot d_d}{\mu_g}; \quad 3.5 \leq \text{Re}_d \leq 7.6 \cdot 10^4; \quad \text{Pr} = 0.7\text{–}380$$

conform to this aim.

With the help of this approach the separated values for the wall-droplets and vapor-droplets heat transfer coefficients may be obtained as it is necessary for the two-liquid two-phase models.

5. Conclusions

The present study is devoted to the reliable wall temperature prediction in the post-crisis region of the heated channels when the inverted annular or the dispersed flow regime occurs. The corrections to the rather simple engineering models for the wall temperature calculation are proposed.

The introduction of the temperature factor $\gamma_{IAFB} = t_w/t_s$ in the original Hammouda's model (Hammouda et al., 1997) (inverted annular flow regime) with new approximations for the influence coefficient θ in the range of the Reynolds numbers Re_g from 0 to 3×10^5 and for the vapor Nusselt number, which were obtained in this work, guarantee a better agreement of the calculated wall temperature values with experimental data.

A modified version of Hein-Köhler's model for heat transfer in the dispersed flow mode region ensures the better coincidence of the calculated temperature curve in the post-crisis region with experimental data than original one due to the use of correction in calculating nonequilibrium factor of two-phase flow.

The wall temperature calculations for the post-crisis flow regimes are represented. The data of Hammouda, Lapperiere, Miropolsky, Bennett and some other investigators were processed. The obtained average error and deviation for IAFB are less than those which take place when special best estimate computer code KORSAR is used. The model by Hein and Köhler enables only the effective heat transfer coefficient $\alpha_w^{\text{H-K,m}}$ from the wall to the dispersed flow to be determined (without dividing in parts: to the vapor and to the liquid droplets). Therefore this model offered to be used for the heat transfer coefficient between the droplets and the wall determination.

Specified results may be applied for the closure correlations improvement which is needed in two-fluid two-phase flow models. Also the conclusion is made that the correct choice of the integration step and the method of the critical quality determination produce effect on the adequate wall temperature prediction.

References

Agafonova, N. D., & Paramonova, I. L. (2013). Heat Transfer in Inverted Annular Mode of Steam–Water Flow. *Thermal Engineering, 60*(3), 176-181. http://dx.doi.org/10.1134/s0040601513030014

Bennett, A. W., Hewitt, G. F., Kearsey, H. A., & Keeys, R. K. F. (1967). Heat transfer to steam-water mixtures flowing in uniformly heated tubes in which the critical heat flux has been exceeded. *United Kingdom Atomic Energy Authority, AERE-R5373.* http://dx.doi.org/10.1243/pime_conf_1967_182_237_02

Chen, Y. (2011). Heat Transfer in Film Boiling of Flowing Water, in: *Heat transfer theoretical analysis, experimental investigations and industrial systems*, edited Aziz Belmiloudi, InTech, 235-260. http://www.intechopen.com/books/heat-transfer-theoretical-analysis-experimental-investigations-and-industrial-systems. http://dx.doi.org/10.5772/14474.

Hammouda, N., Groeneveld, D. C., & Cheng, S. C. (1997). Two-fluid modeling of inverted annular film boiling. *Int. J. Heat Mass Transfer, 40*(11), 2655-2670. http://dx.doi.org/10.1016/s0017-9310(96)00278-5

Kays, W. M., & Leunge, E. V. (1963). Heat transfer in annular passages - Hydrodynamically developed turbulent flow with arbitrarily prescribed heat flux. *Int. J. Heat Mass Transfer, 6*(7), 537-557. http://dx.doi.org/10.1016/0017-9310(63)90012-7

Köhler, W., & Hein, D. (1986). Influence of the wetting state of a heated surface on heat transfer and pressure loss in an evaporator tube. *IAR, NUREG/IA-0003.*

Lapperiere, A., & Groeneveld, D. C. (1984). A Study of Low Quality Film Boiling at High Pressures and Flows. *AIChE Symp. Series 236*(80), 440-445.

Meholic, J. M. (2011). *The development of a non-equilibrium dispersed flow film boiling heat transfer modeling package* (Ph. D. thesis, The Pennsylvania State University, USA). Retrieved from https://etda.libraries.psu.edu/paper/11651

Miropolsky, Z. L. (1975). The Superheated Vapor Heat Transfer with Supply and Removal of Heat. *Teploenergetika, 3,* 75-78 [in Russian].

Nakla, M. E., Groeneveld, D. C., & Cheng, S. C. (2011). Experimental Study of Inverted Annular Film Boiling in a Vertical Tube Cooled by R-134a. *Int. J. Multiphase Flow, 37*(1), 67-75. http://dx.doi.org/10.1016/j.ijmultiphaseflow.2010.08.006

Nilsson, L. (1993). Assessment of RELAP5/MOD3 against twenty-five post-dryout experiments performed at the Royal Institute of Technology. *NUREG/IA-0094. STUDS VIK/NS-90/93*, Sweden.

Obot, N. T., & Ishii, M. (1988). Two-Phase Flow Regime Transition Criteria in Postdryout Region Based on Flow Visualization Experiments. *Int.J. Heat Mass Transfer, 31*(12), 2559-2570. http://dx.doi.org/10.1016/0017-9310(88)90182-2

Petukhov, B. S. (1967). *Heat Transfer and Pressure Drop during Laminar Flow in Tubes.* Moscow: Energiya [in Russian].

Ransom, V. H. (1995). RELAP5/MOD2, Code Manual, Volume 1: Code structure, systems models and solution methods, *NUREG/CR-5535*. http://dx.doi.org/10.2172/105079

Sun, K. H., Gonzales-Santalo, J. M., & Tien, C. L. (1976). Calculations of Combined Radiation and Convection Heat Transfer in Rod Bundles under Emergency Cooling Conditions. *Trans.ASME, J. of Heat Transfer, 98*(3), 414-420. http://dx.doi.org/10.1115/1.3450569

VDI Heat Atlas. (2010). Springer-Verlag Berlin Heidelberg. P. 1585. http://dx.doi.org/10.1007/978-3-540-77877-6

Whitaker, S. (1972). Forced Convection Heat Transfer Correlations for Flow in Pipes, Past Flat Plates, Single Cylinders, Single Spheres and for Flow in Packed Beds and Tube Bundles. *AIChE Journal, 18*(2), 361-371. http://dx.doi.org/10.1002/aic.690180219

4

Determination of Preload of Double-Row Tapered Roller Bearing Used for Supporting Direct-Drive Wind Turbine Rotor

Yunfeng Li[1]

[1] School of Mechatronical Engineering, Henan University of Science and Technology, Luoyang, China

Correspondence: Yunfeng Li, 66#, Xiyan Campus, Henan University of Science and Technology, 48 Xiyan Road, Luoyang 471003, China. E-mail: liyunfeng379@126.com

Abstract

Appropriate axial preload is necessary for the double-row tapered roller bearing used for supporting the rotor of the direct-drive wind turbine; its function is to ensure the rolling motions of the rollers and the long fatigue life of the bearing as far as possible. For this purpose, statics model of the preloaded bearing under the combined action of radial load, axial load and tilting moment load was established firstly; then, for a set of selected preload values which are different proportions of the dynamic equivalent axial load of the external bearing loads, the corresponding loaded roller number, maximum roller load and bearing fatigue life were obtained; thirdly, the effects of different preloads on the calculated indicator values were analyzed, result show that preload can improve the uneven load distribution among the rollers, the preload value also influence the rolling roller number and bearing fatigue life. A preload of 0.5 times of the dynamic equivalent axial load was selected as a trade-off between the rolling roller number and bearing fatigue life.

Keywords: wind turbine, double-row tapered roller bearing, statics analysis, axial preload, rolling motion, fatigue life

1. Introduction

In spite of the global economic depression, the worldwide wind capacity reached 254000 MW by the end of June 2012, out of which 16546 MW were added in the first six months of 2012 (World Wind Energy Association, 2012). The reliability of wind turbines has attract more and more attention from the researcher, bearings are the key components of wind turbines, they need to be pay more attention as they have higher costs associated with repair or replacement (Kotzalas & Doll, 2010). In recent years, some research work of the mechanics problems of slewing bearing used for wind turbine have been conducted by the researchers. Such as the calculation of the static load-carrying capacity of four-contact-point slewing bearings used for the pitching system and yawing system of wind turbine (Aguirrebeitia, Plaza, Abasolo, & Vallejo, 2013), the contact fatigue damage of hardened layer of bearing raceway used for the yaw system in wind turbine, this problem was solved by constructing the nonlinear material constitutive equation of hardened layer (Niu, Yang, & Gao, 2013), and the effect of local structure change on the fatigue life of yaw slewing bearing of wind turbine, this problem was analyzed by the FEM software (Feng, Chen, Huang, & Wang, 2013), all the present work mainly aimed at the slewing bearings used for the pitching system and yawing system of the double-fed induction wind turbine.

Direct-drive wind turbine is a new turbine architecture which has been developed in recent years, this kind of wind turbine eliminate the gearbox and connect the rotor directly to a permanent magnet generator. Direct-drive wind turbine offers significant potential because they eliminate the gear-speed increaser, which is susceptible to significant accumulated fatigue torque loading, related reliability issues, and maintenance costs (Bywaters et al., 2007).

The rotor of direct-drive wind turbine is supported by one set of slewing bearing. Double-row tapered roller bearing is deemed as a more suitable selection to support the rotor of direct-drive wind turbine (Lucas & Pontius, 2005), because this kind of bearing can carry the combined actions of radial load, axial load and tilting moment load; and their design is built around the concept of zero slip which minimizes wear over long periods of operation; especially their load distribution among the rollers can be optimized to avoid load losing or exceeding of some rollers through appropriate axial preload. The cross section of double-row tapered roller bearing is shown in Figure1. In the direct-drive wind turbine, the outer ring of the bearing is connected to the nacelle main frame, and the inner ring of the bearing is connected to the rotor. During the operating process, the bearing sustains the axial load which comes from the aerodynamic load from the rotor, the tilting moment load which comes from the aerodynamic load from the rotor and the weight of the rotor, the radial load form the weight of the rotor mainly.

Axial preload is needed by double-row tapered roller bearing to make the load be well distributed among the two row rollers, its aim is to ensure nearly all rollers to carry load and roll autonomously. As the preload is insufficient, the action of external axial load will result in the unloading of some rollers. Because the normal contact loads between the unloaded rollers and the raceway equal to zero, the raceway surface of the rotating ring can not drag the unloaded rollers to orbit through rolling motion by frictional traction. Slipping frictions appear between the unloaded rollers and the raceway surface; friction heat will lead to the premature failure of the raceway and the cage. While as the preload is excessively large, the combining action of the preload and the external load can result in the exceeding of the roller loads over the rated load, and this will result in heat generation and fatigue failure of the bearing. Therefore, it is necessary to understand deeply the influence laws of preload on the load distribution of the bearing rollers and determine the reasonable value of the preload based on these laws further.

The intend of this paper is to establish the static model of axial preloaded double-row tapered roller bearing used for supporting the rotor of direct-drive wind turbine, and analyze the effect of axial preload on the loaded roller number, maximum roller load and bearing fatigue life; use these analyzed results to determine the appropriate axial preload.

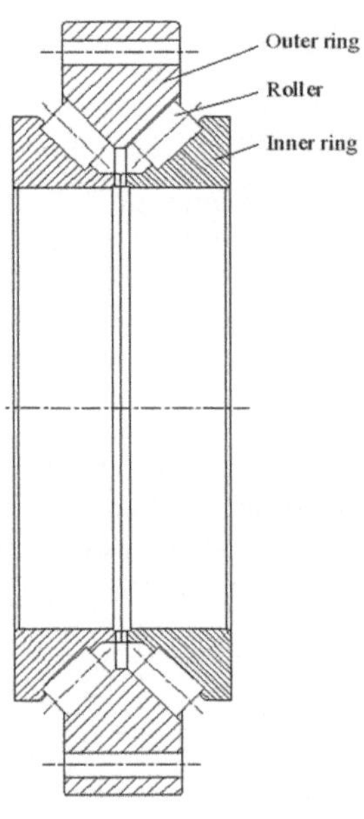

Figure 1. Double-row tapered roller bearing

2. Force Analysis for Tapered Roller

In the tapered roller bearings, in addition to bear the normal loads from the inner raceway and outer raceway, the tapered roller also bear the normal loads from the guide flange of the inner ring. Suppose the acting force between the roller and the inner raceway, outer raceway, guide flange is Q_i, Q_e, Q_f respectively; and the corresponding contact angle is α_i, α_e, α_f respectively, as shown in Figure 2.

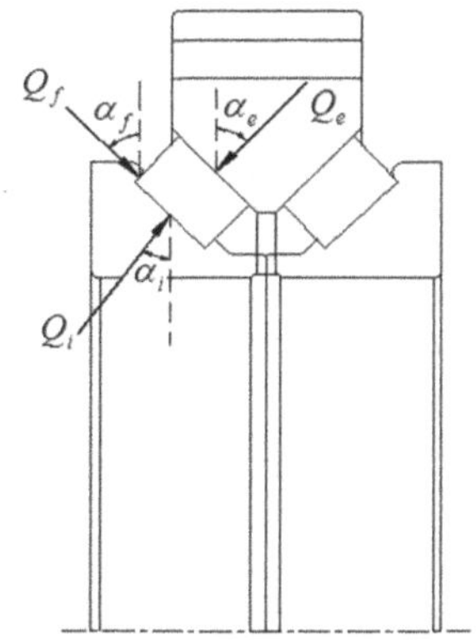

Figure 2. Forces of tapered roller

For the static equilibrium of the tapered roller, the sum of forces in any direction is equal to zero, therefore,

$$Q_e \sin\alpha_e - Q_i \sin\alpha_i - Q_f \sin\alpha_f = 0 \tag{1}$$

$$Q_e \cos\alpha_e - Q_i \cos\alpha_i + Q_f \cos\alpha_f = 0 \tag{2}$$

For the systems of equations composed by Equation (1) and Equation (2), suppose Q_e, α_i, α_e and α_f are known variables, Q_i and Q_f are unknown variables. Then Q_i and Q_f can be represented by other variables through solving the systems of equations, that is

$$Q_i = \frac{\sin(\alpha_e + \alpha_f)}{\sin(\alpha_i + \alpha_f)} Q_e \tag{3}$$

$$Q_f = \frac{\sin(\alpha_e - \alpha_f)}{\sin(\alpha_i + \alpha_f)} Q_e \tag{4}$$

Between the roller and the outer raceway, only normal force Q_e exists, so in the statics analysis of tapered roller bearings, the rollers and the inner ring are looked as an isolated body to consider its equilibrium problem. The rollers and the inner ring together are called "*roller-inner ring*" isolated body in the following.

3. Load-Deflection Relationship

When relative displacement between the inner ring and the outer ring emerges, because the contact angles of the inner raceway and the outer raceway are different, the elastic contact deformation between the roller and the inner raceway is not in the same direction as it between the roller and the outer raceway.

Here, the elastic contact deformation direction between the roller and the outer raceway is looked as a benchmark, project the elastic contact deformation direction between the roller and the inner raceway onto the elastic contact deformation direction between the roller and the outer raceway; the total approach between two raceways under load separated by a roller is the sum of the elastic contact deformations between the roller and each raceway. In this direction, the total load-deflection factor is (Lundberg & Palmgren, 1947)

$$K_{ne} = 8.06 \times 10^4 l^{0.89} \left[1 + c_i^{0.9} \cos(\alpha_e - \alpha_i)\right]^{-1.11} \tag{5}$$

in which

$$c_i = \sin(\alpha_e + \alpha_f) / \sin(\alpha_i + \alpha_f)$$

Then, the normal contact load between the roller and the outer raceway is

$$Q_e = K_{ne} \delta_n^{1.11} \tag{6}$$

4. Equilibrium Model of Bearing Under Axial Preload

After the bearing is mounted on the host machine, the axial preload is applied on the inner ring. As the bearing being analyzed, suppose the outer ring of the bearing is fixed. For the convenience of analysis, subscript "1" is used to represent the row adjacent to the rotor, and subscript "2" represent the row adjacent to the nacelle.

The axial displacement between the rotor side row inner ring and the outer ring is δ_{a0}, the elasticity contact deformation along the normal direction of outer raceway produced by this axial displacement is

$$\delta_{ne}^{0} = \delta_{a0} \sin \alpha_e \tag{7}$$

Substituting Equation (7) into Equation (6), the contact load Q_e between the roller and the outer raceway can be obtained. Then the equilibrium equation of the rotor side row "*roller – inner ring*" isolated body is

$$F_{a0} - ZQ_e \sin \alpha_e = 0 \tag{8}$$

Where, Z is the roller number of each row. From Equation (8), one can obtain the axial displacement of the rotor side inner ring relative to the outer ring under the action of axial preload as follows:

$$\delta_{a0} = \left(\frac{F_{a0}}{ZK_{ne} (\sin \alpha_e)^{2.11}} \right)^{0.9} \tag{9}$$

Similarly, the axial displacement between the nacelle side row inner and the outer ring is δ_{a0} too under the action of axial preload F_{a0}.

5. Equilibrium Model of Bearing Under Combined Loads

Suppose the outer ring of the bearing is fixed, under the combined action of radial load F_r, axial load F_a and tilting moment load M, the corresponding radial displacement, axial displacement and angular displacement of the bearing inner ring is δ_r, δ_a and θ respectively. The "*roller-inner ring*" isolated body sustains the action of external loads F_r, F_a and M, it also sustains the normal loads Q_e acting on the roller from the outer raceway, as shown in Figure 3. Under the actions of these loads, the "*roller-inner ring*" isolated body is in an equilibrium state.

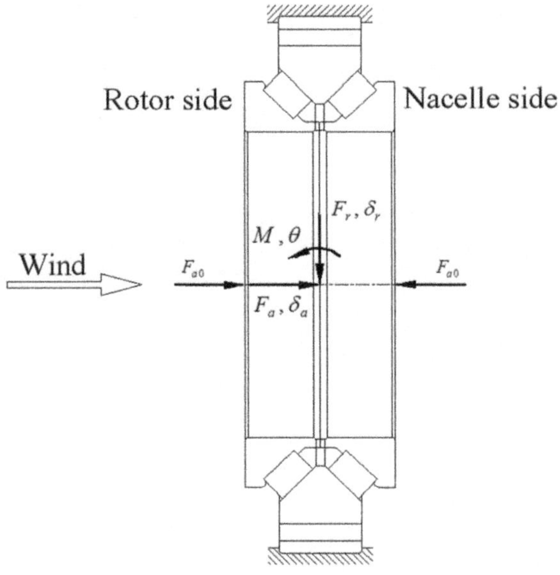

Figure 3. Displacements of inner ring under combined loads

5.1 Displacement of Each Roller Position of Rotor Side Row

Firstly, consider the rotor side row rollers, the axial displacement at the position of *j*th roller produced by the axial displacement δ_a is equal to δ_a, while the radial displacement at the position of *j*th roller produced by the radial displacement δ_r is

$$\delta_{r1(j)}^{r} = \delta_r \cos\psi_{(j)} \tag{10}$$

Where, $\psi_{(j)}$ is the position azimuth angle of the roller along the raceway. The radial displacement at the position of *j*th roller produced by the tilting angular displacement θ is

$$\delta_{r1(j)}^{\theta} = 0.5 d_c \theta \cos \psi_{(j)} \tag{11}$$

Where, d_c is the axial distance between two row roller centers. The axial displacement at the position of *j*th roller produced by the tilting angular displacement θ is

$$\delta^{\theta}_{a1(j)} = 0.5 d_m \theta \cos\psi_{(j)} \tag{12}$$

Where, d_m is the pitch diameter of the bearing. The radial displacement at the position of each roller is produced by the radial displacement δ_r and the tilting angular displacement θ. According to Equation (10) and Equation (11), the total radial displacement at the position of *j*th roller is

$$\delta_{r1(j)} = (\delta_r + 0.5 d_c \theta)\cos\psi_{(j)} \tag{13}$$

For the axial displacement at the position of each roller, in addition to the axial displacement produced by the axial displacement δ_a and the tilting angular displacement θ, the axial displacement produced by the axial displacement δ_{a0} should be also considered. Then, the total axial displacement at the position of *j*th roller is

$$\delta_{a1(j)} = \delta_{a0} + \delta_a + 0.5 d_m \theta \cos\psi_{(j)} \tag{14}$$

Then, the total displacement at the position of *j*th roller along the normal direction of the outer raceway is

$$\delta_{n1(j)} = \delta_{r1(j)} \cos\alpha_e + \delta_{a1(j)} \sin\alpha_e \tag{15}$$

5.2 Displacement of Each Roller Position of Nacelle Side Row

According to the same principle as above, for the nacelle side row rollers, the axial displacement at the position of *j*th roller produced by the axial displacement δ_a is equal to $-\delta_a$.

The total radial displacement at the position of *j*th roller is

$$\delta_{r2(j)} = (\delta_r - 0.5 d_c \theta)\cos\psi_{(j)} \tag{16}$$

Similarly, the total axial displacement at the position of *j*th roller is

$$\delta_{a2(j)} = \delta_{a0} - \delta_a - 0.5 d_m \theta \cos\psi_{(j)} \tag{17}$$

The total displacement at the position of *j*th roller along the normal direction of the outer raceway contact is

$$\delta_{n2(j)} = \delta_{r2(j)} \cos\alpha_e + \delta_{a2(j)} \sin\alpha_e \tag{18}$$

5.3 Equilibrium Equations of "Roller-Inner Ring" Isolated Body

Substituting Equation (15) and Equation (18) into Equation (6) respectively, the expressions of normal load $Q_{e1(j)}$ and $Q_{e2(j)}$ between the two row rollers and the outer raceway can be obtained, where, $j=1,2,3, \ldots, Z$.

In the radial direction of the bearing, the equilibrium equation of the "*roller – inner ring*" isolated body under the actions of external radial load F_r and normal $Q_{e1(j)}$ and $Q_{e2(j)}$ is

$$F_r - \sum_{j=1}^{Z} \left(Q_{e1(j)} + Q_{e2(j)}\right)\cos\alpha_e \cos\psi_{(j)} = 0 \tag{19}$$

In the axial direction of the bearing, the equilibrium equation of the "roller – inner ring" isolated body under the actions of external axial load F_a and normal $Q_{e1(j)}$ and $Q_{e2(j)}$ is

$$F_a - \sum_{j=1}^{Z} \left(Q_{e1(j)} - Q_{e2(j)}\right)\sin\alpha_e = 0 \tag{20}$$

The normal load $Q_{e1(j)}$ acting on the rollers by the outer raceway will produce moment actions on the "*roller-inner ring*" isolated body. One part of them is the moment produced by the axial component of load $Q_{e1(j)}$ on the bearing center, another part of them is the moment produced by the radial component of load $Q_{e1(j)}$ on the bearing center. The equilibrium equation of the "*roller-inner ring*" isolated body under the actions of the two part moments and external tilting moment load M is

$$M - 0.5(d_m \sin\alpha_e + d_c \cos\alpha_e)\left(\sum_{j=1}^{z} Q_{e1(j)} \cos\psi_{(j)} - \sum_{j=1}^{z} Q_{e2(j)} \cos\psi_{(j)}\right) = 0 \tag{21}$$

A system of nonlinear equations can be obtained by using the above three equilibrium Equations (19)-(21). If the design parameters of the bearing are given, the values of unknown variable δ_r, δ_a and θ can be solved corresponding a set of axial preload F_{a0} and external loads F_r, F_a and M, and the roller loads can be calculated further.

6. Example Calculation and Analysis

The design parameters of the rotor bearing of certain type 2.0 MW direct drive wind turbine are listed as in Table 1, and the external fatigue loads are F_r= 451 (kN), F_a= 544 (kN), M=1072 (kNm). According to T. Harris (Harris, Rumbarger, & Butterfield, 2009), the dynamic equivalent axial load of the bearing can be calculated by using these external fatigue loads, the calculated value is as follows: F_{ae}= 1941 (kN). In this paper, the value of F_{ae} is used as a baseline for determining the axial preload of the bearing. As the axial preload equals to the different proportion of F_{ae}, the results calculated are shown in Figure 4-Figure 8.

Table 1. Bearing design parameters

d_m (mm)	d_c (mm)	α_i (°)	α_e (°)	α_f (°)	l_{we} (mm)	D_{we} (mm)	ϕ(°)	Z
2026	116	43	45	46	90	51	1	104

The uneven distribution of roller loads can be attributed to the actions of external axial load and tilting moment load, axial load mainly make the uneven distribution between two rows. Uneven distribution of roller loads can result in the roller slipping, heat generation and fatigue life decrease. Axial preload can improve the uneven load distribution of the rollers. With the increase of axial preload, more and more rollers begin to undertake the bearing external loads, this can be seen in Figure 4 and Figure 5. When the preload is less than $0.6F_{ae}$, the normal contact loads between some rollers and the raceway are equal to zero, these rollers can not roll by themselves, their orbiting are realized by the push of the cage, this is disadvantageous to the bearing. When the preload is equal to $0.6F_{ae}$, the normal contact loads of all rollers of the two rows are nonzero, they all can realize pure rolling, this is advantageous for the bearing to reduce friction and heat generation. When the axial preload is larger than $0.6F_{ae}$, with the increase of the preload, the roller loads begin to increase distinctly; this is disadvantageous to the bearing; because excessive large roller load can result in the premature fatigue failure of the raceways. From the viewpoint of slipping friction and heat generation, the preload of $0.6F_{ae}$ is most favorable for the bearing.

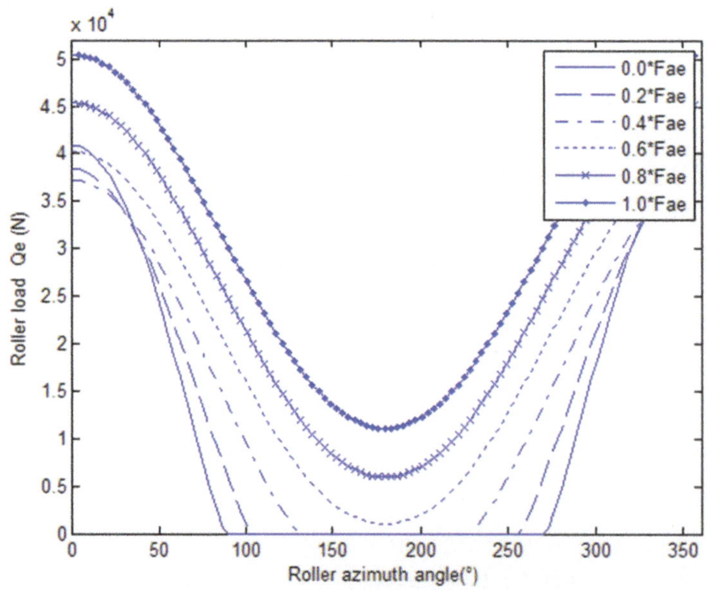

Figure 4. Load distribution of rotor side row

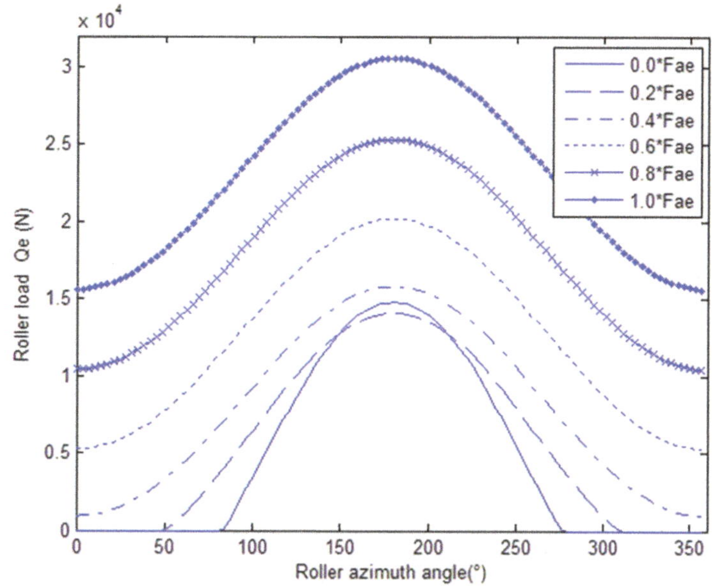

Figure 5. Load distribution of nacelle side row

The effect of the preload on the fatigue life of the bearing is shown in Figure 6. With the increase of the preload from 0 to $0.3F_{ae}$, the bearing fatigue life increase accordingly; this can be attributed to the reason that the load distribution among the rollers become more and more even. When the preload is larger than $0.3F_{ae}$, with the increase of the axial preload, the bearing fatigue life begins decrease distinctly; this can be attributed to the reason that the roller loads begin to increase due to the preload, as shown in Figure 7. Generally speaking, the effect of the roller load increase exceeds the distribution evenness increase through preloading. From the viewpoint of fatigue failure, the preload of $0.3F_{ae}$ is most favorable for the bearing.

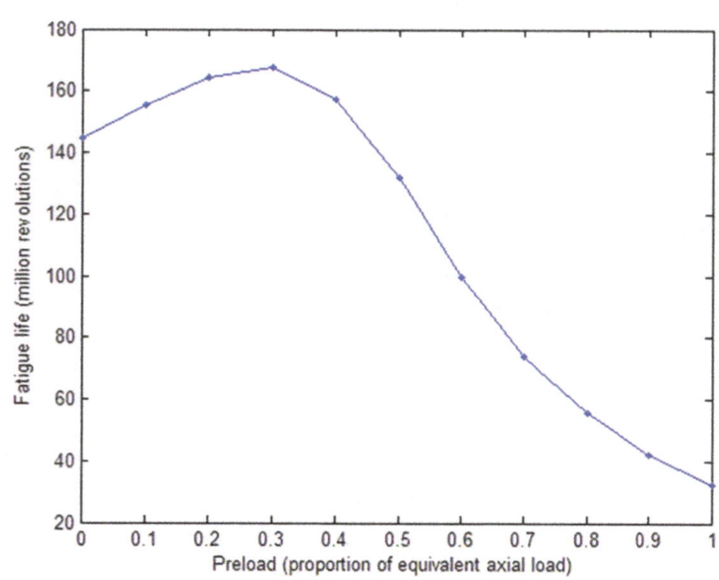

Figure 6. Bearing fatigue life

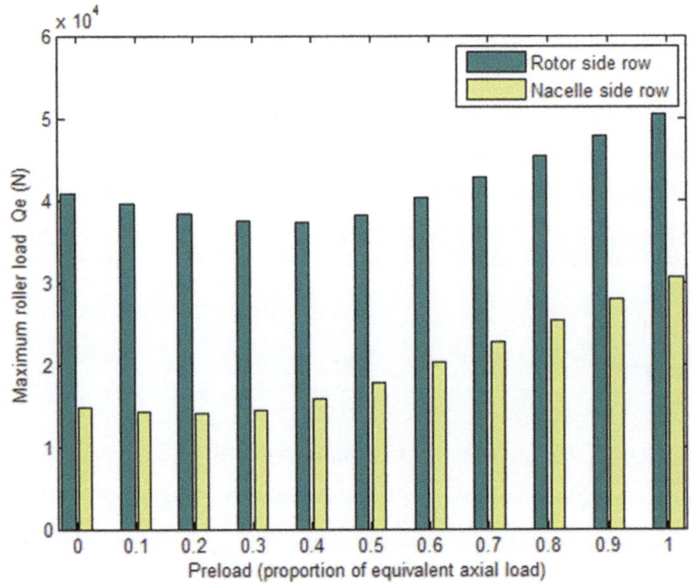

Figure 7. Maximum roller load

Determination of the preload should consider to reduce both roller slipping and raceway fatigue failure. For the former, the proportion of the loaded rollers should be no less than 80 per cent so as to over come the effect of slippage. According to Figure 8, the preload need to be larger than $0.4F_{ae}$. For the latter, if the preload is larger than $0.5F_{ae}$, fatigue life will decrease drastically, as shown in Figure6. Therefore, as a trade-off selection, $0.5F_{ae}$ is selected as the bearing preload.

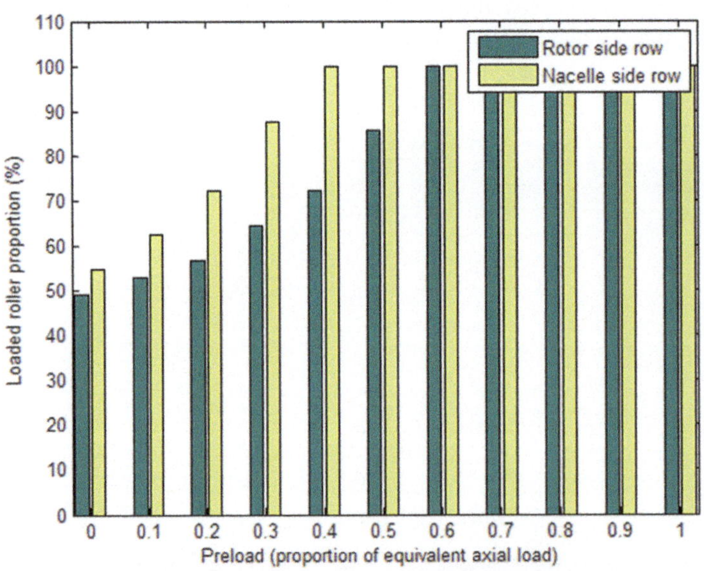

Figure 8. Proportion of loaded rollers

7. Conclusion

Double-row tapered roller bearing is a suitable selection to support the rotor of direct-drive wind turbine, and the appropriate preload is necessary for the bearing to ensure the rolling motions of the rollers and the long fatigue life of the bearing as far as possible. Statics model of the preloaded bearing was established, for a set of selected preload values which are different proportions of the dynamic equivalent axial load, the loaded roller number, maximum roller load and the bearing fatigue life were calculated. With the increase of the preload, the number of loaded rollers becomes more and more, this is advantageous for the pure rolling of the rollers; but excessive large preload will increase the roller loads and decrease the fatigue life of the bearing. Based on these results, a preload of 0.5 times of the dynamic equivalent axial load was determined.

Acknowledgements

The research is financed by the Key Projects of Science and Technology Research of the Education Department of Henan Province (Grant No.14A460008).

References

Aguirrebeitia, J., Plaza, J., Abasolo, M., & Vallejo, J. (2013). General static load-carrying capacity of four-contact-point slewing bearings for wind turbine generator actuation systems. *Wind Energy, 16*(5), 759-774. http://dx.doi.org/10.1002/we.1530

Bywaters, G., Mattila, P., Costin, D., Stowell, J., John, V., Hoskins, S., ... Freeman, B. (2007). *Northern Power NW 1500 Direct-Drive Generator*. US National Renewable Energy Laboratory, USA.

Feng, Y., Chen, J., Huang, X. D., & Wang, W. (2013). Structure fatigue life analysis of a 1.5 MW smart yaw slewing bearing. *Advanced Materials Research, 779*, 664-670.

Harris, T. A., Rumbarger, J. H., & Butterfield, C. P. (2009). *Wind turbine design guide DG03: Yaw and pitch rolling bearing life*. US National Renewable Energy Laboratory, USA. http://dx.doi.org/10.2172/969722

Kotzalas, M. N., & Doll, G. L. (2010). Tribological advancements for reliable wind turbine performance. *Phil. Trans. R. Soc. A, 368*, 4829-4850. http://dx.doi.org/10.1098/rsta.2010.0194

Lucas, D., & Pontius, T. (2005). *Designing large diameter, closely coupled 2-row tapered roller bearings for supporting wind turbine rotor loading*, USA, The Timken Company, USA.

Lundberg, G., & Palmgren, A. (1947). Dynamics capacity of rolling bearings. *Acta Polytechnica Mechanical Engineering Series, 1*(3), Stockholm, Sweden.

Niu, L. K., Yang, J. M., & Gao, J. Y. (2013). Study on contact fatigue damage of hardened layer of yaw bearing raceway in wind turbine. *Acta Energiae Solaris Sinica, 34*(8), 1415-1420.

World Wind Energy Association. (2012). *2012 Half-year Report, WWEA Head Office*, Bonn, Germany.

5

System Hazard Platform: Case Study NASA Field Joint Failure

Kouroush Jenab[1], Josh Herrin[2], Saeid Moslehpour[3] & Sam Khoury[4]

[1] Faculty of College of Aeronautics, Embry-Riddle Aeronautical University, 600 S. Clyde Morris Blvd. Daytona Beach, FL 32114-3900, USA

[2] Graduate Student, Embry-Riddle Aeronautical University, 600 S. Clyde Morris Blvd. Daytona Beach, FL 32114-3900, USA

[3] Faculty of College of Engineering, Hartford University, 200 Bloomfield Avenue, West Hartford, CT 06117, USA

[4] Faculty of College of Business, Athens State University, 300 N. Beaty St., Athens, AL 35611, USA

Correspondence: Kouroush Jenab, Faculty of College of Aeronautics, Embry-Riddle Aeronautical University, 600 S. Clyde Morris Blvd. Daytona Beach, FL 32114-3900, USA. E-mail: jenabk@erau.edu

Abstract

NASA became overconfident with consecutive successful flights with no major failures leading up to Flight 25 of the Space Shuttle Challenger and failed to correctly apply quality assurance to reanalyze the possibilities of failure when extreme cold weather was present for what would become the last Challenger launch. System Hazard Analysis applied correctly to analyze the failure rate patterns of the NASA Space Shuttle Challenger Solid Rocket Booster field joints may have prevented the launch of the tragic Flight 25, where there was a total loss of aircraft and seven astronauts were killed in the accident. The steps of System Hazard Analysis will be explained that if followed may have provided the data necessary for NASA to correct the field joint error prior to instead of after the Challenger explosion.

1. Introduction

Quality Assurance (QA) is a systematic process to inspect and evaluate a product that is being developed and manufactured to ensure it is meeting requirements. It is an effort by the entire production team to prevent mistakes or to discover defects prior to a product's release for use in the field. Every member of a production team should be trained in the importance of applying quality assurance techniques to their product. In the last 100 years, various QA techniques in production have been developed and improved to be used by companies to create a final product with little to no defects. Developing QA in a company involves the correct interpretation of a customer's inputs to create a product that the customer is happy with by creating the proper standards and process control in design, development, and manufacturing. Communication amongst all internal and external stakeholders is also a very important requirement to detect and repair defects in a product. In one of the most memorable tragedies in U.S. history, quality assurance was not applied correctly by NASA management prior to the Flight 25 launch of the Space Shuttle Challenger. On the eve of the launch day of the Challenger the ambient temperature at the Kennedy Space Center (KSC) was forecasted to be below or near freezing for the scheduled launch. NASA and Morton Thiokol-Wasatch, the manufacturer of the Solid Rocket Booster (SRB), became concerned about the performance of the SRBs in cold weather due to problems noticed in past launches. These concerns were specifically related to the field joints of the SRBs based on post analysis of several previous launches, which showed hot combustion gases had charred and eroded the seals of the field joints (Vaughan, 1997). Field joints mate the case segments of the SRB. The SRB consisted of seven segments that were assembled into four segments by the contractor Morton Thiokol-Wasatch. These segments were then shipped to KSC to be mated or stacked vertically using field joints to join and seal each mated segment. Each cylindrical segment had two different ends forming half of each field joint. The bottom of a segment was called the tang which was inserted into a cavity at the top end of another segment called the clevis. Clevis pins were used to hold the joint together. Each field joint consisted of 177 clevis pins installed evenly spaced around the circumference of the SRB case. Shims were used to align each mating tang and clevis. A seal was formed within the field joint by two compressed rubberlike Viton O-rings installed into two channels cut into the clevis which contact the tang. To protect the O-rings, a type of putty made up of asbestos-filled zinc chromate was used to seal the inner wall area between the mated segments (Davis et al., 1990; Vaughan, 1997). Figure 1 shows the details of the mating parts and seals of a completed field joint used on the Challenger.

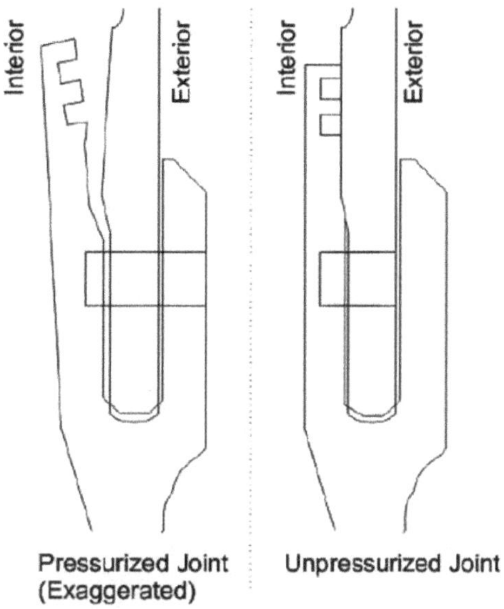

Figure 1. SRB Field Joint

The purpose of the field joint was to contain the hot gases while the SRBs were burning fuel and propelling the shuttle and external tank into space during the first two minutes of the launch. The theory was that the putty in conjunction with the redundant compressed O-rings would seal the joints between segments containing the built up pressure of the hot gases being expelled during launch. During testing prior to the Challenger launch a phenomenon called joint rotation was discovered. Joint rotation was the shifting of hardware within the field joint which could create a gap between the tang and clevis inner wall wide enough for the O-rings to lose their compressed seal and allow hot gases to escape. From this testing effort, decisions were made to make the shims bulkier to improve alignment of the tang and clevis and resist shifting of the tang, clevis, and clevis pins. Also, O-rings with a wider diameter would be used. Through more testing of joint rotation and analysis of field joints from early Space Transportation System (STS) launches, it was agreed upon that the design of the field joint was an acceptable risk (Vaughan, 1997).

As STS missions continued in preparation of the Challenger launch some of the SRB field joints showed signs of hot gases protruding the field joints in post analysis. It was revealed that the primary O-ring was being eroded by the high temperature hot gases escaping the putty seal. The secondary O-ring did its job by maintaining the field joint seal, but this issue raised concern amongst the management of NASA and Morton Thiokol-Wasatch. The approved design of the field joint was still not preventing joint rotation and in extreme cases the second O-ring could also fail. This led to an investigation just a few days before the scheduled launch of the challenger. With the frigid temperatures expected at the scheduled Challenger launch the attention of engineers and management turned to the correlation of O-ring elasticity and behavior at cooler temperatures. On the eve of the Challenger launch, it was suggested by Morton Thiokol-Wasatch engineering that the cooler temperatures may cause the O-rings to harden and not correctly compress in reaction to the initial buildup of pressure once the SRBs had ignited at liftoff (Vaughan, 1997). At this point NASA management was under pressure to go ahead with the scheduled launch of the Challenger after multiple postponements received negative media attention and added strain to an already busy launch schedule. NASA was able to negotiate with Morton Thiokol-Wasatch management to agree to go ahead with the launch based on a lack of extreme failures in past launches even though the cold temperature was an extreme case and had never been tested. "STS 51-L was launched at 11:38 A.M. EST. The ambient temperature at the launch pad was 36°F. The mission ended 73 seconds later as a fireball erupted and the Challenger disappeared in a huge cloud of smoke (Vaughan, 1997).

2. Literature Review

Improvements in quality assurance processes have helped prevent major failures like the tragedy of the Space Shuttle Challenger. Companies large and small have been forced to invest in practices to reduce risk in costly failures later in the life cycle of a product. As far back as the 1920's, pioneers of the quality movement like Walter Shewhart began questioning the quality methods of companies and developing new ways to manage the quality of a product. To reduce the frequency of failures, Shewhart introduced a process of statistical control to distinguish between assignable and chance cause variations (Shewhart, 2011). "His mentoring of other engineers at Western Electric and his groundbreaking work with control charts arguably led a quality revolution and launched the quality profession" (Smith, 2009a). Shewhart's influence on his supervisor George D. Edwards would eventually lead him to become the first president of the American Society for Quality Control, which would combine several local societies formed during World War II into a national society to promote quality education (Smith, 2009b). Another pupil of Shewhart, W. Edwards Deming, successfully learned and applied Shewhart's theories. Deming became a master of quality processes producing his famous 14 point philosophy and his teachings which "affected a quality revolution of gargantuan significance on American manufacturers and consumers" (SkyMark, 2015). He was also the main influence in Japan's economic rise in export-led growth (Best & Neuhauser, 2005). Dodge (1943) was able to use Shewhart's statistical process control to create a sampling inspection plan for products manufactured in high quantities in a continuous process. Eventually Harry Romig would partner with Dodge to create the Dodge-Romig sampling plan. Sampling plans played an important role in quality control systems and "the success of the inspection function has profound effects on the viability of the production process and the profitability of the enterprise" (Jaraiedi & Segall, 1990). Yet another pioneer of quality who worked with Shewhart was Joseph Juran. He was credited with adding the human dimension to the already important statistics of quality. He discovered many human relations problems that were linked to resistance to change or cultural resistance (Phillips-Donaldson, 2004). His level of influence on the global improvement of quality was compared to that of Deming. Deming and Juran's visit to Japan influenced Kaoru Ishikawa to become a leader in quality improvement in Japan. Ishikawa believed in quality through leadership allowing management to improve their employees' work habits based on the quality expected of the external customer by educating and instilling a selfless personal commitment (Watson, 2004). An innovator of quality in the fields of engineering economics, industrial engineering, and statistical quality control and who learned from some of the original developers of improved quality was Eugene Grant. He wrote textbooks and helped teach the techniques to some of the quality experts like Deming, Shewhart, and Dodge (Smith, 2012). Feigenbaum (1956) added a new twist to quality when he discussed the principle of a total quality view, which stated that quality effectiveness started with the design of a product and ended when the product was received by the customer. This principle is considered by some to be the initial theory of Total Quality Management. Another expert in quality is Phillip Crosby who focuses on prevention by stressing an attitude of doing things right the first time and zero defects (Suarez, 1992). His efforts of quality improvement are focused on better planning and analysis of requirements early in development in an attempt to get it right from the start.

A newer quality improvement process called Six Sigma was developed by Bill Smith and Bob Galvin in 1986, which established the idea of improving quality so that the number of defects becomes so few that they are statistically insignificant (O'Farrell, 2015). Keller (2011) provides a step-by-step guide to applying core quality processes to techniques like Six Sigma or Lean processes to improve quality. Another advocate of the Six Sigma technique is Thomas Pyzdek, who developed the Pyzdek Institute to train the Six Sigma technique to improve quality through consultation. Pyzdek feels that Six Sigma cannot be compared to traditional Quality Management and that Six Sigma takes quality control to a new level to improve quality (Noorani, 2014).

A few years after the tragedy of the Challenger, Daniel Goldin was hired at NASA who initiated a revolution to transform the space program by establishing his "faster, better, cheaper" approach to deliver programs of high value without sacrificing safety (Thompson & Davis, 2009). Another major contributor to NASA and quality improvement was Paul Lachance who was responsible for ensuring food safety for the astronauts during missions. In an effort to identify food safety hazards and apply critical control points to reduce threats, Lachance became the developer of the Hazard Analysis and Critical Control Point (HACCP) quality control technique (Valigra, 2012). This technique is a form of the System Hazard Analysis technique of defining hazards of systems, subsystems, and the interaction between the subsystems. This technique, if applied correctly, may have prevented the launch of the Challenger until all hazards had been defined and risks from those hazards had been mitigated.

3. System Hazard Analysis Applied to the STS SRB Field Joint

System Hazard Analysis is used to define the interface effects of subsystems. In the case of the field joint the opposite ends of two SRB segments are being joined together by inserting the tang of one section into the clevis of another section using shims to center the tang in the clevis. High temperature putty is applied to the inner portion of the mated joint to attempt to prevent hot gases from contacting the O-rings. Two grooves or channels are cut into the clevis where the primary and secondary O-rings are seated prior to mating the segments. Once mated a clevis pin is installed which passes through the clevis and tang to lock the segments together. A pressure port is installed to pressure check the seal of the two O-rings after the field joint is complete. Leading up to the launch of the Challenger, historical data had shown that the field joint had minor failures. Investigations of O-rings from past launches showed various levels of erosion of the primary O-ring but the secondary O-ring had always functioned properly and maintained the field joint seal. Based on Figure 2, System Hazard Analysis requires several steps to be completed to efficiently define and mitigate hazards or risks.

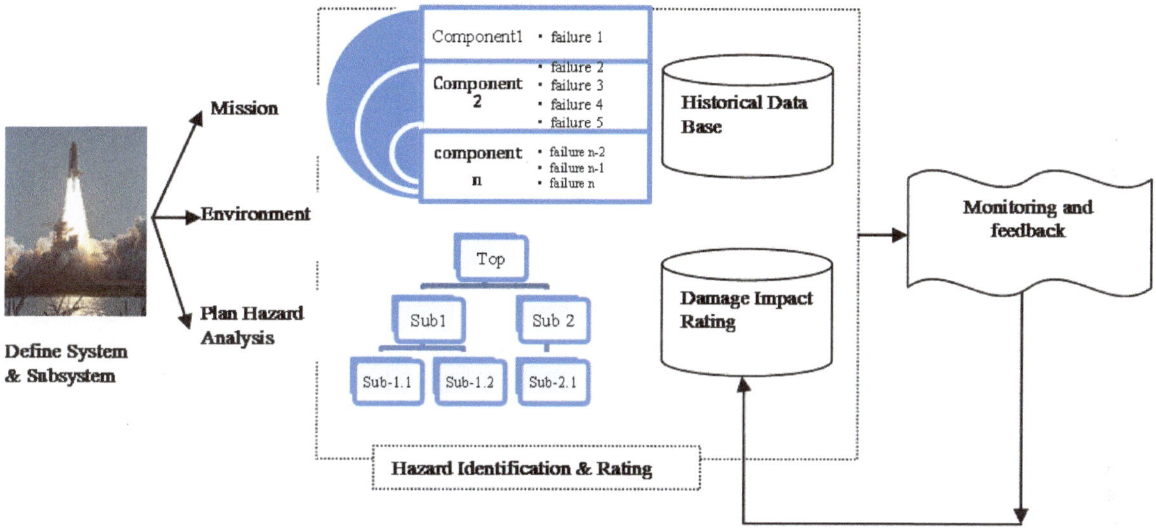

Figure 2. Hazard Analysis

<u>Steps to applying System Hazard Analysis:</u>

- Define System or Subsystem
  - Understand the system design and operation along with mission phases and environments
  - Plan Hazard Analysis
  - Define analysis goals, definitions, and processes
- Select a Team
  - A team leader and an interdisciplinary team will be assembled to apply their expertise and experience to the hazard analysis process
- Acquire Data
  - Team will acquire design and process data along with historical data from test results if available
- Conduct Hazard Analysis
  - Identify items
  - Define parameter list
  - Define Hazards and risk levels
  - Produce necessary documentation/worksheets
  - Conduct necessary meeting to discuss results of analysis
  - Recommend Corrective Action
  - Monitor Corrective Action
  - Track hazards
  - Document hazard process updating information as hazard levels change or new hazards are discovered

3.1 Subsystem Analysis of Field Joint Parts:

Table 1. Field Joint Subsystem Hazard Analysis

Component	Purpose	Failure Mode	Effect	Mitigation	Probability of Occurrence	Risk Level
Tang	Align and mate SRB segments	1.Rotation 2.Bending 3.Separating from Clevis inner wall	1.Hot gas from SRB propellant escapes	1. Test different materials. 2. Structure testing to determine possibility of range of movement. 3.Consider Redesign.	Low	II
Clevis	Align and mate SRB segments	1.Rotation 2.Bending 3.Separating from tang	1.Hot gas from SRB propellant escapes	1. Test different materials. 2. Structure testing to determine possibility of range of movement. 3.Consider Redesign.	Low	II
Clevis Pin	Pin or bolt the tang and clevis	1.Rotation 2.Bending	Loss of integrity of field joint	1. Test different materials. 2. Structure testing to determine possibility of range of movement. 3.Consider Redesign.	Low	II
Shim	Align tang and clevis	1.Rotation 2.Bending	Loss of integrity of field joint	1. Test different materials. 2. Structure testing to determine possibility of range of movement. 3.Consider Redesign.	Low	II
Primary O-Ring	Maintain seal between tang and clevis	Erosion due to escaped hot gas	1.Hot gas from SRB propellant escapes 2.Possible loss of aircraft	1. Test different materials. 2. Structure testing to determine possibility of loss of compression. 3.Consider Redesign.	Medium	I
Secondary O-Ring	Maintain seal between tang and clevis	Erosion due to escaped hot gas	1.Hot gas from SRB propellant escapes 2.Possible loss of aircraft	1. Test different materials. 2. Structure testing to determine possibility of loss of compression. 3.Consider Redesign.	Medium	I
Putty	Protect O-rings from hot gas escaping inner wall of joint	Leaking due to joint rotation or separation	1.Hot gas from SRB propellant escapes	1. Test different materials. 2. Structure testing to determine possibility leaks. 3.Consider Redesign.	Medium	I

Table 2. Historical Data of Field Joint Failures from previous Challenger launches:

# O-rings fail	3	1	1	1	0	0	0	0	0	0	1	1	0	0	0	0	2	0	0	0	0	0	0
Launch Temperature °F	53	57	58	63	66	67	67	67	68	69	70	70	70	70	72	73	75	75	76	76	78	79	81

Table 2 shows historical data from 23 previous launches of the Space Shuttle Challenger before the loss of aircraft of flight 25. One data point was lost due to not being able to recover the SRBs for one of the missions. If the data is split at the 70˚F mark, it shows 6 out of 12 launches had O-ring failures at temp [53, 70]. The other half of the data shows 1 out 11 launches had O-ring failures at temp [70, 81]. Although the data trend is somewhat obvious when observing the failures with respect to temperature, using a binomial distribution could be used to provide further analysis of O-ring reliability.

$R(s) \quad = \sum_{x}^{N} \binom{N}{x} R^x (1-R)^{N-x}$ for colder temperatures where N = 23, x = 6, and R = 6/12

$R(s) \quad = \sum_{x}^{N} \binom{N}{x} R^x (1-R)^{N-x}$ for colder temperatures where N = 23, x = 6, and R = 1/11

Formulas can be used to analyze the probability of failures with respect to temperature in a given amount of launches. The field joint is not considered completely failed until the putty seal and both the primary and second O-rings fail. So the complete failure probability is:

$R(s) \quad = \sum_{x}^{N} \binom{3}{1} R^1 (1-R)^{3-1} + \sum_{x}^{N} \binom{3}{2} R^2 (1-R)^{3-2} + \sum_{x}^{N} \binom{3}{3} R^3 (1-R)^{3-3}$ using the probability of failure based on temperature from earlier.

4. Post-Challenger Improvements

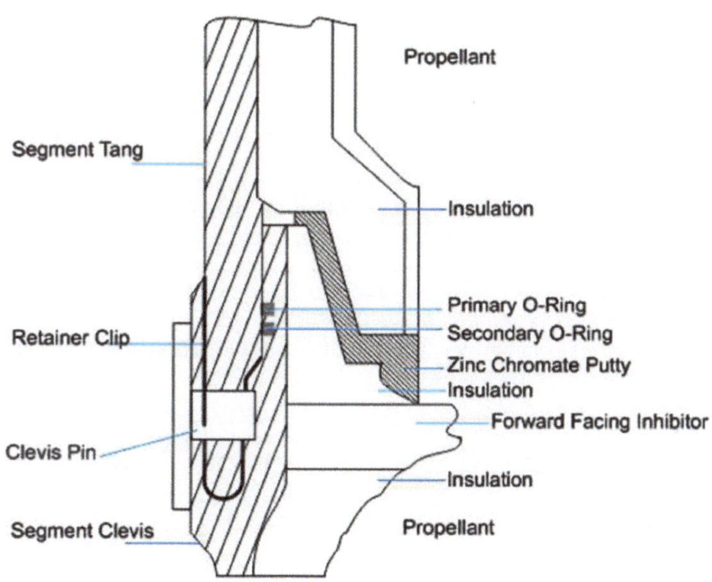

Figure 3. Field Joint Redesign

The tragedy of the Challenger, "triggered an in-depth, detailed review of analytical, design, manufacturing, assembly, and testing methods for solid rocket motor field and factory assembly joints" (Harkins, 1999). NASA overhauled their quality assurance processes to improve hazard analysis and the process to attain safety of flight approval. A new independent organization was created to oversee critical flight safety matters, working directly with teams investigating issues but retaining its separate identity as final arbiter reporting directly to the NASA Administrator (NASA, 1987). Once the organization was established the first priority was to focus on the redesign and testing necessary to repair the SRB field joints. The redesign featured a new tang capture latch that included a third O-ring seal that helped control movement of the mated tang and clevis. External heaters were added to heat the field joints prior to launch to ensure O-ring compression and reaction to structural loads and pressure build was optimal. Also insulation design improvements were implemented to prevent seal leakage from the inner mating walls of the field joint. Figure 3 shows a diagram of the new design of the field joint.

5. Conclusion

NASA's breakdown in communication and ignorance of a lingering issue led to a historical tragedy causing the loss of seven lives. There was already a great amount of public interest in the Challenger launch with the selection of a school teacher to be launched into space. Combined with the extra attention and the busy launch schedule of that year put extra stress on NASA management. They had received negative media attention, because of the many delays prior to the Challenger launch. On the eve of the Challenger launch issues were brought to the attention of NASA management about possible failures due to extreme cold conditions, but they chose to chance the risk of failure because they were under extraordinary pressures to launch the Challenger (Vaughan, 1997). NASA management chose to ignore their quality standards and to reassess the possible elevated risk. Following a detailed process of System Hazard Analysis may have proven to NASA management that the risk was too high for safety of flight approval. The added pressure to delay another flight would have been welcomed in comparison to the stress of a failure investigation that showed a lack of communication amongst NASA officials and the lack of applying quality assurance processes to reevaluate the level of risk in a last minute issue that affected human lives. For future works, application of virtual reality in Hazard Analysis would effectively pinpoint the risk elements.

References

Best, M., & Neuhauser, D. (2005). W Edwards Deming: Father of Quality Management, Patient and Composer. *BMJ Quality and Safety Journal*, *14(1)*, 310-312. http://dx.doi.org/10.1136/qshc.2005.015289

Davis, R. C., Bowman, L. M., Hughes IV, R. M., & Jackson, B. J. (1990). *A Parametric Shell Analysis of the Shuttle 51-L SRB Aft Field Joint.* Hampton: NASA. Retrieved from http://ntrs.nasa.gov/archive/nasa/casi.ntrs.nasa.gov/19910005316.pdf

Dodge, H. (1943). A Sampling Inspection Plan for Continuous Production. *The Annals of Mathematical Statistics*, *14(3)*, 264-279. http://dx.doi.org/10.1214/aoms/1177731420

Feigenbaum, A. V. (1956). Total Quality Control. *Harvard Business Review*, *34(6)*, 93-101.

Harkins, W. (1999). *Public Lessons Learned Entry: 0759.* NASA MSFC. Retrieved from http://www.nasa.gov/offices/oce/llis/0759.html

Jaraiedi, M., & Segall, R. (1990). Mathematical Modelling of Dodge-Romig Sampling Plans for Random Incoming Quality. *Applied Mathematical Modelling*, *14(5)*, 264-270. http://dx.doi.org/10.1016/0307-904x(90)90017-y

Keller, P. (2011). *Statistical Process Control Demystified.* New York: McGraw-Hill. Retrieved from http://accessengineeringlibrary.com.ezproxy.libproxy.db.erau.edu/browse/statistical-process-control-demystified#Preface03

NASA. (1987). *Implementation of the Recommendations of the Presidential Commission on the Space Shuttle Challenger Accident.* NASA. Retrieved from http://ntrs.nasa.gov/archive/nasa/casi.ntrs.nasa.gov/19930010801.pdf

Noorani, M. (2014). *Thomas Pyzdek-Taking Quality Management to the Next Level.* Retrieved from http://project-management.com/thomas-pyzdek-taking-quality-management-to-the-next-level/

O'Farrell, R. (2015). Who Started Six Sigma? *Houston Chronicle.* Retrieved from http://smallbusiness.chron.com/started-six-sigma-16865.html

Phillips-Donaldson, D. (2004). Gurus of Quality: 100 Years of Juran. *Quality Progress*, *37(5)*, 25-39.

Shewhart, W. (2011). Giants of Quality - Wiley & Sons. *Quality and Reliability Engineering International*, 979. Retrieved from http://onlinelibrary.wiley.com.ezproxy.libproxy.db.erau.edu/doi/10.1002/qre.1283/epdf

SkyMark. (2015). Dr. W. Edwards Deming. Pittsburgh, PA, USA. Retrieved from http://www.skymark.com/resources/leaders/deming.asp

Smith, J. L. (2009a). Remembering Walter A. Shewhart's Contribution to the Quality World. *Quality Magazine*. Retrieved from http://www.qualitymag.com/articles/85973-remembering-walter-a-shewhart-s-contribution-to-the-quality-world

Smith, J. L. (2009b). The History of Modern Quality. *Peoria Magaizines*. Retrieved from http://www.peoriamagazines.com/ibi/2009/jul/history-modern-quality

Smith, J. L. (2012). The Legacy of Eugene Grant. *Quality Magazine*. Retrieved from http://www.qualitymag.com/articles/88833-the-legacy-of-eugene-grant

Suarez, J. G. (1992). *Three Experts on Quality Management.* Arlington: Dept. of the Navy, TQL Office. Retrieved from http://dtic.mil/dtic/tr/fulltext/u2/a256399.pdf

Thompson, E., & Davis, J. (2009). *Daniel Saul Golding: NASA Administrator, April 1, 1992 - November 17, 2001*. Retrieved from history.nasa.gov: http://history.nasa.gov/dan_goldin.html

Valigra, L. (2012). Former NASA Food Coordinator Pioneered the HACCP System. *Food Quality & Safety*. Retrieved from http://www.foodquality.com/details/article/1721927/Former_NASA_Food_Coordinator_Pioneered_the_HACCP_System.html?tzcheck=1&tzcheck=1

Vaughan, D. (1997). *Challenger Launch Decision : Risky Technology, Culture, and Deviance at NASA*. Chicago: University of Chicago Press. Retrieved from http://site.ebrary.com.ezproxy.libproxy.db.erau.edu/lib/erau/reader.action?docID=10417007

Watson, G. (2004). The Legacy of Ishikawa. *Quality Progress, 37(4),* 54-57. Retrieved from http://search.proquest.com.ezproxy.libproxy.db.erau.edu/docview/214762859/abstract?accountid=27203

6

Laws of Radiation From Large Gas Volumes and Claculation of Heat Transfer in Steam Boiler Boxes

A. N. Makarov[1]

[1] Tver' State Technical University, Russia

Correspondence: A. N. Makarov, Tver' State Technical University, Russia. E-mail: tgtu_kafedra_ese@mail.ru

Abstract

The progress in the development of methods for calculating heat transfer in torch furnaces, fireboxes, and combustion chambers is analyzed. In the XIX-XX centuries calculations of radiative heat exchange were followed the methods based on the laws of radiation from black body, disclosed in the XVII-XX. It is substantiated that there is crisis of methods for calculating heat transfer in torch furnaces, fire boxes and combustion chambers. In the early XXI century the author of this manuscript disclosed the laws of radiation from torch gas volumes, described in the article.

Keywords: laws for radiation, heat exchange, torch, fire box, steam boiler, discovery.

1. Introduction

Radiant heat transfer is the main kind of heat transfer and accounts for 90–98% of the total heat transfer in steam boiler fireboxes (Blokh, Juravlev, & Ryzkov, 1991), torch heating and melting furnaces (Telegina, 1993; Krivandin & Egorov, 1989; Makarov & Svenchanskii, 1992).

Since the late 19th century and throughout the 20th century, heat transfer in torch furnaces, fireboxes, and combustion chambers was calculated based on the law of radiation from black body, particularly on the law, that was experimentally established by Y. Stefan in 1879 in studying radiation from solid bodies, which was then theoretically substantiated by L. Boltzmann in 1884 in studying radiation from solid bodies analytically.

In the late 19th – early20th century, solid lumped fuel (coal, blacks, peat, wood) was fired in furnaces on fire grates, and the first descriptions of heat transfer processes were essentially descriptions of calculation problems of radiant heat transfer between two arbitrarily located hard surfaces (a fuel bed and a heating surface) on the basis of Stefan–Boltzmann's law.

2. Retrospective View of Calculation Procedure of Heat Transfer

The first descriptions of heat transfer processes included bulk fuel burning on fire grates, the only method applied to industrial furnaces, solid fuel furnace. The authors of these descriptions assumed, that burning is completed within fuel bed with overall constant temperature,and gases filling furnace chamber are diathermic and aren't involved in radiative heat exchange between two arbitrarily-spaced surfaces (a fuel bed and a heating surface) using Stefan-Boltzmann law. In 1924, M.V. Kirpichev gave an analysis of methods for solving this problem that had been developed by different researchers (Kirpichev, 1924). Solution of this problem is presently given as follows:

$$q = c_S \varepsilon_{\mathrm{red}} \left[\left(\frac{T_1}{100} \right)^4 - \left(\frac{T_2}{100} \right)^4 \right] \phi_{12}, \tag{1}$$

where q is the density of heat flux radiated from fuel to the heating surface; c_s is the black body emis sivity factor; ε_{red} is the reduced emissivity; T_1, T_2 are the temperatures of fuel bed and heating surface, respectively; ϕ_{12} is the view factor for radiation from the fuel bed on the heating surface.

In 1931, V.N. Shreter also recommended to use Stefan–Boltzmann's law (1) for torch furnaces in the case of firing fuel oil, with the torch was regarded as an isothermal gray body filling the entire space of the chamber (Shreter, 1931). Despite the use of oversimplified assumptions and conceptions about heat transfer processes in chambers that were rather far from reality, the results of calculation carried out for pulverized coal and oil fired furnaces with small sizes of the heat absorbing surface and highly extended lining turned to be close to measured results.Subsequent investigations in such furnaces and fireboxes showed how great importane the lining in heat flux flattering and generating dominating radiation from gray solid body is. However, with the advent of waterwall fireboxes, such calculations turned to be unsatisfactory.

In 1932, L.K. Ramzin developed an analytical method for calculating heat transfer that uses the temperature field in the chamber, correlated to a first approximation with the fuel combustion conditions instead the mistaken assumption of constant temperatures in the furnace.

In 1934–1940, V.N. Timofeev continued the works of L.K. Ramzin that proposed to take the torch emissivity equal to the emissivity of triatomic gases contained in it.

The equation for calculating radiation from gas to the wall is presently formulated as follows:

$$q = c_S \varepsilon_{ew} \varepsilon_{\Gamma} \left[\left(\frac{T_g}{100} \right)^4 - \left(\frac{T_w}{100} \right)^4 \right], \qquad (2)$$

where ε_{ew} - effective emissivity factor (emissivity) of the wall; ε_g - emissivity factor (emissivity) of gas; T_g, T_w -gas and wall temperatures respectively. In order to calculate gas radiation to the heating surface by equation (2), gas emissivity must be known. Basic sources of radiation from nonluminous torch are triatomic gases CO_2 and H_2O. Radiation of triatomic gases is selective, that is gases radiate and absorb heat energy in the wavelength range, aligned in multispectral regions. Many experiments were carried out on determining the absorption and emission bands for gas and it should be noted, that different researchers obtained different data (Nevskii, 1971). This may possibly be attributed to the fact, that the width of the bands depends on the gas temperature, and beside that, as the gas layer thickness increases, the radiation from spectrum regions with low emissivity is amplified in the neighborhood with radiating bands (Shreter, 1931).

Schack was the first, who attempted to develop a method for calculating emissivity factors of CO_2 and H_2O in 1933 (Shreter, 1931). Design formulas deduced by Shack turned out to be cumbersome and not suitable for account. The difficulties involved with developing analytical methods for calculation emissivity factors of CO_2 and H_2O made many researchers follow the course of graphical calculations. Egbert and Houttel nomograms, plotted in 1939-42, received widespread use. The graphs are presented in two basic diagrams, connected emissivity factors CO_2 and H_2O with temperature and absorbing power on semilogarithmic ccordinates (Shreter, 1931).

Experimental findings of Barr, Fishenden, Houttel ,Egbert, Mangelsdorf, Eckert and other researchers (Mitor, 1963; Blokh, 1984; Nevskii, 1963) for emissivity factors CO_2 and H_2O are conflicting, the difference in results are from 5 to 70 %. Eckert data departure from Houttel and Mangelsdorf data by 15 – 50%. Houttel and Eckert don't offer a satisfactory explanation for the reasons, caused such significant experimental data gaps.

In 1947–1952, specialists of the All Union (presently All Russia) Thermal Engineering Institute (ARTEI) carried out the work aimed at analyzing all these investigations for selecting the most reliable results with a view to construct the calculated radiation graphs for CO_2 and H_2O. In 1952, graphs for carbon dioxide and steam emissivity factors were constructed on the basis of these data in standard method for thermal design of boiler unit (ND, 1952).

The gas temperature is plotted on the abscissa ,the emissivity factor is plotted on the ordinate versus the reduced beam lengths $x_{red} = pl$. Updated and specified uniform time standards of thermal design for boiler units were published in 1957, 1973, 1998 (Energiya, 1973; ND, 1998).

Method for calculating gas self- radiation and gas radiation on the heating surfaces, created in the 50–s of 1910 had took the complete form by 1960, and during the subsequent years was developed, but hasn't changed significantly.

3. Modern Methods for Calculating Radiant Heat Transfer

The range of modern methods for calculating radiant heat transfer is fairly wide, starting from calculations at the microlevel of photon emissions by atoms and ending with obtaining a macropattern of radiation in large objects. In calculations of radiation in the microworld, the molecular kinetic theory of elementary radiation processes is developed, and with regard of macroworld, the theory of radiation in process macrosystems is being sophisticated. The theory of radiant heat transfer in fireboxes, furnaces, and combustion chambers is a theory of radiating technological macrosystems, in which the radiation in microlevel objects and in the macrosystem as a whole has to be calculated.

Let us consider modern methods for calculating radiative heat transfer. The Monte-Carlo statistical modeling method is based on object physical model, on the use of homogeneous Markov chains (Blokh, Juravlev, & Ryzkov, 1991). It is applicable for solving transfer integrable equations as well. The Monte-Carlo method enables us to bring the amount of simulated photons into millions or more for maximum object approach. Complex radiating volumes, the distribution of thermogas dynamic parameters by them can be modeled, applying the Monte-Carlo method. However, because this method is numerical, it doesn't allow us to obtain analytic dependences correlating radiant fluxes and the data given even for unsophisticated physical models.

The Shwartzshild–Shuter and Eddington method for solving transfer equations is available for short-cut, evaluative calculations owing to averaging of directional intensity, assumption of isotropic dissipation and isothermality of medium and low accuracy of calculations. The use of the Shwartzshild–Shuter method is justified in research on the effect of the parameters and pattern of disperse systems heat radiation at the level of preliminary, qualitative estimates.

The iteration method of Chandrasekar (Chandrasekar, 1993), Nobel Prize winner in physics of 1983, allows to calculate integral and angular parameters of radiation for nonisothermal areas with any distribution of radiative properties by volume and different boundary conditions, for this reason the method has found a wide utility. However, the disadvantages of the method lie in complexity realizations for non-one-dimensional geometry and iteration of computational process in solving finite difference set of equations. Further development of the discrete coordinate method in two-dimensional geometry is iterative method of characteristics, or Vladimirov's method. It is commonly of high precision (more precisely than Carlson method), particularly at heavy optical thickness. The main weakness of this method is spatial checkout fixation , a variety of quadrature formula coefficients, that makes multidimensional problem calculations difficult (Adzeriho, 1975).

The routine method for solving transport equation is the spherical harmonics method, which numerical solution allows to approximate step by step integro- differential equations system by finitely difference equations, followed by matrix isolating technique. In work (Smelov, 1972) it is underlined, that calculation using the matrix factorization method doesn't lead us to round-off accumulation and being stable.

The method of zonal mathematical modeling has received the widest use for calculations of heat transfer in steam boiler boxes.

The basic idea of the zonal method is that the radiating surfaces and gas volumes are subdivided into zones, and that the continuous distributions of temperatures and physical characteristics are replaced by their piecewise continuous approximations consisting of a finite number of radiating macrosystems, homogeneous zones. In consequence of this, the integral-differential equations describing radiant heat transfer are replaced by an approximating system of algebraic equations, and the sought temperatures of zones and heat fluxes between them are determined by solving this system of equations:

$$\sum_{i=1}^{m+n} a_{ij}T_i^4+\sum_{i=1}^{m'} V_{ij}c_i t_i-c_j t_j\sum_{i=1}^{m} V_{jl}+\sum_{i=1}^{n'}\alpha_{kij}F_{ij}\left(t_i-t_j\right)+\sum_r V_{ojr}c_{or}t_{or}=0\ , \tag{3}$$

$$j=1\div m$$

$$\sum_{i=1}^{m+n} a_{ij}T_{ij}^4+\sum_{i=1}^{n'}\alpha_{kij}F_{ij}\left(t_i-t_j\right)+k_i F_j\left(t_n-t_j\right)\ ,\quad j=1\div n \tag{4}$$

where a_{ij} – is the radiation heat-transfer coefficient; *m, n, m', n'* –the number of volume and surface and neighboring zones respectively ; V_{ij} – the torch intensity from the *i*-th to the *j*-tha zone ; c_{or}, c_i – the heat capacity; α_{kij} – the heat transfer coefficient by convection; F_{ij} – the contact area of volume – surface zones; V_{ojr} – the actuiating flow to the *j*-th zone.; k_j – the heat transfer coefficient; F_j – zone area; T_i, t_n – the temperature.

Calculation of Heat Transfer in a Steam Boiler Box by the Zonal Method

In Blokh, Juravlev, and Ryzkov (1991) heat transfer is calculated by the zonal method taking as an example the BKZ-320-140 PT steam boiler box. Furnace space was subdivided into 7 calculated tiers, 14 volume and 40 surface zones. Fig 1,a shows a schematic design of the furnace, Figure 1,b presents distribution of incident radiant fluxes at vertical symmetry axis of the front wall calculated using the zonal method.

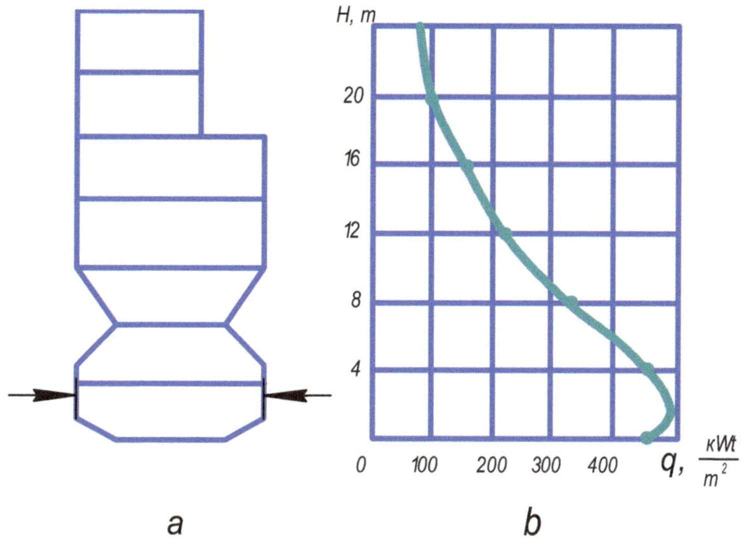

Figure 1. Subdivision of steam boiler box into calculated tiers (a) and distribution of incident radiation fluxes along the front waterwall axis (b)

The arrangement of burners in the furnace is arrowed in Figure 1,a. The maximal heat fluxes falling on the waterwall surfaces are observed in the area of the burners and account for about 500 kW/m^2.

Radiation fluxes falling from the torch decrease along the furnace height to 230 kW/m^2 at a height of 12 m and to 75 kW/m^2 in the underceilng zone of the furnace.

The calculated results of distribution for heat fluxes on the front wall vertical symmetry axis are experimentally confirmed.

The difference between heat fluxes along the front wall height obtained by calculation and by measurements reaches 15-20% (Blokh, Juravlev, & Ryzkov, 1991).

4. Crisis of the Existing Methods for Calculating Heat Transfer in Torch Furnaces, Fireboxes, and Combustion Chambers

Despite the fact that the zonal method has been constantly improved for the last 50 years and has been an engineering calculation method, it still has certain shortcomings:

—The differential and algebraic equations are not analytical expressions which crown the development of any theory if it is possible to solve differential equations;

—In calculating radiation from gas, Stefan–Boltzmann's law is used, which was derived for calculating radiation from solid bodies;

—Radiation and absorption from macrolevel systems, volume and surface zones are calculated, whereas radiation in microlevel systems, radiating particles, and elementary (infinitesimal) radiating volumes is not taken into account;

—The volume radiation from the gas that fills fireboxes, furnaces and combustion chambers is replaced in calculations by surface radiation.

The zonal method for calculating heat transfer in torch furnaces and fireboxes that is based on Stefan–Boltzmann's law does not allow to reveal the distribution of heat fluxes over all heating surfaces.

The zonal method was used to calculate the distribution of radiant heat flux in the steam boiler firebox chambers over the frontal wall height along the wall's vertical symmetry (Blokh, Juravlev, & Ryzkov, 1991; Blokh, Juravlev, & Ryzkov, 1991; Alekhnovich, 2007). No data were available on calculating the density of radiant heat fluxes over the perimeter and height of the frontal and lateral walls, and at the periphery of the walls of steam boiler fireboxes. An analysis of the distribution for radiant flux densities over the wall perimeters and heights observed during the operation of steam boiler fireboxes showed, that this distribution has a significantly nonuniform pattern (Makarov, 2012a).

Modern science is not a dogma, and obvious need arose to use a fundamentally approach to solving the problem of heat transfer in steam boiler boxes.

At the IV Minsk International Forum on heat-mass exchange (MIF-2000), the existing radiative heat exchange design procedure in furnaces, fire boxes with gaseous ,fuel, pulverized flame that flame is simulated by equivalent gas hemisphere , isothermal volume or bulk bands with *i*- parameters were criticized in series of reports (Germann, 2000).

At the I International Symposium on radiant heat exchange in 1995 it was noted, that there isn't sufficiently safe and effective radiative heat exchange procedure, every existing method has its imperfection and restricted range of application(Germann, 2000).It is concluded in (Germann, 2000), that configuration of radiating gas volume and radiant heat absorbing surface of furnace chamber are disregarded explicitly in the existing method for calculating radiative heat transfer in furnaces and fireboxes. The size, pattern of the radiating volume exert significant determining influence on heat transfer between the heating surfaces.

Radiating rectangular parallepiped, sphere, cylinder follow different laws for radiation and the accuracy of radiative heat transfer calculations in furnaces, fireboxes, combustion chambers is influenced on which set of volume bodies the torch will be modeled. Torch burner length may be different, the torch may be directed to the heating surfaces at different angle, different results of heating can be obtained when it depends on these geometrical parameters of the constant torch power. It has been verified experimentally and analytically, that the determining influence on radiative heat transfer in furnaces and fireboxes exert the torch length and the relative positions of the torch and the heating surfaces (Kutateladze, 1990).

By now, quite a number of facts have been revealed that point to the need of adjusting the existing methods for calculating radiative heat transfer in torch furnaces,fireboxes, and combustion chambers.

Gaseous radiation is characterized by a slightly less dependence from temperature than radiation from gray solid bodies . Radiation from carbonic acid vapour is proportional to the 3.5 power of the temperature, a water vapour radiation is proportional to the third power of the temperature (Kutateladze, 1990). For simplicity, in practical calculations we assume, by convention, that gas radiation is in proportion to its fourth temperature with presenting temperature corrections to the gas emissivity factors (Kutateladze, 1990).

Rise in torch temperature without increase in its power doesn't lead us to the development of furnace capacity. The determining influence on the incident radiation flux densities from the torch on the heating surfaces exerts it's power rather than temperature. At the same temperature of the torch and different power releasing in it, the radiation fluxes falling on the heating surface and the resulting fluxes will be different.The distribution of incident radiation fluxes on the heating surfaces depends on the distribution of power along the torch length.

The torch power can be increased by heating the air supplied. For example, with air heated to 600°C, the torch power increased by 17%, and its temperature rose from 1300°C to 2000°C, i.e., by a factor of 1.5 (Kutateladze, 1990). According to expression (2), the density of total heat flux from the torch to the calculated zone should increase by a factor of 5, and the heating rate should also increase by a factor of 5, which is in contradiction with the energy conservation law. Under the real conditions of heating furnace operation, with air subjected to preheating and with the torch power increased by 17%, the heat flux density and the heating rate increase by 12–15%, i.e., in direct proportion to the growth of torch power and not to the fourth power of temperature (Kutateladze, 1990).

The use of zonal method for torch modeling by rectangular parallepipeds or radiant areas and determination of the total radiation fluxes on the heating surfaces depending on gas-radiation coefficient and the difference in the fourth power of temperature between parallepipeds, areas, by which a flame is simulated and heating surfaces lead us to quick approximate calculation results. This is because gas structure, geometry, pattern, attitude of a flame and heating surfaces is neglected and Stefan–Boltzmann's law for calculating heat transfer between solid surfaces is used. It must be admitted that path of heat transfer research in furnaces, fire boxes, combustion chambers chosen in 1910-20 investigation and being improved throughout the 20th century has proven to be stalemate.

5. Laws for Radiation From Large Gas Volumes

The laws for heat radiation, disclosed in the late 19th – early 20th century, characterize radiation from black bodies, i.e. solid bodies. The fact, that black body radiation is a quartic function of its absolute temperature was first determined experimentally by Stefan in 1879, then theoretically by Boltzmann in 1884 (Stefan-Boltzmann law).

Planck's law (the Nobel prize of 1918) characterizes spectral irradiance distribution of black body along the wavelength in the radiation spectrum against the body temperature.

Wavelength,at which density of black body radiation attains its maximum value, arises from Wien displacement law (the Nobel prize of 1911).

Kirchhoff's law suggests, that ratio between surface density of real body radiation and absorption body coefficient is equal for all the bodies at the same temperature and equals surface density of black body radiation at the temperature given

According to Lambert's law(1760) the intensity of surface radiation in any direction, making an α angle with the normal to a surface is directly proportional to cosine of the angle:

Absorption and scattering of radiation flux is characterized by Bouguer-Lambert-Beer law, that surface radiation fallout in its passage through unit layer absorber material is in proportion to material attenuation coefficient and beam path length in this layer.

In the 17th– 20th centuries in the scientific world there was a good tradition to name the law after the author, discovered it. It is evident nowadays, that the scientific ethic binds the researchers to mention the author's name when they follow his law in their studies. The point of the discovery is thus the following.

Solid fuel, fired on fire grates, was the main type of fuel in the 19th – early 20 century, and calculations of heat transfer, made on the base of black body radiation satisfactorily agreed with the measured data of heat surface temperatures. In the 20th and 21st centuries liquid, gaseous, pulverized fuel was used in furnaces, fire boxes, combustion chambers with the flaring formation of volume radiant body - gas radiative torch of a few kW to some hundreds, sometimes several thousands MW in one or a score of burners. (Figure 2, Figure 3).

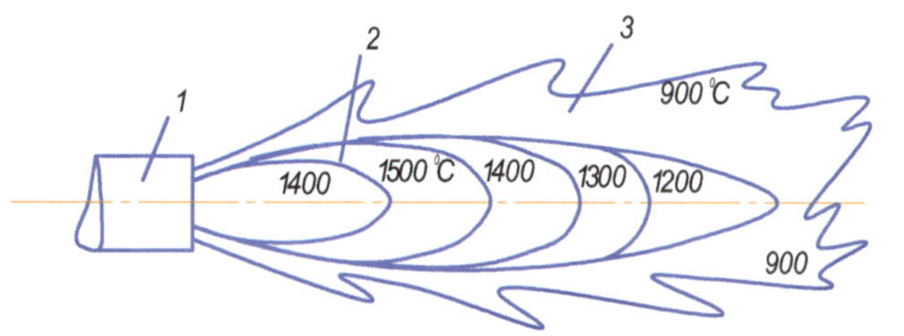

Figure 2. Fuel torch shape and structure with distribution of isotherms over the volume.

1 – burner; 2 – torch; 3 – combustion products

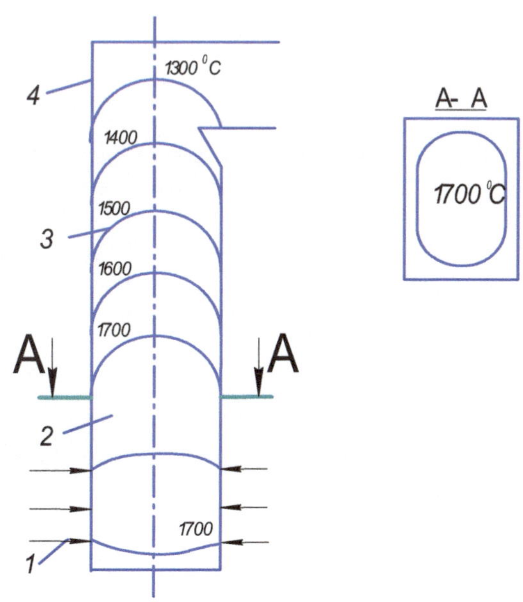

Figure 3. Torch and distribution of isotherms in the steam boiler firebox.

1 – Burner; 2 – firebox; 3 – torch; 4 – waterwall surfaces.

Laws of blackbody radiation are not applicable to heat transfer calculations of large gas volumes.

In flaring fuel on a single burner, the torch represents gas radiant volume in the form of spheroid of 0, 3 to 6 meters in length, with a diameter of major circle from 0,1 to 3,0 meters (Figure 2). Torch, generated by few tens of burners in steam boiler boxes represents gas radiant volume in the form of cylindroids of 10 to 50 meters height (Figure 2), the major axis of ellipse is of 3 to 20 meters in length. Thus, in flaring liquid, gaseous, pulverized fuel , we obtain large radiant gas volumes and in calculating heat transfer, it is necessary to calculate volume radiation rather than surface radiation, considering radiation of a number of gas layers, a number of radiative atoms, gas electrons. For solving this problem, we need triply integrate by gas volume, that is an intricate problem.

The laws of radiation from gas volumes, torch gas layers and electric arc burning in metal vapors of electric arc and plasma arc steel melting furnaces disclosed in 2001 make it possible to calculate volume radiation of torch gas layers to the heating surface. The laws of radiation from torch gas layers were established by the author of this article in 2001 (Makarov, 2003a, 2003b), and during the subsequent decade, this scientific disclosure was verified and approbated, and works were carried out on experimentally confirming and developing the geometrical, physical, and analytic models of torch as a source of heat radiation, and on elaborating the techniques for calculating heat transfer in torch furnaces, fireboxes, and combustion chambers (Makarov & Dunaev, 2004; Makarov & Krivnev, 2002; Makarov, 2000; Makarov, Chernyshov, & Voropaev, 2006; Makarov & Voropaev, 2004). After carrying out these works, the author of this article filed an application for a scientific disclosure, and in 2011 the relevant patent was granted (Makarov, 2012b).

Radiant heat transfer is common for torch and electric arc furnaces.

The essence of the laws for heat radiation from gas volumes consists in the following. Cylinder gas volumes may be inscribed in gas radiating volumes in the form of ellipsoid and elliptic cylinder, formed in fuel flaring, that is radiation from gas ellipsoid, elliptic cylinder may be approximated by radiation from gas cylinder volumes.

From the laws of radiation from coaxial cylinder gas layers, volumes, radiation from all the layers of large cylinder gas volumes may be equivalently replaced by radiation from central cylinder gas volume of a small diameter. Make certain of justification for these arguments. Let us assume, that isothermal cylinder gas body of 10 meters height, 4.9 meters in diameter is formed in steam boiler box during fuel firing (Figure 4). Let us assume that 18 kg of fuel oil is burnt in furnace every second and hereby all the 30×10^{30} chemical atoms, involving in the combustion process, implicate in photon energy emission. The power of 700 MWth, which is steadily distributed along the full cylinder gas volume, releases hourly in a furnace. Let us divide isothermal cylinder radiative and absorbing gas volume by three equal in gas cylinder volume, layers 1-3. Every layer has equal 10×10^{30} atoms. We have calculated angular radiation coefficients, average beam path lengths, radiation fluxes density on dF calculated platform for every cylinder, layer (Makarov, 2012c, 2012d). As a result of calculations, we have disclosed four laws for heat radiation from isochoric coaxial cylinder gas volumes (Makarov's regularities) (Makarov, 2014e).

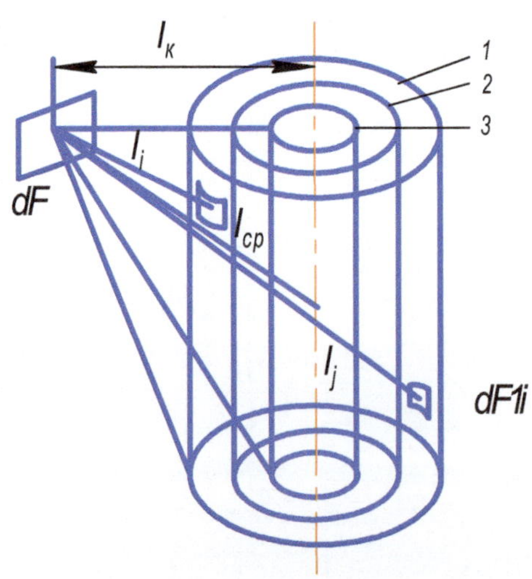

Figure 4. To the calculation of radiant fluxes from coaxial cylinder gas volumes 1-3. l_i, l_j, l_{av} – the distance from dF_{1i} surface elements and arithmetic mean distance to dF calculated area.

5.1 The First Law:

"Elementary view factors of radiation from coaxial cylindrical gas layers are equal to each other ":

$$\phi_{F_1 dF} = \phi_{F_2 dF} = \phi_{F_3 dF} = \phi_{F_i dF} \tag{5}$$

where $\phi_{F_1 dF}$, $\phi_{F_2 dF}$, $\phi_{F_3 dF}$, $\phi_{F_i dF}$ – elementary view factors of radiation from the first, second , third, *i*-th coaxial cylinder layers on the calculated area *dF*.

5.2 The Second Law:

"The average path length of beams from coaxial cylindrical gas layers to the calculation area is equal to the arithmetic mean distance from the symmetry axis of cylindrical layers to the calculation area":

$$l_{av} = l_1 = l_2 = l_3 = l_i = \left(\sum_{j=1}^{n} l_j / n \right), \tag{6}$$

where, l_1, l_2, l_3, l_i, l_{av} – the average beam path length from the first to the *i*- th cylinder gas layer , respectively, the arithmetic mean distance from the summetry axis to the calculation area dF.

The third law: "The densities of radiation fluxes falling from isochoric coaxial cylindrical gas volumes the calculation area are equal to each other" :

$$q_{F_1 dF} = q_{F_2 dF} = q_{F_3 dF} = q_{F_i dF} \tag{7}$$

5.3 The Fourth Law:

"The total density of radiation fluxes falling on the calculation area from a few radiating and absorbing cylindrical gas volumes (layers) forming the flame is equal to the density of radiation flux from a small diameter coaxial cylindrical gas volume falling on the calculation area at the radiation power releasing in a small diameter cylindrical gas volume and equal to the total radiation power releasing in all coaxial cylindrical gas volumes radiating on the calculation area,"

$$q_{F_3 dF} = \sum_{i=1}^{n} q_{F_i dF} \tag{8}$$

5.4 The Fifth Law:

"In modeling a flame and an arc by isothermal coaxial cylindrical gas layers and calculating radiative heat transfer, the results of calculations include the volume radiation and absorption in all flame layers and the transfer of heat between them and all heating surfaces."

It follows from five laws of heat radiation, that radiation from isothermal coaxial cylindrical gas volumes is invariant, i.e., the parameters of their radiation, properties are equal to each other.Hence, the radiation from any large gas volume of a high power may be equivalently replaced by radiation from coaxial cylinder gas volumes of a small diameter equal in power .The laws for heat transfer from gas volumes spare us from triple , quadruple integration in radiative heat transfer calculating and make it possible to estimate parameters for radiation from cylinder gas volumes by single integrating of geometric and trigonometric dependences of coaxial cylinder volume of small diameter or , as it is often determined, linear radiation source. Integrating geometrical relations between linear radiation sources and heating surfaces at their any attitude, the author of this article has practically solved the problems of determinating radiation parameters of mentioned bodies and surfaces (Makarov, 2014e).

6. Calculation of Heat Transfer in the Firebox of a Tgmp-314 Steam Boiler

Let us calculate heat transfer in the fire box of a TGMP-314 steam boiler with reference to distribution of isotherms and power along the torch height which is the case for opposite arranged burners (Abrutin, Antonov, & Usman, 1981; Usman, Eniakin, & Filatov, 1985; Zroichikov, Eniakin, Glusker, 2002). The firebox of a 300 MW power unit has the shape of a rectangular parallelepiped with height H_f = 35 m, width a = 14 m, and depth b = 7 m. The boiler operates on fuel oil and is equipped with 16 burners with a total throughput capacity of 67 t/h , which are installed in opposite directions in two tiers,8 in each tier at 3 and 6 m height from the bottom surface (Figure 5).

The air excess factor in the furnace $\alpha = 1.03$, and the gas recirculation ratio $r = 0.16$ (Abrutin, Antonov, & Usman, 1981). With these values of air excess factor and gas recirculation ratio the average values : particle diameter d_p= 0,278 mcm, density $\rho = 2\times10^3$ kg/m^3, and concentration μ= 0,06 g/m^3 are determined in Blokh (1984). The medium attenuation coefficient k = 0,162.

Figure 5 shows the distribution of isotherms along the height of the TGMP-314 steam boiler box with opposite directed burners.

The torch fills the entire firebox chamber along the firebox height and perimeter and has the shape of a The isotherms divide the straight elliptical cylinder into six volume bodies.

Six volume bodies with ellipses in their bases and vertices and with a parabolic generatrix (the torch vertical parts) have the shape of elliptical paraboloids resting on the seventh volume body having the shape of a truncated ellipsoid of revolution (the horizontal part of torch). Ellipsoid of revolution is spaced in the interior of a truncated ellipsoid, which is formed by isotherm of 1900 ^{0}C in the shape of volume zone with maximal temperature in the firebox of a TGMP-314 steam boiler.

For increasing the calculation accuracy , we inscribe four straight circular cylinders in a torch and simulate it by seven tiers 1-7 of vertical cylinder sources (tier of fours horizontal cylinder sources),and by two tiers 8-9 of horizontal cylinder sources (tier of fours horizontal cylinder sources) , while they are spaced in the course of burner axes (Figure 5).

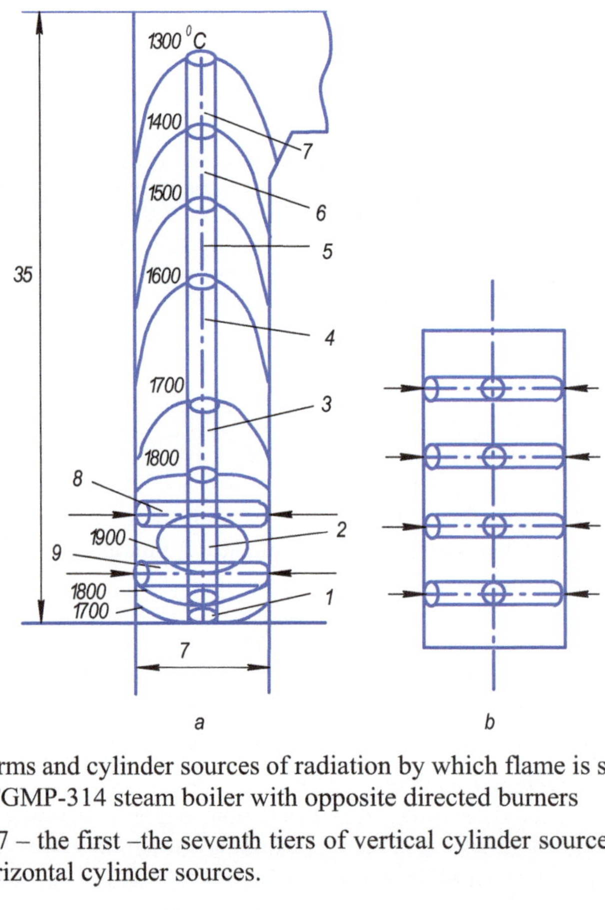

Figure 5. Distribution of isotherms and cylinder sources of radiation by which flame is simulated in the firebox of a TGMP-314 steam boiler with opposite directed burners

a – side view; b – top view; 1-7 – the first –the seventh tiers of vertical cylinder sources, respectively; 8, 9 – the eighth and the ninth tiers of horizontal cylinder sources.

The power releasing in the torch is determined from the expression:

$$P_f = Q_H^P B_k \tag{9}$$

where Q_H^P is the fuel heating value equal to 41 MJ/kg, and B_k is the fuel flow rate equal to 67 t/h; P_f = 770 MW.

For calculating the power distribution along the torch height, let us make a proportion which composed of temperatures and zone volumes of the torch. For any flame a proportion for determining the power releasing in each of volume zones may be written:

$$P_1 : P_2 : P_3 \ldots P_i = T_1^3 V_1 : T_2^3 V_2 : T_3^3 V_3 \ldots T_i^3 V_i \ , \tag{10}$$

where $P_1 - P_i$ – the powers releasing in the first- *i*-th volume zones of a torch; $T_1 - T_i$ – the temperatures of the first – *i*-th volume zones respectively.

The product obtained during calculation the proportion (10) torch power fraction, releasing in the torch volume zone into the torch power, gives the heating power , releasing in the volume zone given.

According the Equation (10),we obtained the powers releasing in volume bodies, torch zones and the power, releasing in every linear radiative source of 1-9 tiers : P_1 = 15 MW; P_2 = 24 MW; P_3 = 24 MW; P_4 = 31 MW; P_5 = 14 MW; P_6 = 14 MW; P_7 = 7,5 MW; P_8 = 33 MW; P_9 = 33 MW .

In the lower volume zone, restricted by isotherms of 1700 ^{0}C and made up the third part of the firebox space 66 % of torch power is released. In the lower one third part of the firebox space at a distance of 0 to 11 m over the bottom height 66 % of fuel is burned that causes here the torch power concentration and rise of volume zone with the maximal temperature of 1900 C (Figure 5). It is known, that volume burning velocity increases with the temperature increase and vice versa, with the increase in the volume burning velocity,volume temperature increases that causes the power concentration and high temperatures in the lower volume zone of the firebox. The third height part of torch volume between isotherms of 1700 and 1500 ^{0}C, accounts for 23 % of fuel rate and torch power and 11 % of fuel is burning and 11 % of torch power is releasing in the one third higher part of torch volume torch power .

We calculate the incident integral heat flux density from the torch on the following surfaces: on the frontal wall over the wall's vertical symmetry axis; on the frontal wall over the vertical line 4.7 m off the frontal wall vertical symmetry axis or 2.3 m off the parallepiped side edge (firebox); on the sidewall along the vertical symmetry axis; on the sidewall along the vertical line 2.0 m off the sidewall vertical symmetry axis or 1.5 m off firebox side edge. Calculation of the incident integral radiation flux density from every cylinder source of horizontal and vertical volume zone on the *i*-th wall surface elemental area was made by Equation (Makarov, 2014e):

$$q_{inj} = \frac{\phi_{ji} P_j e^{-kl}}{F_i} \ , \tag{11}$$

where ϕ_{ij} is the local view factor of radiation from the *j*-th cylindrical radiation source on the *i*-th area, which is determined from the analytical expressions given in (Makarov, 2014e); P_j is the power of the *j*-th cylindrical source, MW; F_i is the surface area of the *i*-th elemental area,m^2; and l is the average beam path length, m.

The density of incident integral radiation flux from the torch on the *i*th elemental area is determined as the sum of integral heat flux densities from all cylindrical sources:

$$q_{iqi.fl} = \sum_{j=1}^{36} q_{inj} . \tag{12}$$

The density of incident integral radiation flux on the *i*th area caused by reflection of torch radiation from the bottom, ceiling, walls as well as the density of incident integral radiation flux from the radiant surfaces of the bottom, ceiling, walls and convective flux density is not determined due to the fact, that they add up to no more than 5 % of incident integral heat flux density from the torch to the *i*th elementary area p is determined as the sum of integral heat flux densities from all cylindrical sources if the refractory coat is absent (Makarov, 2014e).

Figure 6 shows the calculation results of incident integral radiation fluxes on the waterwall surfaces of the firebox of a TGMP-314 steam boiler with opposite directed burners.

Graph 2 characterizes the distribution of integral radiation flux density along the frontal wall vertical symmetry axis. In consequence of symmetry of the firebox, it also characterizes the distribution of integral radiation flux density along the rear wall vertical symmetry axis. According the measurement results the distribution of integral radiation fluxes along the frontal and side vertical symmetry axis is the same and is characterized by curve *1* (Abrutin, Antonov, & Usman, 1981). The measurement results (curve *1*) did not vary more than10 per cent from calculation results along the frontal wall vertical symmetry axis, testifying that the developed firebox of a TGMP-314 steam boiler adequately reflects the real conditions.

The maximal integral radiation fluxes are observed on the front and rear walls at a height of 12-16 m (4-5 m above the top tier of burners). The waterwall surfaces at a height of 12-16 m experience the maximal radiation from the horizontal volume zone and rear walls at a height of 12-16 m (4–5 m above the top tier of burners). The waterwall surfaces at a height of 12–16 m experience the maximal radiation from the horizontal volume zone and

Zone of maximum heat release at height of 2-5 m from the back bottom is characterized by the maximal integral radiation flux densities from the flame to the walls equaled to 680 kW/m^2.Heat release in torch decreases along the torch height, as integral radiation flux densities from the torch decrease on the waterwall surfaces along the wall height.

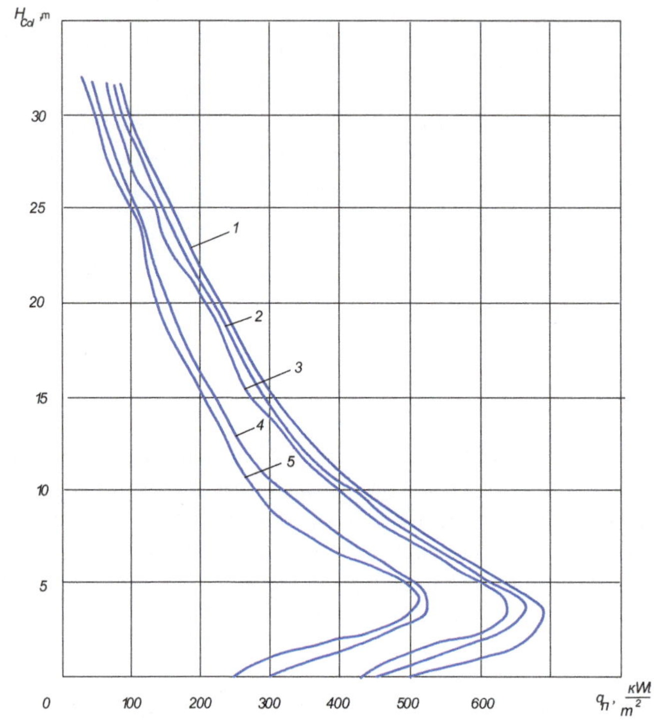

Figure 6. The distribution of integral radiation flux density along the frontal and side wall height of the firebox of a TGMP-314 steam boiler with opposite directed burners

1 – the measurement results of the distribution of integral radiation flux density along the frontal and side walls vertical symmetry axis (Abrutin, Antonov, & Usman, 1981); 2 – the calculation results of the distribution of integral distribution flux density along the frontal wall vertical symmetry axis ; 3 – the same is along the side wall vertical symmetry axis; 4, 5 – the calculation results of distribution of the integral flux density along the frontal wall vertical symmetry axis ; calculation results of distribution of the integral radiation flux density along the height of lateral wall respectively at a distance of 2 m from lateral wall vertical symmetry axis and along front wall at a distance of 4.7 m from front wall vertical symmetry axis.

The surface zone of frontal wall is located at a height of 4.5 m from the surface bottom opposite the volume zone of the maximal heat release and torch temperature which is characterized by the maximal nonuniformity of integral radiant heat flux densities from the torch (Graph 1). The maximal density of integral radiant flux of 680 kW/m^2 is observed in the center of the waterwall surfaces, the density of integral radiant fluxes decreases by 230 kW/m^2 to the wall periphery. The maximal nonuniformity of integral radiant heat flux densities is observed on the wall surfaces opposite the zone of the maximal heat release from torch: the maximal density of integral radiant heat flux in the center of the frontal wall three times higher than the density of integral radiant heat flux at the periphery of the frontal wall.The distribution of integral radiant flux densities along the lateral wall surfaces are characterized by similar nonuniformity.

It is known, that with increasing the density of incident integral radiation flux on the waterwall surfaces, increases the density of deposits inside the tubes (Davidzon, 2001).

The nonuniformity of integral radiant heat flux densities along the perimeter decreases at a height of 15–25 m from the bottom, the difference in the maximal in the center and minimal at the wall periphery densities of integral radiant fluxes doesn't amount to more than a factor of 2. The calculated data of distribution of integral radiant heat flux densities along the frontal wall width of steam boiler box with opposite arranged burners agree closely with the results from measurements of deposits inside the tubes.

The deposits are distributed ununiformly over the tube waterwall surfaces: the maximal number are observed in the zone of the largest heat flow near the burners (along the furnace height) and in the center of each wall (Kozlov, Zroichikova, & Belov, 2003), the minimal – at the furnace wall periphery.

7. Conclusions

Thus, theory for radiative heat exchange based on the laws for heat radiation from surfaces of black body radiation is replenished by five laws for heat radiation from gas volumes. The disclosed laws for heat radiation from gas volumes allow us to develope the torch model in the form of radiative gas volume and modern method for calculating heat transfer in torch furnaces, fireboxes. Volume torch pattern, consisted of a set of coaxial radiating and absorbing cylinder gas volumes, layers is used in radiative heat exchange calculation of torch furnaces (Makarov, Chernyshov, & Voropaev, 2006; Makarov, 2012d), steam boiler boxes (Makarov, 2003; Makarov, 2003; Makarov, 2003; Makarov, & Voropaev, 2004), combustion chambers of gas-turbine plants (Makarov, Chernyshov, & Voropaev, 2006; Makarov & Voropaev, 2004). Calculation results agree closely with measured data of heating fluxes and temperatures in furnaces, fire boxes, combustion chambers (Kozlov, Zroichikova, & Belov, 2003). The discovery enabled us to design new torch furnaces, fire boxes, combustion chambers. On the basis of disclosed laws for radiation from large gas volumes, it was developed 18 innovative steam boiler boxes, torch heating and electric arc melting furnaces, combustion chambers, ways of heating and melting the metal, providing the improvement in the quality of production, reduction in time of heating and melting the metal, improving the efficiency of plants, decrease in fuel and electric energy consumption, unit pollutant emissions (Makarov, 2012d; Makarov, 2014e; Abrutin, Antonov, & Usman, 1981; Usman, Eniakin, & Filatov, 1985; Zroichikov, Eniakin, Glusker, 2002).

The developments are protected by 18 patents (Makarov, & Shevchenko, 2010; Makarov & Voropaev, 2012; Makarov, Kruglov, & Rybakova, 2014a; Makarov, Kruglov, Rybakova, 2014b).

References

Abrutin, A. A., Antonov, A. Y., & Usman, Y. M. (1981). Pecularities of heat exchange in powerful oil-fired unit with bottom arrangement of burners. *Electricheskiye stancii, 9*, 27-30.

Adzeriho, K. S. (1975). *Lectures on Radiation transfer theory/ under the editorship of M.A. Elyashevich*. Minsk: BSU publishing.

Alekhnovich, A. N. (2007). The thermal efficiency factor of furnace waterwalls as applied to the standard method for thermal calculation of boilers. *Therm. Eng., 9*, 23-29.

Blokh, A. G. (1984). *Heat Transfer in Steam Boiler Fireboxes*. Energoatomizdat, Leningrad.

Blokh, A. G., Juravlev, Y. A., & Ryzkov, L. N. (1991). *Radiative heat exchange: Reference*. M.: Energoatomizdat.

Chandrasekar, S. (1993). *Radiation transfer*. M.: IL.

Davidzon, M. I. (2001). The effect of heat flux density on the formation of in tube deposits. *Therm. Eng., 1*, 72-73.

Energiya, M. (1973). *Thermal Design of Boiler Units (a Standard Method)* (in Russian).

Germann, M. L. (2000). Engineering method for calculating temperature condition of shell boilers with terminal furnace. In M. L. Herman, V. A. Borodylia, E. F. Nogotov, & G. I. Palchenok (Eds.), *The IV Minsk International Forum on heat-mass exchange (MIF-2000): Proceedings of the Forum.P.2. Radiative and combined heat transfer* (pp. 21-31). Minsk: ANK publishing. A.V. Luikov Heat and Mass Transfer Institute of the National Academy of Sciences of Belarus.

Kirpichev, M. V. (1924). About heat transfer in boilers. *Usp.Tepl. Tekhn., 2*, 17–35.

Kozlov, Y. V., Zroichikova, T. V., & Belov, V. A. (2003). Ways of improving the reliability of furnace waterwalls. *Electricheskiye stancii, 5*, 17-19.

Krivandin, V. A., & Egorov, A. V. (1989). *Thermal performance and constructions of iron and steel furnaces: textbook*. M.: Metallurgiya.

Kutateladze, S. S. (1990). *Heat Transfer and Flow Friction: Handbook.* Energoatomizdat, Moscow. (in Russian).

Makarov, A. N. (2000). Determining the view factors for radiation from a line source and from the fireball of steam boiler furnaces. *Therm. Eng., 8*, 63-66.

Makarov, A. N. (2003a). Heat-flux distribution in steam boiler box TGMP – 204. *Electric Power Plants, 1*, 20-25.

Makarov, A. N. (2003b). Heat exchange in electric arc and flame furnaces, steam-boiler boxes: Tutorial, Tver, TSTU.

Makarov, A. N. (2012a). *Laws of Heat Transfer for Electric Arc and Torch in Metallurgical Furnaces and Power Installations*. TvGTU, Tver (in Russian).

Makarov, A. N. (2012b). A Regular correlation between the parameters characterizing radiation from isothermal coaxial cylindrical gas layers generated during flame combustion of fuel and during the burning of electric arc in metal vapors at atmospheric pressure (Makarov's regularities): A Diploma No. 417. In V. V. Pototskii (Ed.), *Scientific Discoveries* (RAEN, Moscow, pp. 33–37).

Makarov, A. N. (2012c). Regularities accompanying the burning of electric arc and flame in metallurgical furnaces. Part 1. Geometrical physical and analytical model of flame and arc. *Electric Metallurgy, 7*, 22-32.

Makarov, A. N. (2012d). Regularities accompanying the burning of electric arc and flame in metallurgical furnaces. Part 2. Use of scientific discovery for decreasing the power consumption in metallurgical furnaces. *Electric Metallurgy, 8*, 28-35.

Makarov, A. N. (2014e). Regularities of heat transfer in the gas layers of a steam boiler furnace flame. Part II. Gas layer radiation laws and the procedure for calculating heat transfer in furnaces, fire boxes, and combustion chambers developed on the basis of these laws. *Thermal Engineering, 61*(10), 717-723. http://dx.doi.org/10.1134/S0040601514100073

Makarov, A. N., & Dunaev, A. Y. (2004). Calculation of heattransfer in a regenerative heating well. *Prom. Energ., 1*, 49-53.

Makarov, A. N., & Krivnev, E. I. (2000). Procedure for calculating rational attitude of a flame for Heat transfer optimization in in industrial furnace. *Prom. Energ., 2*, 36–42.

Makarov, A. N., & Krivnev, E. I. (2002). Calculation of heatfluxes in the TGMP-204 steam boiler furnace. *Prom. Energ., 2*, 38–42.

Makarov, A. N., & Shevchenko, M. N. (2010). A firebox for firing gas–oil fuel. *RF Patent No. 2400668 (RU 2400668 C1) F23C3/00, Izobret., No. 27.*

Makarov, A. N., & Svenchanskii, A. D. (1992). Optimum Thermal Conditions of Electric Arc Furnaces. Energoatomisdat,Moscow.

Makarov, A. N., & Voropaev, V. V. (2004). Simulation of the torch by radiative cylinders and calculation of heat transfer in the TGMP-314 steam boiler furnace. *Therm. Eng., 8*, 48-52.

Makarov, A. N., & Voropaev, V. V. (2012). A Regenerative Soaking Pit. *RF Patent No. 2457262 (RU 2312907 C1) C21D9/70, Izobret., No. 35.*

Makarov, A. N., Chernyshov, D. V., & Voropaev, V. V. (2006). Calculation of heat transfer in the combustion chamber of a stationary gas turbine unit. *Prom. Energ., 1*, 31–36.

Makarov, A. N., Kruglov, E. V., & Rybakova, V. V. (2014a). DC Arc –steel melting furnace. *RF Patent No. 2516896 (RU 2312907 C1) C21D9/70, Izobret., No. 35.*

Makarov, A. N., Kruglov, E. V., Rybakova, V. V. (2014b). Reheating furnace with ring bottom. *RF Patent No. 2517079 (RU 2517079C1) C21D9/70, Izobret., No. 35.*

Mastrukov, B. S. (1972). *Thermotechnical calculations of industrial furnaces: textbook*. Metallurgy, Moscow.

Mitor, V. V. (1963). *Heat Exchange in Steam Boiler Boxes*. M: Mashgiz.

ND. (1952). *Regulations for the Thermal Design of a Boiler Unit*. Gosenergoizdat, Moscow, Leningrad (in Russian).

ND. (1998). *Thermal Design of Boiler Units (a Standard Method)*. NPO TsKTI, St. Petersburg (in Russian).

Nevskii, A. S. (1963). *Heat Transfer in Open Hearth Furnaces*. Metallurgizdat, Moscow (in Russian).

Nevskii, A. S. (1971). *Heat Exchange by radiation in Furnaces and Fire Boxes*. M: Metallurgiya.

Pinar Mengus, M. (Ed.) (1995). *Radiative Transfer-1. Proceeding of the First International symposium on Radiation Transfer*. Kusadasi, Turkey.

Shreter, V. N. (1931). Steam Boilers in Calculations and Figures. Gostekhizdat, Moscow (in Russian).

Smelov, V. V. (1972). *Lectures on neutron transfer theory*. M.: Atomizdat.

Telegina, A. S. (Ed.) (1993). *Heat engineering calculations in metallurgical furnaces: textbook*. M.: Metallurgiya.

Usman, Y. M., Eniakin, Y. P., & Filatov, A. V. (1985). Comparative studies of gas, oil burning and its mixture in the TGMP-314 boiler furnace with bottom arrangement of burners. *Electricheskiye stancii, 1*, 33-37.

Zroichikov, N. A., Eniakin, Y. P., Glusker, B. N. (2002). Developing the TGMP-314 boilers aided by cyclone primary furnaces for decreasing the irritants and increasing the reliable performance of burner devices and heating surfaces. *Therm. Eng., 1*, 17-21.

The Effect of Position of Heated Rod in Tube Banks on the Heat Transfer Coefficient

Ehsan Fadhil Abbas[1]

[1] Department of Refrigeration and Air Conditioning, Kirkuk Technical College, Foundation of Technical Education, Iraq

Correspondence: Ehsan Fadhil Abbas, Department of Refrigeration and Air Conditioning, Kirkuk Technical College, Foundation of Technical Education, Iraq. E-mail: ehsanfadhil@gmail.com

Abstract

Heat transfer in flow across a bank of tubes is of a particular importance in the design of heat exchangers. Heat exchangers are used in numerous services and industrial applications. Experimental studies were performed, carried out in cross-flow tube banks. They contain copper heated rod and 17 Aluminum rods arranged in the form of staggered and at arrange of Reynolds number 2700 to 21530. The experimental results indicated that the location of the heated rod in the work section affects the value of heat transfer coefficient thus this value is increased when the heated rod location changes both the y and x-directions. The experimental data revealed good agreement with Zukauskas correlation in four columns.

Keywords: tube banks, cross-flow heat exchanger, convection from tube banks, heat transfer coefficient

Nomenclature

A : Surface area.	m^2
A_e: Effective surface area.	m^2
C : Column.	
C_p: Specific heat capacity of copper.	$J/kg.^oC$
d : Diameter.	m
h : Heat transfer coefficient.	$W/m^2\ .^oC$
k : Thermal conductivity of copper	$W/m.^oC$
k_a: Thermal conductivity of air.	$W/m.^oC$
m : Mass of copper.	kg
Nu_d: Nusselt number based on rod diameter.	
Pr : Prandtl number.	
Re_d: Reynolds number based on rod diameter.	
S_D: Diagonal pitch.	m
S_L: Longitudinal distance between two consecutive tubes	m
S_T: Transverse distance between two consecutive tubes	m
T : Temperature of the surface at any time.	^oC
T_∞ : Ambient temperature.	^oC
T_o : Initial temperature.	^oC

1. Introduction

A tube bundle is one of the simplest but more effective geometries in heat transfer applications. The bundle consists of a multiple cylindrical bars or circular tubes. The bundle tube heat exchangers are found in numerous industrial applications, ranging from low- tech applications such as domestic water heaters to high-tech applications such as hypersonic jet engine. Their size similarly varies extensively, from massive power plant boilers to extremely small chip coolers. In general, for convenience, tube bundles are installed such that the shell-side stream flows perpendicularly to the tubes (Jeong et al., 2011). Tube banks are usually arranged in-line or in staggered manner and are characterized by the dimensionless transverse, longitudinal, and diagonal pitches. Typically, one fluid moves over the tubes, while the other fluid at a different temperature, passes through the tubes (Kan, Culhum, & Yovanovich, 2006). Below is a brief summary of a number of researchers who experimentally, analytically and numerically have developed and determined the rate of heat transfer and flow structure around circular cylinder placed in a tube bank heat exchanger.

Brevoort & Trifford (1942) have presented an experimental investigation of the details of the flow of fluids across tube banks. Information that clarifies the picture of the flow has been obtained by surveys of total dynamic and static pressure by visualization of flow through the use of titanium tetrachloride smoke. Martin & Gnielinsk (2000) have discovered a new type of analogy between pressure drop and heat transfer, that may be used in chevron-type plate heat exchanger in packed bed, and tube bundles in cross-flow for both inline and staggered arrangements. The new method is to calculate heat and mass transfer from pressure drop in tube bundles with a single adjustable parameter of fraction of total pressure drop due to friction, which has been found to be 0.5 in all cases recommended especially for the staggered tube bundle. However this method in case of inline bundles gives slightly less accuracy than the older one in the law of Reynolds number range. Khan et al. (2006) proposed analytical method to investigate the heat transfer from tube banks in cross-flow under isothermal boundary conditions in both cases i.e in-line and staggered arrangements. The results showed that the models for in-line and staggered arrangements are applicable for use over a wide range of Reynolds and Prandtl numbers as well as longitudinal and transverse pitch ratios. Horvat, Leskovar, and Mavko (2006) compared heat transfer coefficient in tube bundle cross-flow heat exchanger for different tube shapes. Numerical analysis of heat transfer was performed for the heat exchanger segments with the cylindrical, the ellipsoidal and the wing-shaped tubes in the staggered arrangement. This comparative study showed that the wing-shaped which gives the best result based on these average values, drags the coefficient and Stanton number with respect of Reynolds number. Takemoto, Kawanishi, and Mizushima (2010) studied heat transfer in the flow through a bundle of heated tubes and transitions of the flow and discuss the effect of transition of flow upon the heat transfer characteristics numerically. The results of this study indicated that the physical quantities such as the Nusselt number and pressure exhibit discontinued jumps with continuous change in the Reynolds number. Ji et al. (2011) presented and investigated the effects of the evaluation method on the average heat transfer coefficient for a mini-channel tube bundle. The study was performed with a tube diameter of 1.5 mm and shell side Reynolds number of $3000 \leq Re \leq 7000$. The average convection heat transfer coefficient can be estimated by a surface temperature method based on Newton's cooling law and the LMTD method. The result showed that the convection heat transfer coefficient by LMTD method was 22.6% smaller than that yielded by the surface temperature method. Mehrabian (2007) evaluated experimentally the value of heat transferred coefficient directly between a cylindrical copper element and the air flowing past it in specially designed test rig. The comparison of experimental results with predications of a standard correlation shows that the two sets of data are in close agreement; also the experimental data gives a linear relationship between upstream velocity and pressure drop across the tube bank. Abd (2012) studied the effect of the external shape of a heated rod (cylindrical, square and triangular) on the convection heat transfer coefficient when the body is subjected to free or forced convection heat transfer by cross flow air stream. The results were in good agreement with the standard correlation.

The objective of this paper is to study experimentally the effect of heated rode position on the value of heat transfer coefficient in cross flow heat exchanger in a range of Reynolds number 2700 to 21530 and compare the test results with two standard correlation methods.

2. Equipment Description

The TE93 cross-flow heat exchanger apparatus, manufactured by (TecQuipment) was used in this study. A constant speed (2500 rpm) electric centrifugal fan (1hp) draws ambient air through duct having a square cross section of 12.5*12.5 cm (the working section) is transparent and includes holes for insertion of aluminum rods arranged as staggered form at right angle to the airflow. The air flow rate over rods regulated by a throttling valve installed to the discharge end. The base of the working section consists of two static pressure tapping used to measure the pressure drop across the rods. The air velocity measure directly by pitot tube as shown in Figure (1). The components and all dimensions of work section are shown in Figure 2. A special heated rod is placed at a selected position in the working section, and it is made of several parts; the main part heating rod is a pure copper, machined and drilled to allow for insertion of a thermocouple, and the two ends of the rod are closed by

Teflon. The Teflon ends are for thermal insulation. The schematic diagram of the heated rod shown in Figure 3. And technical specification of heating rod a shown in Table 1. The apparatus consists of a control and instrumentation unit, which is included the electrical control elements such as main switch and on/off control for the electric fan and heater compartment for heated rod. Also included digital measuring devices for each of the rod surface temperature, inlet and outlet air temperatures measured by thermocouples type K, pressure drop across working section which was measured by tube adapters fit into the ends of the pipe work. All collected data were shown directly through a small display. This unit can be connected to PC by VDAS-B hardware to record all required data instantaneously during a test.

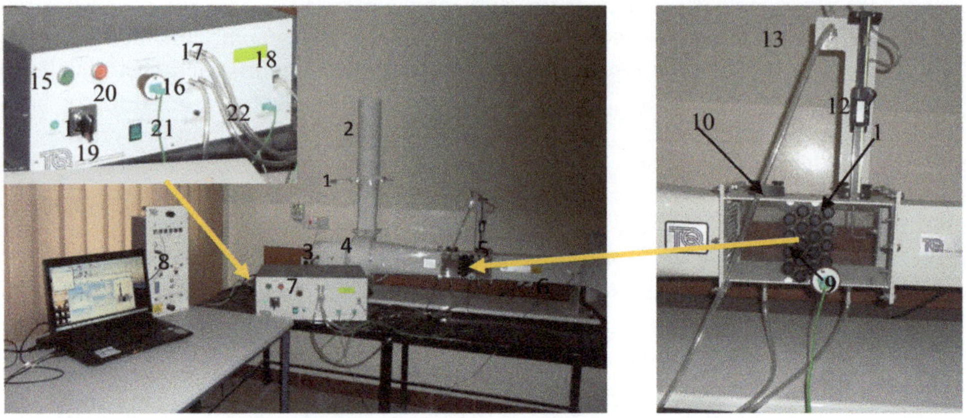

Figure 1.The experimental apparatus

1	Throttle valve	2	Discharge pipe	3	Electric motor	4	Centrifugal fane
5	Work section	6	Suction pipe	7	Control and instrument unit		
8	VDAS hardware	9	Heated rod	10	Aluminum rods	11	Pitot tube
12	Digital indicator	13	Pitot assembly	14	Main switch	15	Fan switch
16	Downstream hose	17	Upstream hose	18	Display monitor	19	Heater switch
20	Heater compartment and heated rod	21	Zero pressure button	22	Ambient thermocouple		

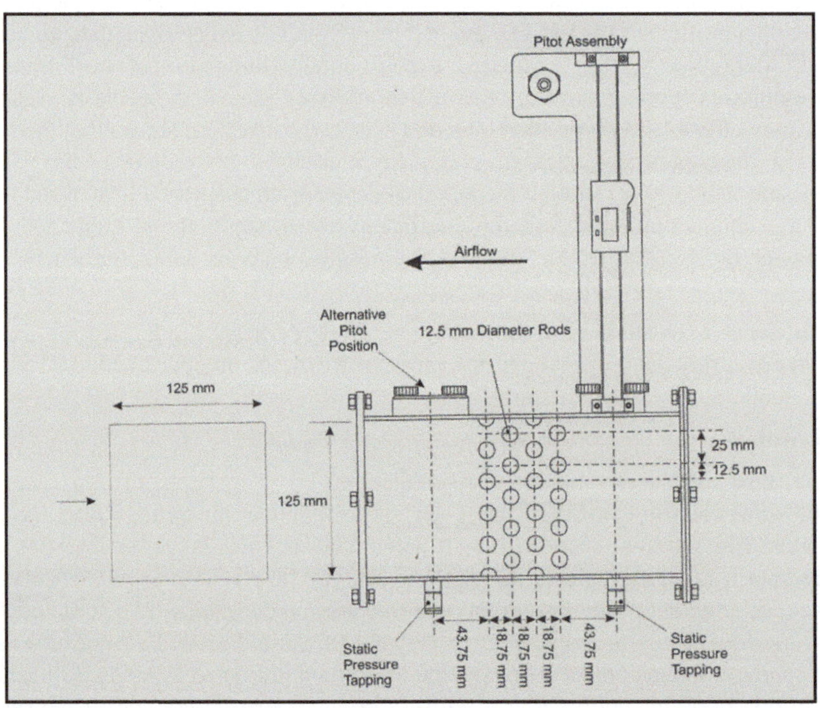

Figure 2. Schematic diagram of the working section (TE93 Cross-Flow Heat Exchanger, User Guide 2009)

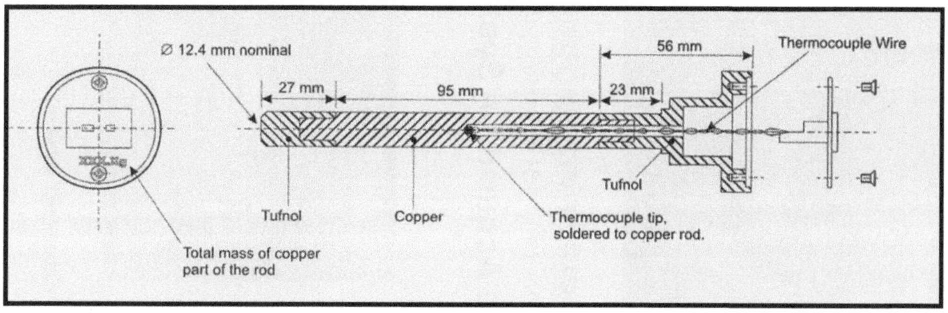

Figure 3. Schematic diagram of the heated rod (TE93 Cross-Flow Heat Exchanger, User Guide 2009)

Table 1. Specification of heated rod (TE93 Cross-Flow Heat Exchanger, User Guide 2009)

Item	Specification
Material	Copper
Mass	152 g
Diameter	0.0124 m
Length	0.095 m
Effective Length	0.1034 m
Surface area	0.0037 m^2
Effective Area	0.00403 m^2
Specific Heat capacity	380 J/kg.°C

3. Experimental Procedure

The experimental procedure in the present study was developed to obtain the value of Nusselt number under various air flow rate regulated by throttling valve. Before starting the experimental work all aluminum rods fit into the working section, except the hole that was allocated for heated rod. The pitot assembly is in the upstream position, adjusted so that its probe points upstream, in the middle of the working section and then connected to the socket of the differential pressure input on the control and instrumentation unit. After these procedures, the test rig becomes ready to proceed with the experiment as follows:-

a. The control and instrumentation unit was set up and the heated rod is inserted into heater compartment, and heated to a maximum value of about 75°C.
b. The throttle valve was open fully (100%).
c. Heater rod was removed after it reached to set temperature and fit it into position C1R1.
d. The fan was switched on, while the software stared up to record required data every ten seconds until the surface temperature of heated rod reached the ambient temperature
e. The experimental procedure (a to d) was repeated with throttle valve open at 90%, 80% , 70%, 60%, 50%, 40%, 30%, 20%, and 10%.
f. The experiment was repeated for different location of heated rod in each of the other rows in the first column (C1R2, C1R3, C1R4, and C1R5), and same procedure repeated in other three columns sequentially.

4. Computational Method

a) Assumptions

- All heat generated in the heated rod is transferred to the surrounding by convection.
- Heat transferred by radiation is neglected.
- Temperature variations within the mass can be neglected in comparison with the temperature difference between the mass and the surrounding fluid and surface-convection resistance, is large as compared with the internal-conduction resistance (Lumped-capacity system). Such analysis may be expected to yield reasonable estimates when the following condition is met (Holman, 1997):

$$\frac{h(V/A)}{k} < 0.1 \qquad (1)$$

b) Problem Formulation

Energy balance:

Change of internal energy of the solid= Net heat transfer from heated rod (Copper rod) to the air stream

$$\rho C_p V \frac{dT}{d\tau} = -hA(T - T_\infty) \quad (2)$$

When $\boldsymbol{T}$ is greater than T_∞ , the mass is losing heat to the surroundings. Thus, the rate of addition is the negative of the heat loss rate.

$$\text{Let } \theta = T - T_\infty \quad (3)$$

Differentiating the Equation 3, the equation becomes:

$$d\theta = dT \quad (4)$$

Substitute Equations 3 and 4 into the Equation 2, we have

$$\rho C_p V \frac{d\theta}{d\tau} = -hA\theta \quad (5)$$

Rearrange Equation 5

$$\frac{d\theta}{\theta} = -\frac{hA}{\rho C_p V}.d\tau \quad (6)$$

Since this is a first-order differential equation, when we integrate it, there will be an arbitrary constant in the solution. To evaluate it, we'll need an initial condition on the temperature of the mass (T_o).

At $\tau = 0$ $\quad\theta = \theta_o$

By integrating both sides of Equation 6

$$\int_{\theta_o}^{\theta} \frac{d\theta}{\theta} = -\frac{hA}{\rho C_p V}\int_0^{\tau} d\tau \quad (7)$$

The result yields:

$$\ln(\theta - \theta_o) = -\frac{hA}{\rho C_p V}\tau \quad (8)$$

$$\ln\left(\frac{\theta}{\theta_o}\right) = -\frac{hA}{\rho C_p V}\tau \quad (9)$$

$$\ln\left(\frac{T - T_\infty}{T_o - T_\infty}\right) = -\frac{hA}{\rho C_p V}\tau \quad (10)$$

A is replaced into A_l because the effective length of the heater rod is actually slightly longer than the 95mm length exposed copper part. The Teflon ends are not perfect insulator and conduct a small amount of heat. This gives a larger effective length. Test on lengths of solid copper rod of the diameter has shown that the heated rod has the same properties as a solid rod that is 8.4mm longer. This gives a nominal effective length of 103.4 mm (TE93 Cross-Flow Heat Exchanger, User Guide 2009)

$$\ln\left(\frac{T - T_\infty}{T_o - T_\infty}\right) = -\frac{hA_e}{\rho C_p V}\tau \quad (11)$$

Equation 11, giving a chart of time against temperature should yield a straight line of slope M,

$$M = -\frac{hA_e}{\rho C_p V} \quad (12)$$

Since the value of heat transfer coefficient ***h*** can be found from Equation 12, as other factors are known. Where the value of ***h*** can be found from

$$h = -\frac{mC_p}{A_e}M \tag{13}$$

where

$$m = \rho V \tag{14}$$

5. Sample of Calculation

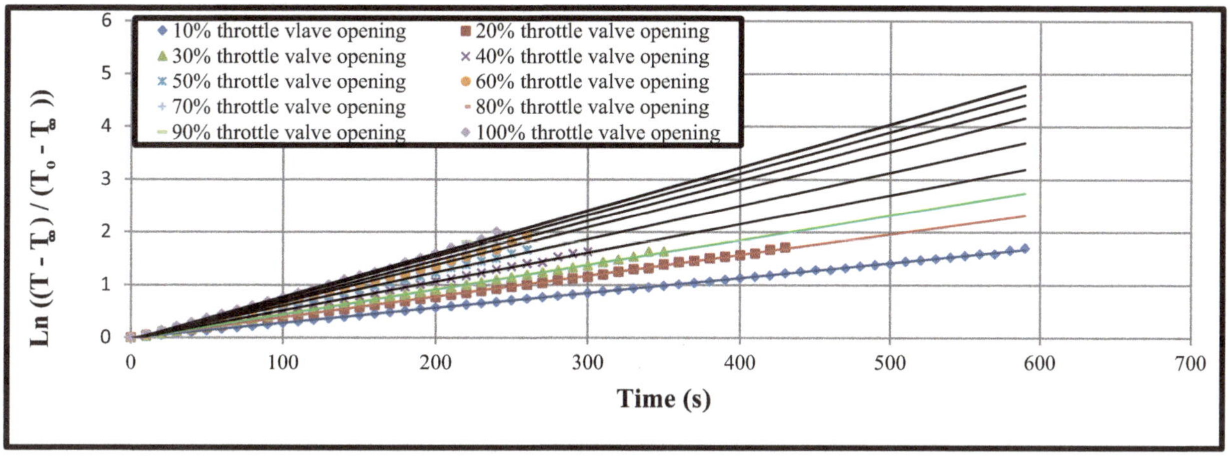

Figure 4. Relationship between $\ln\frac{T-T_\infty}{T_o-T_\infty}$ and time for different air flow rate

From the data recorded according to the time variation by VDAS-B unit, which were air stream velocity, air temperature, surface temperature of heated rod, upstream and downstream pressure. Below is the calculation of C1R1 of location of heated rod in work section. The value of heat transfer coefficient calculated from eq.(13), and the value of M for each flow rate of air can be obtained from Figure (4), and it is equal to the slope of straight line. Table (2) shows the value of M in each percentage of throttle valve opening, coefficient of heat transfer (h) and Nusselt number by using the following relation

$$\text{Nu} = \frac{h*d}{k} \tag{15}$$

Table 2. Sample calculation for C1R1

Throttle valve opening	Re_d	M	h $W/m^2.°C$	Nu
10%	2691.21	0.002815	40.24134	18.27812
20%	5382.419	0.003869	55.31639	25.12539
30%	8073.629	0.004514	64.53685	29.31344
40%	9541.695	0.005286	75.57952	34.32916
50%	12110.44	0.006151	87.94251	39.94458
60%	14801.65	0.006772	96.822	43.97776
70%	17492.86	0.007131	101.9511	46.30745
80%	18838.47	0.007587	108.4712	49.26896
90%	20184.07	0.00775	110.7961	50.32498
100%	21529.68	0.007828	111.9233	50.83697

Table 3. Properties of air based on film temperature (Holman, 1997)

Properties of air	Value
Density ρ_a	1.0877 kg/m^3
Thermal conductivity k_a	0.028135 W/m.°C
Viscosity μ_a	1.9606×10^{-5} kg/m.s
Specific heat capacity C_{pa}	1.00735 kJ/kg.°C

Table 4. The minimum and maximum values of Nu for each row in four columns

Column	Row1		Row2		Row3		Row4		Row5	
	Min.	Max.	Min.	Max.	Min.	Max.	Min.	Max.	Min.	Max.
1	25	83	25.4	85.9	26.9	89.4	28.6	96.6	30.9	104.5
2	28.7	102.6	29.9	106.7	30.9	110.2	32.1	114		
3	29	103	31	110	32.3	115.3	35.7	127.4	36.5	130.4
4	33.4	116.8	34.4	119.6	36.5	127.1	40.5	141		

6. Results and Discussion

1) Effect of heated rod location on Nusselt number

A set of experimental data were obtained from changing the position of heated rod through C1R1 to C4R4 and within a Reynolds number range of 2700 – 21530 for each location. Data obtained from all tests in this research proved that the value of Nusselt number increased when heated rod was moved in two directions such as in y and x-directions in work section. And also the value of Nusselt number increased when increasing Reynolds number. Table 4 shows as the maximum and minimum value of Nusselt number for each row in the four presented columns. This fact can be observed through Figures 5 to 8, which describe a relationship between Nu with Re in each row from column 1 to column 4 respectively.

2) Comparison of the result of present study with standard correlations(Yovanovich, Kan, |& Culhum, 2005)
The result of experimental procedures was compared with two standard correlations of (Zukauskas and Grimison) for staggered arrangement tube bank in cross flow (. The mathematical expression of Zukauskas, 1972 correlation is shown as follows:

$$\mathrm{Nu_d} = \mathrm{FCRe_d^n Pr^m} \tag{16}$$

Where this correlation works for $N_L \leq 16$, and $10^3 \leq Re_d \leq 2\times10^5$. N_L is number of columns, F is correction factor. The constants C, n *and* m vary depending on the Reynolds number and tube bank geometry, as shown in appendix.

The mathematical expression of Grimison is shown below:

$$\mathrm{Nu_d} = \mathrm{FCRe_d^n Pr^{1/3}} \tag{17}$$

This correlation is valid for $N_L \leq 16$ where F is correction factor and (C, n) are constants that depend on tube bank geometry, as shown in appendix.

In the present study the average value of Nusselt number in each column can be calculated from the following relation:

$$\overline{\mathrm{Nu}} = \frac{\sum_{i=1}^{i=n} \mathrm{Nu_i}}{\mathrm{n}} \tag{18}$$

where n is the number of rows in each column.

The values of average Nusselt number calculated from these correlations are compared with the results of the present study for Reynolds number ranging from 2700 to 21530 as shown in Figures 9, 10, 11 and 12 for first column to fourth column respectively. The results of the four columns in present study are in close agreement with Zukauskas correlation.

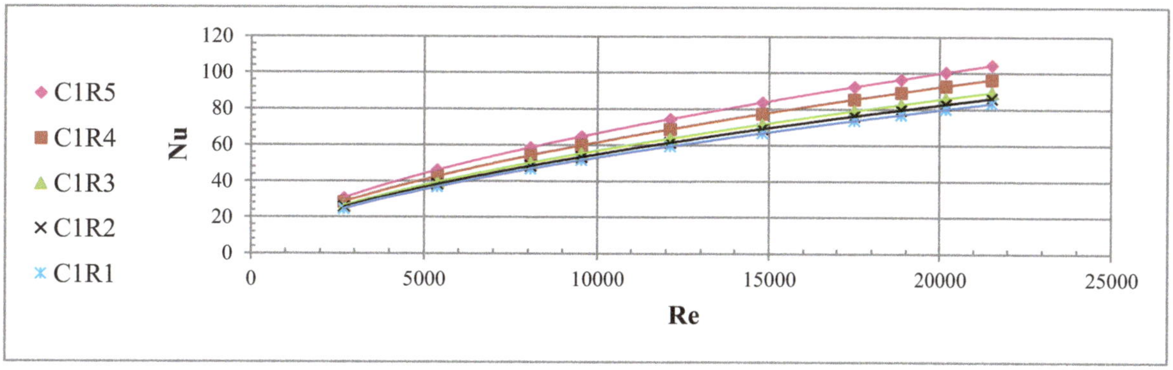

Figure 5. The effect of heated rod position on the value of Nu with respect of Re in column 1

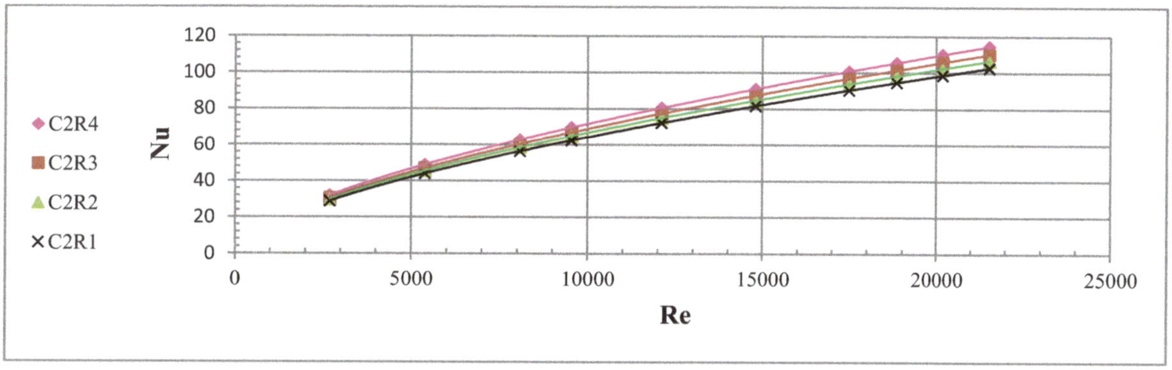

Figure 6. The effect of heated rod position on the value of Nu with respect of Re in column 2

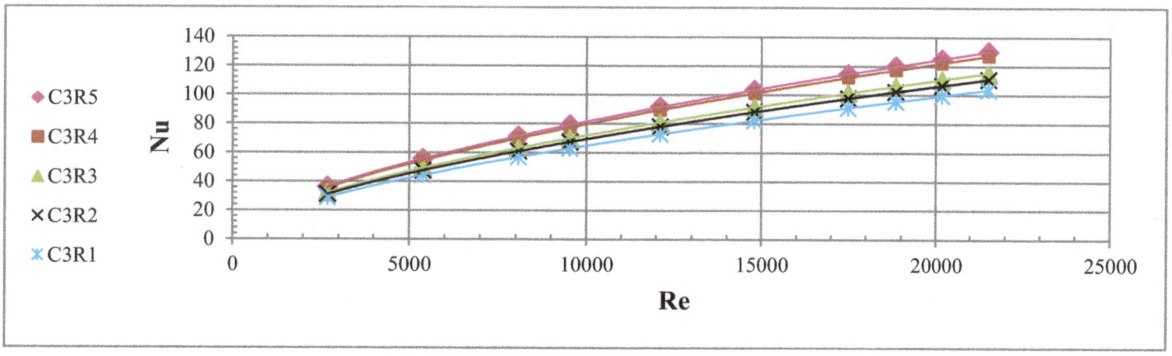

Figure 7. The effect of heated rod position on the value of Nu with respect of Re in column 3

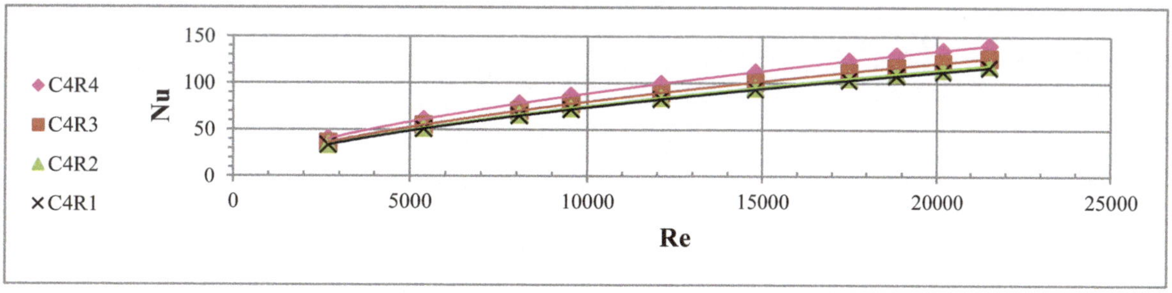

Figure 8. The effect of heated rod position on the value of Nu with respect of Re in column 4

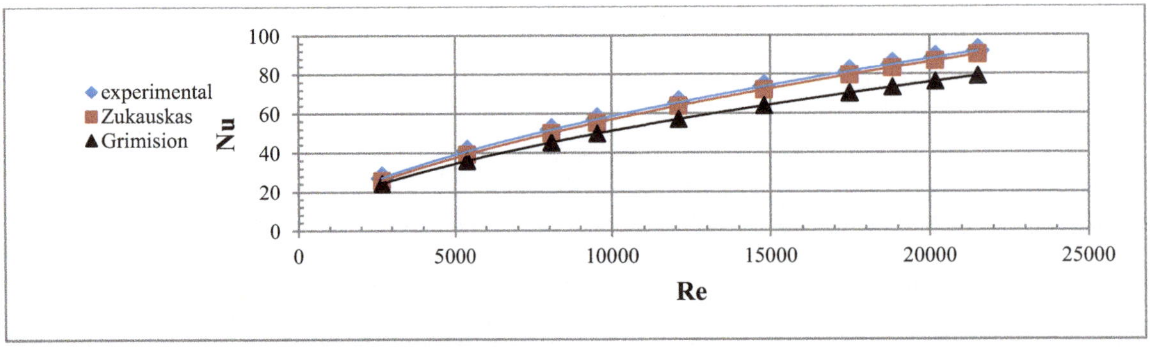

Figure 9. Comparison of experimental Nu and standard correlations for column1

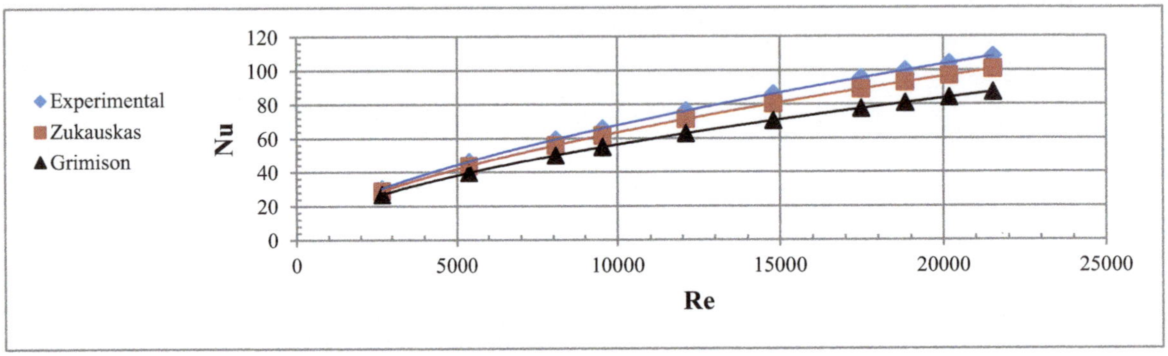

Figure 10. Comparison of experimental Nu and standard correlations for column2

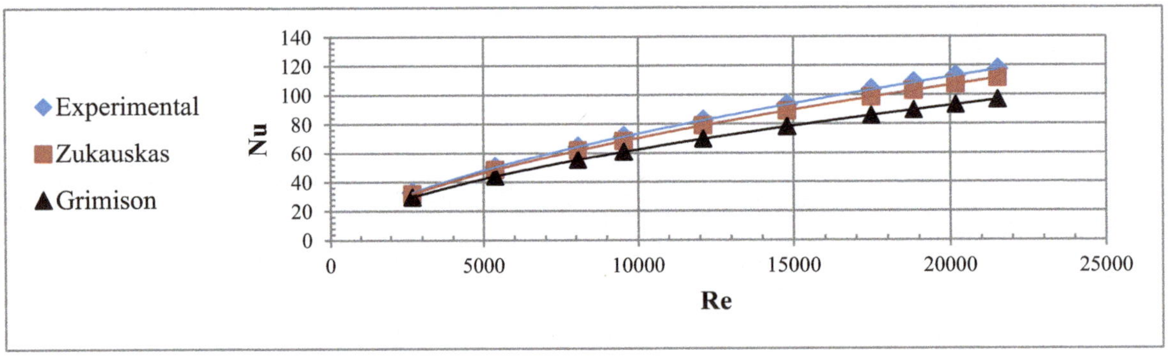

Figure 11. Comparison of experimental Nu and standard correlations for column3

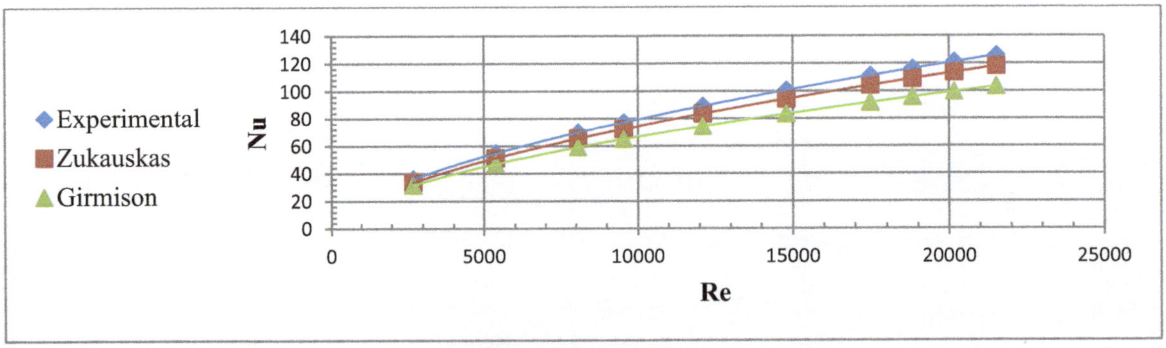

Figure 12. Comparison of experimental Nu and standard correlations for column4

7. Conclusion

It is clear from the experimental results that the value of heat transfer coefficient increases when the location of heated rod is changed in x and y-direction with respect of C1R1 in work section. Also this value increased with increasing Reynolds. The results of all relationships between Nusselt number and Reynolds number in the four columns showed a good agreement with Zukauska correlation.

References

Abd, S. H. (2012). Experimental Study on the Impact of External Geometrical Shape on Free and Force Convection Time Dependent Average Heat Transfer Coefficient during Cooling Process. *Al-Khwarizmi Engineering Journal, 8*(3), 74 – 89. Retrieved from http://repository.uobaghdad.edu.iq/ArticleShow.aspx?ID=1979

Brevoort, M. J., & Tifford, A. N. (1942). An Experimental Investigation of Flow Across Tube Banks. *National Advisory Committee for Aeronautics Wartime Report, L-232.* Retrieved from http://ntrs.nasa.gov/archive/nasa/ casi.ntrs.nasa.gov/19930093156.pdf

Holman, P. H. (1997). *Heat Transfer* (8th ed.). New York, NY: McGraw-Hill.

Horvat, A., Leskovar, M., & Mavko, B. (2006). Comparison of Heat Transfer Conditions in Tube Bundle Cross-Flow for Different Tube Shape, *International Journal of Heat and Mass Transfer, 49*(5-6), 1027-1038. Retrieved from http://www.sciencedirect.com/science/article/pii/S001793100500668X

Jeong, J. H., Nam, K. W., Min, J. K., Kim, K. S., & Ha, M. Y. (2011). The Effects of the Evaluation Method on theAverage Heat Transfer Coefficient for Mini-Channel Tube Bundle. *International Journal of Heat and Mass Transfer, 54*(25-26), 5481-5490. Retrieved from http://www.sciencedirect.com/science/article/pii/S0017931011004297

Khan, W. A., Culham, J. R., & Yovanovich M. M. (2006). Convection Heat Transfer from Tube Banks in Cross-Flow, Analytical approach, *International Journal of Heat and Mass Transfer, 49*(25-26), 4831-4838. Revised from http://www.sciencedirect.com/science/article/pii/S0017931006003796

Khan, W. A., Culham, J. R., & Yovanovich M. M.(2006). Analytical Model Force Convection Heat Transfer from Tube Banks. *Thermophysics and Heat Transfer, 20(4),* 720-727. http://arc.aiaa.org/doi/abs/10.2514/1.15453

Martin, H., & Gnielinsk, V. (2000). Calculation of Heat Transfer from Pressure Drop in Tube Bundles. *Paper presented at the third European Thermal Science Conference.* Retrieved from http://digbib.ubka.uni-karlsruhe.de/volltexte/documents/2579184

Mehrabian, M. A. (2007). Heat Transfer and Pressure Drop Characteristics of Cross low of Air Over Circular Tube in Isolation and/or in a Tube Bank. *The Arabian Journal for Science and Engineering, 32*(2B), 365-376. Retrieved from http://ajse.kfupm.edu.sa/articles/322b_p.14.pdf

Takemoto, Y., Kawanishi, K., & Mizushima, J. (2010). Heat Transfer in the Flow Through a Bundle of Tubes and Transitions of the Flow. *International Journal of Heat and Mass Transfer, 53*(23-24), 5411-5419.

TE93 Cross-Flow Heat Exchanger, User Guide. (2009). TecQuipment Ltd, Nottingham NG102AN, UK.

Yovanovich, M. M., Khan, W. A., & Culham, J. R. (2005). *Convection Heat Transfer from Tube Banks in Cross Flow: Analytical Approach,* 43rd Aerospace Science Meeting and Exhibit, Reno, Nevada. Retrieved from http://www.mhtlab. uwaterloo.ca/pdf_presentations/aiaa0105-2.pdf

Appendix

Available Correlations (Tube Banks)

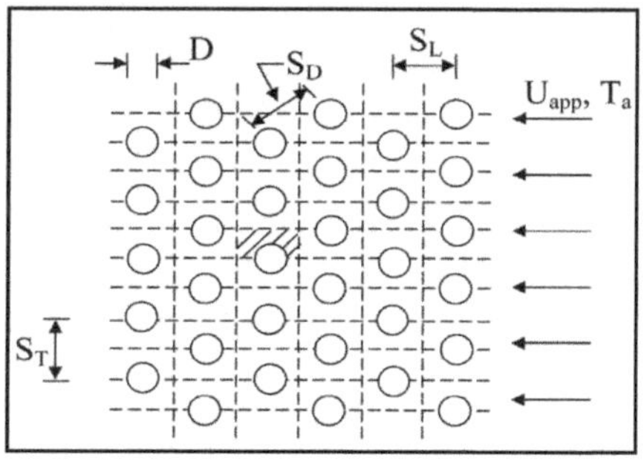

1) Zukauskas, 1972: $Nu_D = FCRe_D^n Pr^m$

Geometry	C	n	m	Conditions
Staggered				
	1.04	0.4	0.36	$10 \le Re_{Dmax} \le 500$
	0.35 $(S_T/S_L)^{0.2}$	0.6	0.36	$S_T/S_L < 2$ $10^3 \le Re_{Dmax} \le 2 \times 10^5$
	0.4	0.6	0.36	$S_T/S_L > 2$ $10^3 \le Re_{Dmax} \le 2 \times 10^5$
	0.022	0.84	0.36	$Re_{Dmax} \le 2 \times 10^5$

F = Correction factor for $NL \le 16$

N_L	1	2	3	4	5	7	10	13	16
staggered	0.64	0.76	0.84	0.89	0.93	0.96	0.98	0.99	1.00

2) Grimision, 1937: $Nu_D = FCRe_D^n Pr^{1/3}$

S_T/D →	1.25		1.5		2		3	
$S_{L/D}$ ↓	C	n	C	n	C	n	C	n
Staggered								
0.600	-	-	-	-	-	-	0.213	0.636
0.900	-	-	-	-	0.446	0.571	0.401	0.581
1.000	-	-	0.497	0.558	-	-	-	-
1.125	-	-	-	-	0.478	0.565	0.518	0.560
1.25	0.518	0.556	0.505	0.554	0.519	0.556	0.522	0.562
1.5	0.451	0.568	0.460	0.562	0.452	0.568	0.488	0.568
2.000	0.404	0.572	0.416	0.568	0.482	0.556	0.449	0.570
3.000	0.310	0.592	0.356	0.580	0.440	0.562	0.421	0.574

F = Correction factor for $NL \le 10$

N_L	1	2	3	4	5	6	7	8	9	10
staggered	0.68	0.75	0.83	0.89	0.92	0.95	0.97	0.98	0.99	1.00

8

Effects of Various Geometric Designs on the Flow Characteristics of a Triangular Rotary Engine

Chiu-Fan Hsieh[1] & Hao-Yu Cheng[1]

[1] Department of Mechanical and Computer-Aided Engineering, National Formosa University, Taiwan

Correspondence: Chiu-Fan Hsieh, Department of Mechanical and Computer-Aided Engineering, National Formosa University, 64 Wunhua Road, Huwei, Yunlin, Taiwan. E-mail: cfhsieh@nfu.edu.tw

Abstract

Although rotary engines can potentially be used as hydrogen fuel engines, the contact condition between the casing and the apex seal on the triangular rotor impacts sealing performance, which then directly affects leakage problems. In this paper, therefore, a rotary engine's internal gas flow characteristics are investigated by using the CFD package FLUENT to construct a fluid analysis model. For comparative convenience, assuming that the simulated gas is air, and then, using a set circular radius outside the casing, analyze three different cases with various geometric design parameters (the K factor) and their effects on the internal pressure, streamlining, and leakage. The results indicate that the K factor design produces different rotor profiles with different working chamber volumes: the lower the K factor, the larger the working chamber volume. Nevertheless, although this design can improve the combustion and compression efficiencies, it may lead to increased internal pressure variation and raised pressure. In addition, when the clearance is small, it may result in a larger leakage problem, negatively affecting rotary engine performance.

Keywords: rotary engine, sealing property, leakage, flow characteristics, CFD

1. Introduction

The Wankel engine, commonly called a rotary engine, is a type of internal combustion engine that instead of using reciprocating pistons uses an eccentric rotary design to convert pressure into a rotating motion. In addition to apex seals, an eccentric shaft, and sun gears, the rotary engine consists of two important parts: a specially designed chamber and a triangular rotor. The movement of this latter is such that its tips are always in contact with the chamber walls so that at any instant, the chamber is divided into three regions. Each of these three areas houses a different part of the combustion process, so that each of the rotary engine's four different jobs – intake, compression, combustion, and exhaust – occurs in its own section of the housing. It is thus equivalent to having a dedicated cylinder for each of the four jobs, with the piston moving continually from one to the next.

The relevant research on rotary drives includes an early study of trochoid envelopes, which outlined a planetary motion that can be applied in engines or pumps (Colbourne, 1974). A more recent paper proposed a very general mathematical relation for a generating arc traveling on a hypotrochoidal path and then compared the flow rate and pocket displacement (Beard, Hall, & Soedel, 1991). Subsequent work then developed kinematic equations and performed a dynamic force analysis of the apex seal on the rotary drive (Beard & Pennock, 1997). Other kinematic analyses of the epicycloidal and hypocycloidal mechanisms for a rotary engine identified differences in their compression ratios and transmission angle (Shih, 1993) and the influence of flowrate irregularity, compression ratio, and pressure angle using various geometric design parameters (Bonandrini, Mimmi, & Rottenbacher, 2009). Recently, a new method for deriving the deviation function was developed for designing a rotary engine with apex seal profiles that can be arc or non-arc curves (Rose & Yang, 2013, 2014).

For the use of hydrogen fuel in the Wankel engine, related studies can be traced back to Mazda in the 1990s, since when several drawbacks of hydrogen engines have been identified (Salanki & Wallace James, 1996). A very recent study, however, shows that adding hydrogen to gasoline can improve both thermal efficiency and power output (Amrouche et al., 2014), meaning that the rotary engine has the potential to be a "green" hydrogen fuel engine. Nevertheless, if this goal is to be realized, many problems remain to be solved, including the rotary engine's combustion efficiency and the strength and sealing properties of its rotor. It is for this reason that the rotary engine is not currently prevalent in industry.

Because the key to the rotary engine's performance is geometric design, in this paper, we investigate the flow characteristics of various geometric designs for the rotary engine by developing a fluid analysis model of a triangular rotary engine. The following sections outline the mathematical model of the casing profile and triangular rotor and then discuss fluid modeling and the flow characteristics inside the rotary engine.

2. Mathematical Model of the Rotary Engine Geometry

The theoretical curve of the rotary engine housing is most commonly epitrochoidal, as shown in Figure 1. Assuming that the rolling circle Σ_r with radius ρ_r follows another fixed circle Σ_b with radius ρ_b rolling by an internal tangent, then when $\rho_r > \rho_b$, $\rho_r = 1.5\rho_b$, and $\alpha = 3\beta$, the housing curve is formed by the movement trajectory of point p. The equation can then be represented as

$$\mathbf{r}_1 = [x_1 \quad y_1 \quad 1]^T \tag{1}$$

where

$$x_1 = c\cos\alpha + \frac{3Kc^2}{\rho_r}\cos\beta \tag{2}$$

$$y_1 = c\sin\alpha + \frac{3Kc^2}{\rho_r}\cos\beta \tag{3}$$

Here, c is the eccentric distance, K is defined as ℓ/c. x_1 and y_1 are the coordinate positions of p, α is the revolution angle of the rolling circle Σ_r around the fixed circle Σ_b, and β is the revolution angle of the rolling circle Σ_r around the its own axis.

Based on the operational principle of the rotary engine, the apexes of the triangular rotor should always stay in contact with the housing, and the triangular rotor and housing should never collide during operation. Hence, to obtain the triangular rotor profile, circle Σ_r should be fixed (see Figure 1). Specifically, circle Σ_b rotates along circle Σ_r and then forms a series of housing curves with circle Σ_b. The inner envelope of the housing curves is the triangular rotor profile. According to the condition of a common tangent on the geometrical contact point, the equation of the triangular rotor curve can be derived as follows:

$$\mathbf{r}_2 = [x_2 \quad y_2 \quad 1]^T \tag{4}$$

$$x_2 = c\cos 2\gamma + c\cos(\alpha+\gamma) + \frac{3Kc^2}{\rho_r}\cos(\frac{\alpha}{3}+\gamma) \tag{5}$$

$$y_2 = -c\sin 2\gamma + c\sin(\alpha+\gamma) + \frac{3Kc^2}{\rho_r}\sin(\frac{\alpha}{3}+\gamma) \tag{6}$$

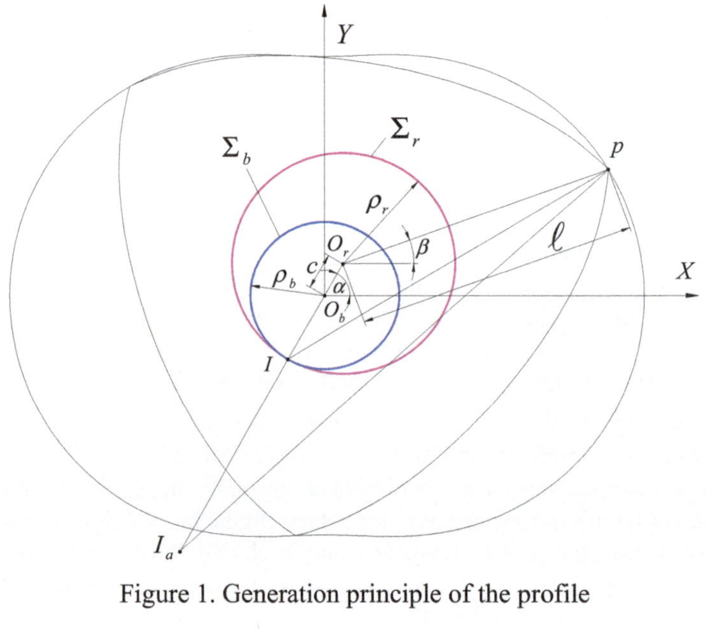

Figure 1. Generation principle of the profile

The contact point can then be obtained by considering

$$f = \frac{\partial x_2}{\partial \alpha} \cdot \frac{\partial y_2}{\partial \gamma} - \frac{\partial x_2}{\partial \gamma} \cdot \frac{\partial y_2}{\partial \alpha} = 0 \tag{7}$$

and then

$$f = -\sin\frac{2}{3}\alpha + \frac{Kc}{\rho_r}\sin(\alpha + 3\gamma) + \sin(\frac{1}{3}\alpha + 3\gamma) = 0 \tag{8}$$

An actual rotary engine structure, however, contains apex seals whose arc radius is *e* and whose center is the peak of the triangular rotor. Hence, the actual housing profile must be offset outward by distance *e*, as represented by

$$\mathbf{R}_1 = \begin{bmatrix} x_1 + n_{x1}e & y_1 + n_{y1}e & 1 \end{bmatrix}^T \tag{9}$$

where

$$\mathbf{n}_1 = \frac{\frac{\partial \mathbf{r}_1}{\partial \alpha} \times \mathbf{k}}{\left|\frac{\partial \mathbf{r}_1}{\partial \alpha}\right|} = \frac{\begin{bmatrix} t_{y1} & -t_{x1} & 0 \end{bmatrix}^T}{\sqrt{t_{x1}{}^2 + t_{y1}{}^2}} = \begin{bmatrix} n_{x1} \\ n_{y1} \\ 0 \end{bmatrix} \tag{10}$$

$$\frac{\partial \mathbf{r}_1}{\partial \alpha} = \begin{bmatrix} t_{x1} \\ t_{y1} \\ 0 \end{bmatrix} = \begin{bmatrix} -\frac{Kc^2}{\rho_r}(\frac{Kc}{\rho_r}\sin\alpha + \sin\frac{\alpha}{3}) \\ \frac{Kc^2}{\rho_r}(\frac{Kc}{\rho_r}\cos\alpha + \cos\frac{\alpha}{3}) \\ 0 \end{bmatrix} \tag{11}$$

The actual rotor profile equation can then be expressed as

$$\mathbf{R}_2 = \begin{bmatrix} x_2 + n_{x2}e & y_2 + n_{y2}e & 1 \end{bmatrix}^T \tag{12}$$

where

$$\mathbf{n}_2 = \frac{\frac{\partial \mathbf{r}_2}{\partial \alpha} \times \mathbf{k}}{\left|\frac{\partial \mathbf{r}_2}{\partial \alpha}\right|} = \frac{\begin{bmatrix} t_{y2} & -t_{x2} & 0 \end{bmatrix}^T}{\sqrt{t_{x2}{}^2 + t_{y2}{}^2}} = \begin{bmatrix} n_{x2} \\ n_{y2} \\ 0 \end{bmatrix} \tag{13}$$

$$\frac{\partial \mathbf{r}_2}{\partial \alpha} = \begin{bmatrix} t_{x2} \\ t_{y2} \\ 0 \end{bmatrix} = \begin{bmatrix} -\frac{Kc^2}{\rho_r}\left[(1+3\gamma')\sin(\frac{\alpha}{3} + \gamma) + (\frac{\rho_r}{Kc} + \frac{\rho_r}{Kc}\gamma')\sin(\alpha + \gamma) + 2\frac{\rho_r}{Kc}\gamma'\sin 2\gamma\right] \\ \frac{Kc^2}{\rho_r}\left[(1+3\gamma')\cos(\frac{\alpha}{3} + \gamma) + (\frac{\rho_r}{Kc} + \frac{\rho_r}{Kc}\gamma')\cos(\alpha + \gamma) - 2\frac{\rho_r}{Kc}\gamma'\cos 2\gamma\right] \\ 0 \end{bmatrix} \tag{14}$$

$$\gamma' = -\frac{1}{9}\left[1 - \frac{2Kc(\cos\frac{2}{3}\alpha - \frac{\rho_r}{Kc}\cos(\alpha + 3\gamma)}{\rho_r\cos(\alpha + 3\gamma) + \cos(\frac{1}{3}\alpha + 3\gamma)}\right] \tag{15}$$

For comparative convenience, when assessing the various rotor and housing profile designs and their effects on a rotary engine's flow characteristics, we assume a set circular radius outside the casing of 92.75 mm. Table 1 shows the design parameters for the three cases, including the three K factors used to explain inter-design differences. Figure 2 shows the design results for the rotary engine: when the K factor is small, the working chamber volume between the casing and the triangle rotor is large.

Table 1. Parameter values (Unit: mm)

	e	ℓ
K = 10 (Case 1)	8.432	84.318
K = 7.5(Case 2)	10.912	81.838
K = 6 (Case 3)	13.25	79.5

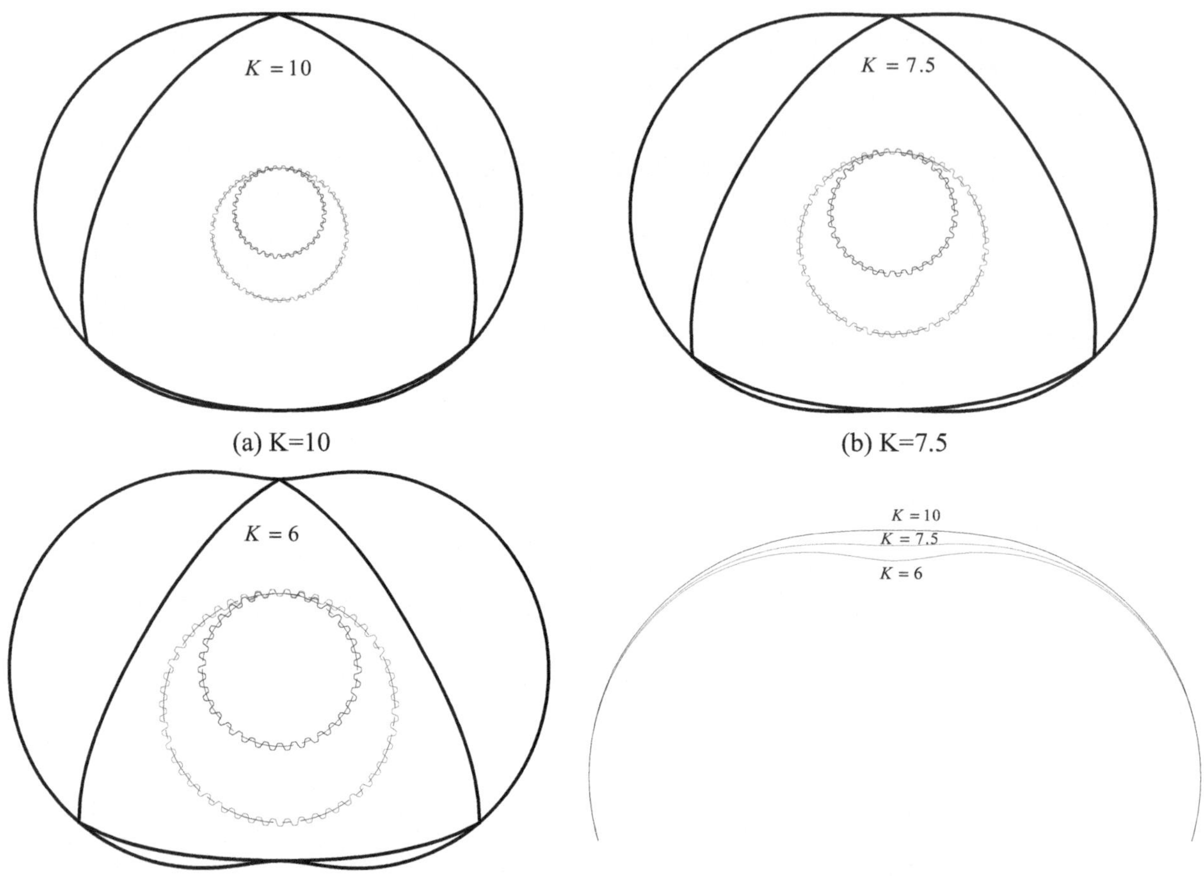

(a) K=10 (b) K=7.5

(c) K=6 (d) Casing profiles

Figure 2. Design results for the rotary engine

3. Fluid Analysis and Discussion

3.1 Governing Equations

The CFD (Computational Fluid Dynamics) package FLUENT used in this study solves conservation equations of mass and momentum using a finite volume approach. The continuity equation can be expressed as

$$\frac{\partial \rho}{\partial t} + \nabla \cdot \rho \vec{V} = 0 \tag{16}$$

The momentum equation is then

$$\rho \left[\frac{\partial (\vec{V})}{\partial t} + (\nabla \cdot \vec{V})\vec{V} \right] \nabla p + (\nabla \tilde{\tau}) = \rho \vec{g} + \vec{f} \tag{17}$$

where p is the static pressure, $\tilde{\tau}$ is the stress tensor, and $\rho \vec{g}$ and $\vec{f}$ are the gravitational body force and surface forces, respectively.

Turbulence is accounted for using a standard *k*-ε two-equation model, chosen because its convergence is better than any other model type with no appreciable difference in flow structure:

$$\frac{\partial}{\partial t}(\rho k)+\nabla\cdot(\rho k V)=\nabla\cdot\left[\left(\mu+\frac{\mu_t}{\sigma_k}\right)\nabla k\right]+G_k+G_b-\rho\varepsilon-y_m+S_k \tag{18}$$

$$\frac{\partial}{\partial t}(\rho\varepsilon)+\nabla\cdot(\rho\varepsilon V)=\nabla\cdot\left[\left(\mu+\frac{\mu_t}{\sigma_\varepsilon}\right)\nabla\varepsilon\right]+C_{1\varepsilon}\frac{\varepsilon}{k}(G_k+C_{3\varepsilon}G_b)-C_{2\varepsilon}\frac{\varepsilon^2}{k}+S_\varepsilon \tag{19}$$

Here, G_k represents the generation of turbulent kinetic energy due to the mean velocity gradients, G_b is the generation of turbulent kinetic energy due to buoyancy, and y_m represents the contribution of the fluctuating dilatation in compressible turbulent flow to the overall dissipation rate. $C_{1\varepsilon}$, $C_{2\varepsilon}$, and $C_{3\varepsilon}$ are constants, σ_k and σ_ε are the turbulent Prandtl numbers for k and ε, respectively, and S_k and S_ε are user-defined source terms.

3.2 Assumptions and Boundary Conditions

For simulative convenience, the fluid used in this research is air (density is 1.293 kg/m^3 and viscosity is 1.81E-05 kg/m-s). The fluid analysis also makes the following important assumptions:

(1) The fluid is analyzed using a turbulent model (k–ε model).

(2) Flow is unsteady, compressible and turbulent.

(3) Conjugate heat transfer and radiation effects are neglected.

(4) The fluid is initially stationary; the flow is two-dimensional.

The input conditions are as follows: inlet pressure = two atmospheric pressure (202,650 Pa), outlet pressure = one atmospheric pressure (101,325 Pa), shaft speed = 1800 rpm.

During operation, the effect on the rotary engine of imbalance gas pressure and rotation inertia or centrifugal force changes the contact force on the apex seal causing it to move, which leaves a tiny clearance between the apex seal and the housing surface that can lead to leakage problems. Hence, for simulative convenience, we assume a set clearance of 0.15 mm in each case and then compare the leakage condition in various designs.

3.3 Grid Generation

Taking the case of K=6 as an example, the fluid model of the rotary engine is in 2D and has four regions: the inlet, outlet, rotor, and casing. The rotor's swaying motion, which changes the volumetric displacement between rotor and casing, can be discretized using an unstructured grid (e.g., triangular cell) generated by the FLUENT GAMBIT package. The resulting mesh in the case of K=6 is shown in Figure 3. This simulation includes 6 simulation cycles of the rotation shaft with a total of 20,000 steps, each counted as 1E-05 sec. In the dynamic grid model, the internal pressure in the interval 0.01~0.05s is calculated as in Figure 4. As Figure 4(a) shows, the volume is largest on the right side, with the apex seal separated from the inlet, which is ready for compression. In Figure 4(b), the volume is lessening and starting to compress, and the internal pressure rises to about 7 atmospheric pressures. Figure 4(c) then shows the condition after compression, when the pressure drops off because of leakage; Figure 4(d) illustrates the depression after compression, when volume gradually increases; and Figure 4(e) depicts the major depression for exhausting gas to the outlet.

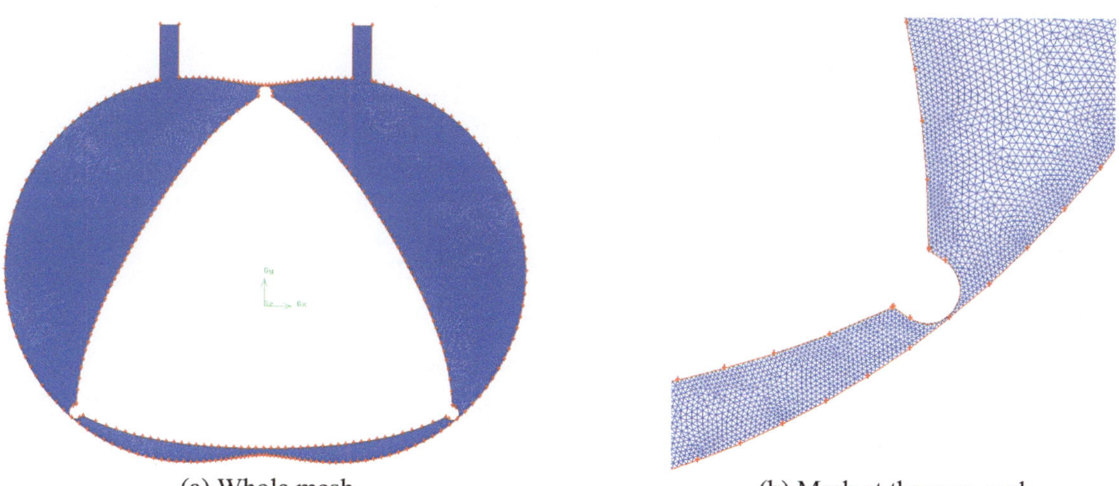

(a) Whole mesh

(b) Mesh at the apex seal

Figure 3. Resulting mesh

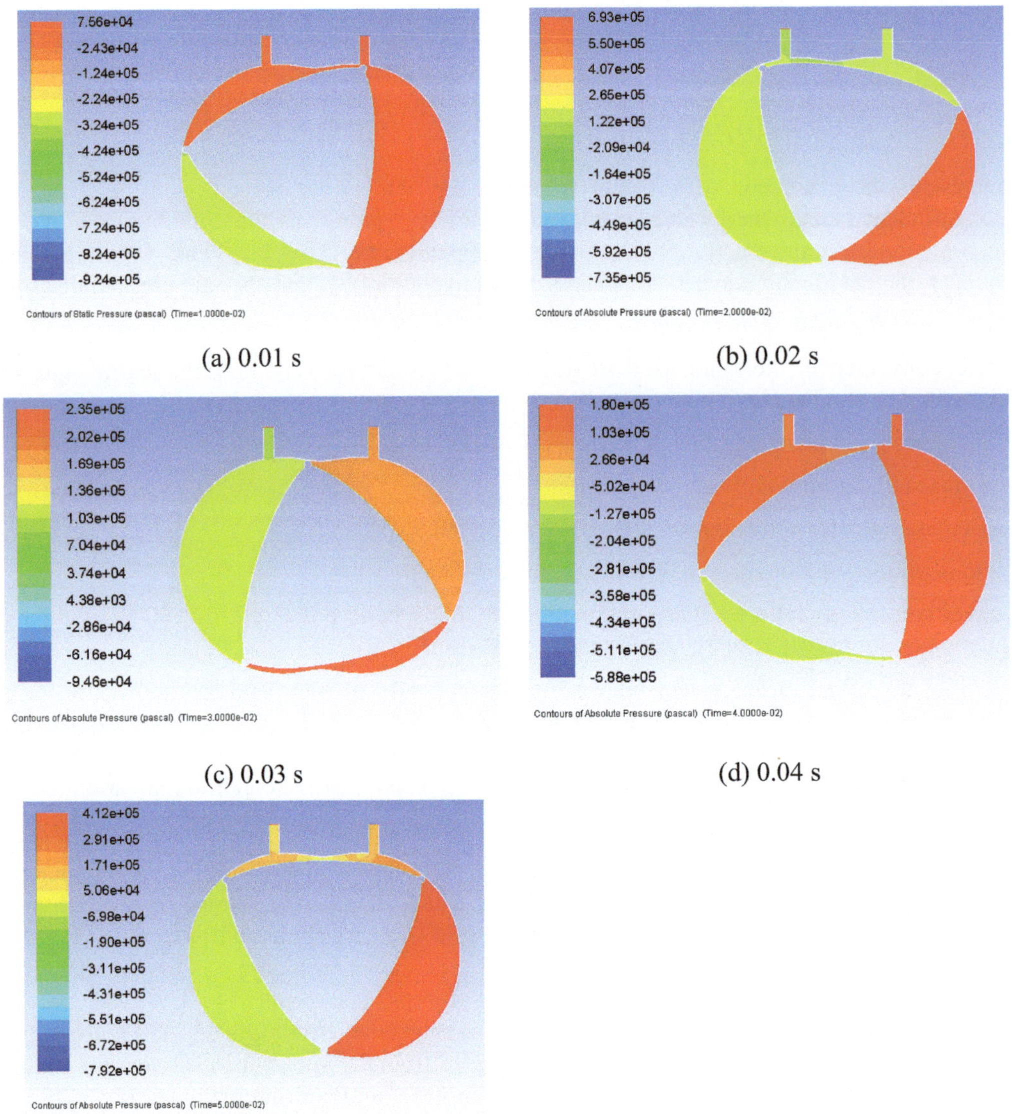

(a) 0.01 s (b) 0.02 s

(c) 0.03 s (d) 0.04 s

(e) 0.05 s

Figure 4. Pressure calculation for the rotary engine

Based on the calculation mode used in this paper, the grid information for the three cases is as shown in Table 2.

Table 2. Grid information

	Cells	Faces	Nodes
Case 1 (*K*=10)	193,386	291,957	98,571
Case 2 (*K*=7.5)	237,668	358,328	120,660
Case 3 (*K*=6)	275,812	415,494	139,682

3.4 CFD Results and Discussion

The working chamber volume of the rotary engine affects both compression efficiency and the fluid flow characteristics. For example, when *K* equals 6, Case 3 has the largest volume, followed by Case 2 and then Case 1. Setting particular survey points in the three cases (see Figure 5) throws more light on the effect of different rotary engine geometric designs on the internal flow field, particularly flow variations and leakage in the rotor. The CFD results reported here are based on one revolution of the triangular rotor.

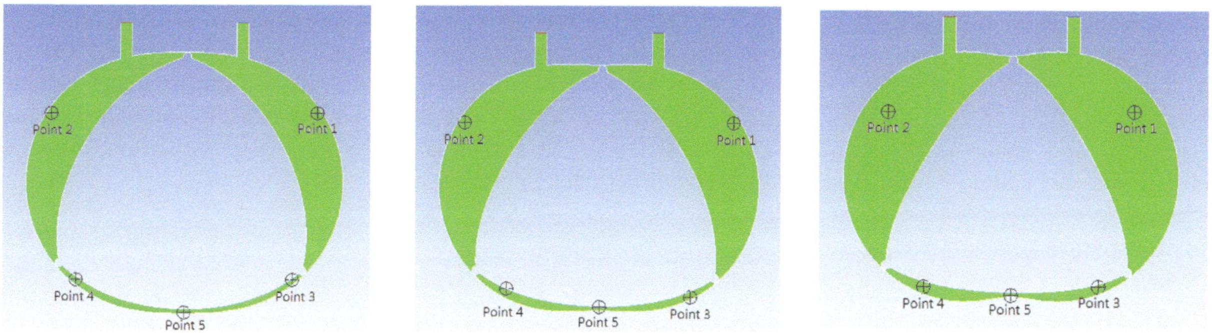

(a) Case 1 (b) Case 2 (c) Case 3

Figure 5. Survey points in the three cases

Because the survey points are covered during rotor operation, their pressure curves may be subject to a discontinuous phenomenon that leads to missing data. Even so, it is still possible to calculate most pressure values and analyze the differences. The pressure results for points 1 and 2 (see Figure 6) show no obvious differences among the three cases. Hence, near the inlet and outlet port, the rotor profile design has no obvious influence. Entering the compression area at point 3, however, there is a significant rise in pressure as the pressure curve changes with variation in the working chamber volume (Figure 7). The larger the change in volume, the greater the pressure variation, with the largest change occurring in Case 3 (Figure 7(a)). Nearing the compression area at point 4, the larger working chamber volume produces more obvious compression and an increased pressure (Figure 7(b)). At point 5, which is located in the position of being ready to leave the compression, all three cases have significant depression (Figure 7(c)).

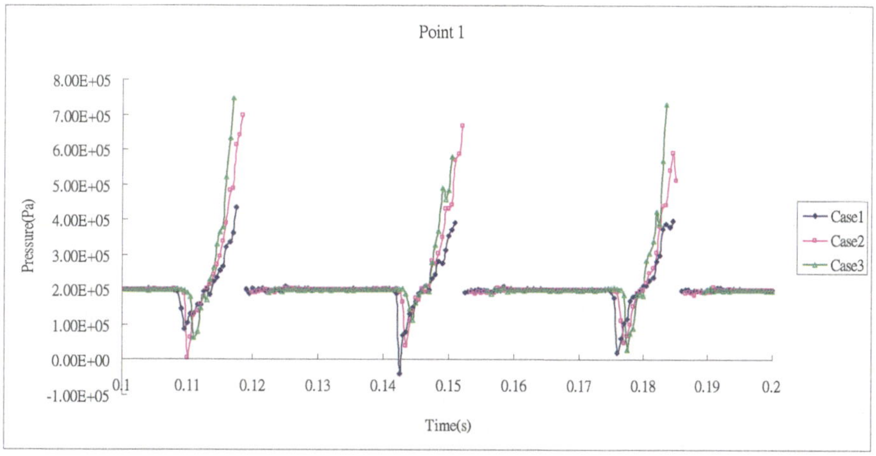

(a) Point 1

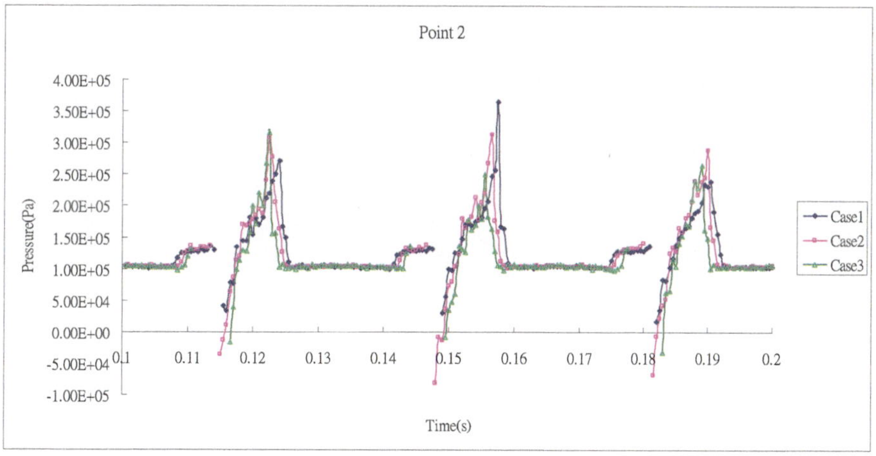

(b) Point 2

Figure 6. Pressure on points 1 and 2

No matter the case, from point 3 to point 5, the pressure increases during compression but at the moment of leaving the compression area, all pressure lowers, even the larger negative pressure. During the process, Case 3 has the largest positive and negative pressure peaks, so when K is small, it has a larger working chamber volume. Overall, the rotary engine produces large volume changes during operation, which enlarges internal pressure differences and leads to better compression efficiency. Nevertheless, there may also be larger pressure fluctuation, which could lower engine stability and affect power performance.

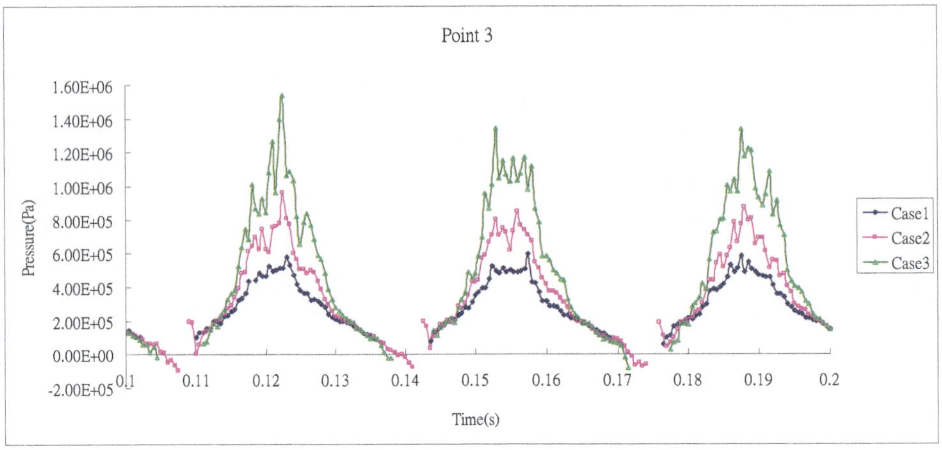

(a) Point 3

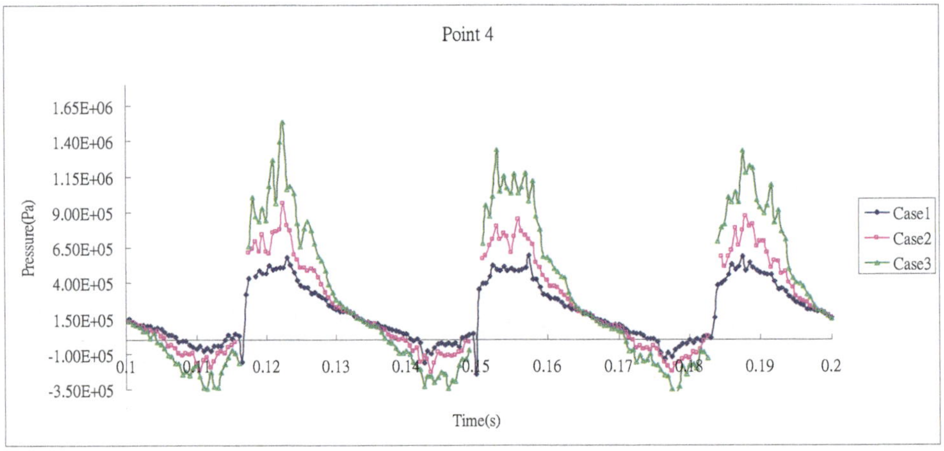

(b) Point 4

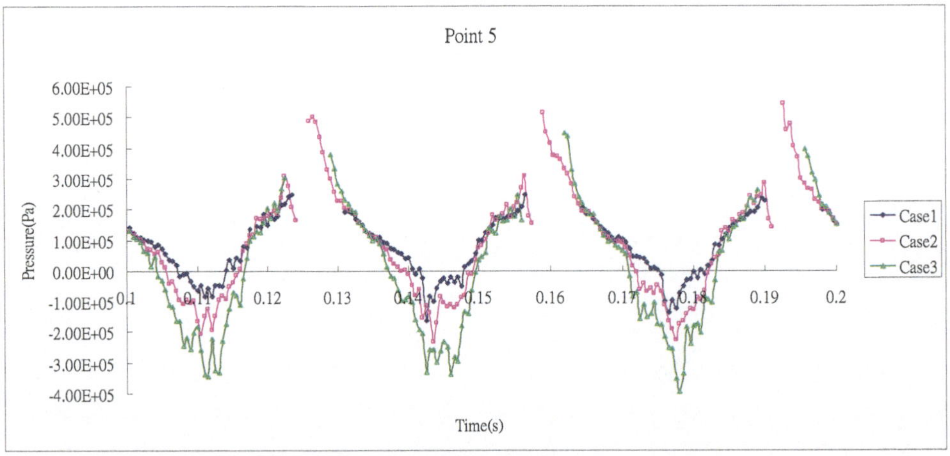

(c) Point 5

Figure 7. Pressure on points 3–5

As regards variation in the inlet and outlet pressures, it is affected by changes in the working chamber volume, which is at its largest as gas moves into the chamber from the inlet. After that, the inlet pressure decreases because of the *K* factor: the smaller the *K* factor, the larger the working chamber volume, meaning a large pressure drop in the inlet port. The magnitude of this pressure fluctuation in the inlet port ranks as follows (see Figure 8): Case 3 > Case 2 > Case 1. Changes in the working chamber volume during operation also affect the outlet pressure in the low pressure area, resulting in pressure drops and fluctuations in certain positions. A comparison of the outlet pressure fluctuations among the three cases, however, shows no significant differences. The results for the inlet and outlet ports suggest that the *K* value is a key factor for the inlet port, one that causes large fluctuations in the inlet pressure. Specifically, the smaller the *K* value, the greater the fluctuation.

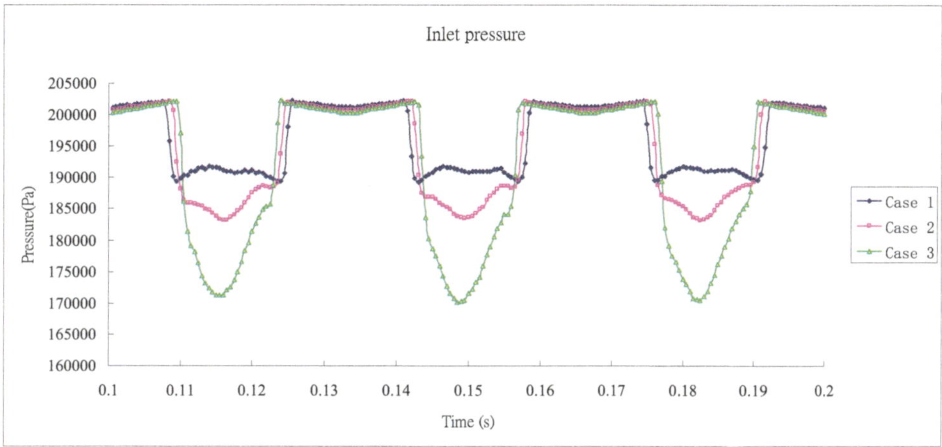

Figure 8. Pressure at the inlet

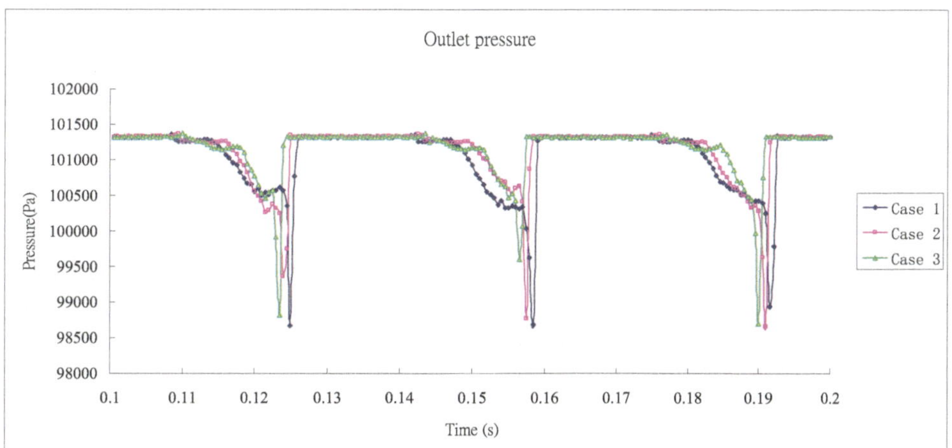

Figure 9. Pressure at the outlet

Streamlining of the three cases also reveals that when gas enters the inlet port, a vertex occurs between the casing and rotor (see Figure 10). These turbulence flows are also likely to affect the rotary engine's leakage level.

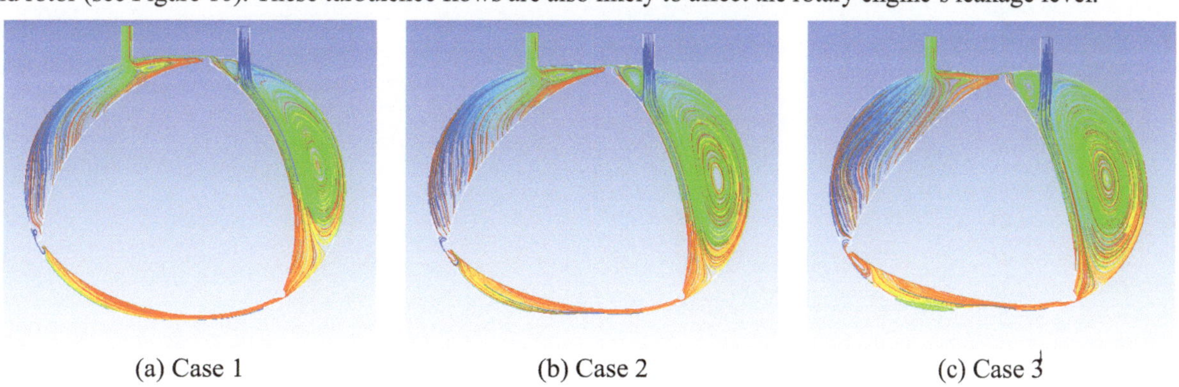

(a) Case 1 (b) Case 2 (c) Case 3

Figure 10. The three cases streamlined

Assuming the existence of a tiny clearance between the casing and apex seal, the leakage situation is represented by streamlining the flow velocity of all three cases (see Figure 11), which demonstrates that fluid flows from the high pressure area to the low pressure area. Figure 11(a) illustrates the situation in Case 1, in which the leakage position is concentrated in the clearance, with a flow velocity up to 189 m/s compared to 218 m/s in Case 2 (Fig, 11(b)). The most severe leakage occurs in Case 3 (Figure 11(c)), with a flow velocity of up to 293 m/s. These observations indicate that when internal pressure rises, the leakage level increases: the highest internal pressure in Case 3 results in both the largest pressure difference and the greatest leakage magnitude. The analysis thus clearly demonstrates that the design of the rotary engine profile directly affects leakage during engine operation and has a significant effect on engine performance.

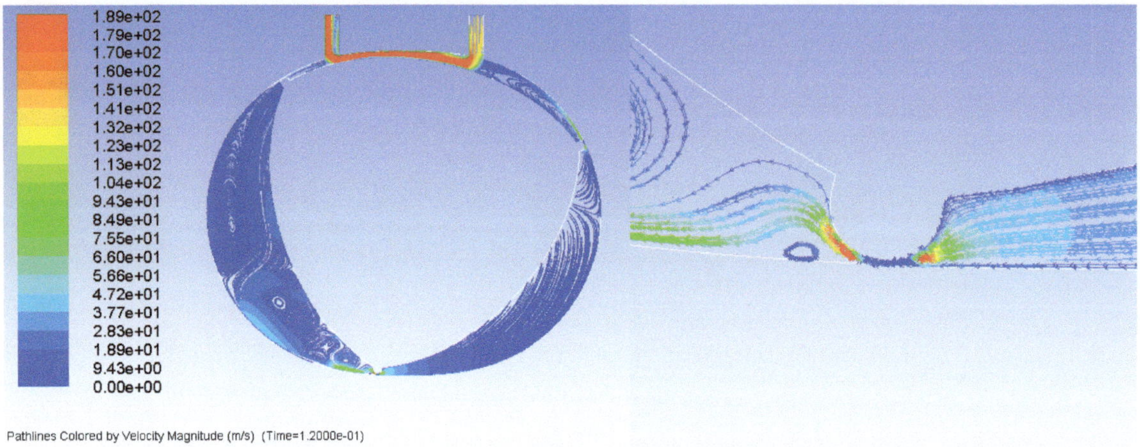

(a) Case 1

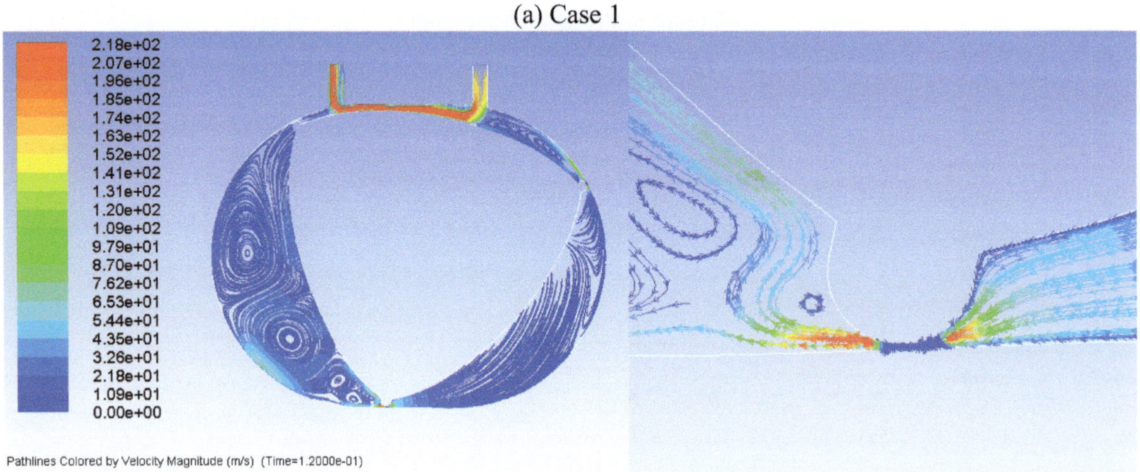

(b) Case 2

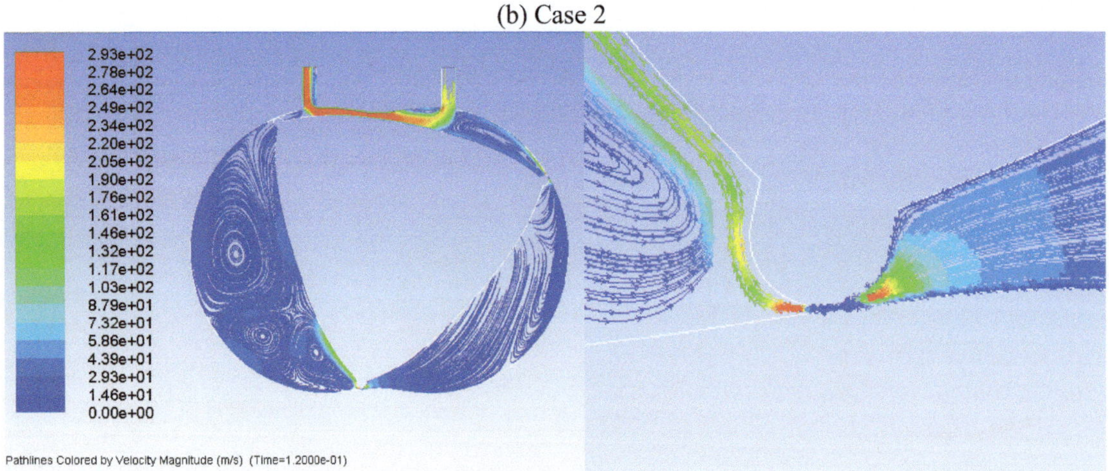

(c) Case 3

Figure 11. Velocity plot in the three cases

4. Conclusions

This study analyzes the effects on rotary engine flow characteristics of different design parameters; here, three variations in the curve (K) factor. Within the three different curve designs, a larger working chamber volume and higher compression efficiency with a lower K factor leads to higher leakage, larger internal pressure fluctuation, and unstable inlet pressure. Conversely, a high K factor produces low leakage, small internal pressure fluctuation, and stable inlet pressure, indicating a greater design advantage. Hence, when designing a rotary engine, designers must consider the interactions between profile curves and fluid flow characteristics. This paper can serve as a useful reference for such consideration.

References

Amrouche, F., Erickson, P., Park, J., & Varnhagen, S. (2014). An experimental investigation of hydrogen enriched gasoline in a Wankel rotary engine. *International Journal of Hydrogen Energy, 39*, 8525-8534. http://10.1016/j.ijhydene.2014.03.172

Beard, J. E., & Pennock, G. R. (1997). Force analysis of the apex seals in the Wankel rotary compressor including the influence of fluctuations in the crankshaft speed. *Mechanism and Machine Theory, 32*, 349-361. http://10.1016/S0094-114X(96)00051-1

Beard, J. E., Hall, A. S., & Soedel, W. (1991). Comparison of hypotrochoidal and epitrochoidal gerotors. *ASME J. Mech. Design, 113*, 133-141. http://dx.doi.org/10.1115/1.2912761

Bonandrini, G., Mimmi, G., & Rottenbacher, C. (2009). Theoretical analysis of internal epitrochoidal and hypotrochoidal machines. *Proceedings of the Institution of Mechanical Engineers, Part C: Journal of Mechanical Engineering Science, 223*, 1469-1480. http://dx.doi.org/10.1243/09544062JMES1163

Colbourne, J. R. (1974). The geometry of trochoid envelopes and their application in rotary pumps. *Mechanism and Machine Theory, 9*, 421- 435. http://dx.doi.org/10.1016/0094-114X(74)90025-1

Rose, S. W., & Yang, D. C. H. (2013). Design of rotary engine from the apex seal profile (Abbr.: Rotary engine design by apex seal). *Mechanism and Machine Theory, 64*, 200-209. http://10.1016/j.mechmachtheory.2013.01.015

Rose, S. W., & Yang, D. C. H. (2014). The deviation function method of rotary engine design by geometric parameters. *ASME J. Mech. Design, 136*, 051004. http://10.1115/1.4026867

Rose, S. W., & Yang, D. C. H. (2014). Wide and multiple apex seals for the rotary engine (Abbr.: Multi-apex-seals for the rotary engine). *Mechanism and Machine Theory, 74*, 202-215. http://10.1016/j.mechmachtheory.2013.12.011

Salanki, P. A., & Wallace James, S. (1996). Evaluation of the hydrogen fuelled rotary engine for hybrid vehicle application. *SAE paper, no. 960232*. http://dx.doi.org/10.4271/960232

Shih, A. J. (1993). Analysis and comparison of epicycloidal and hypocycloidal internal combustion engine mechanisms. *ASME J. Mech. Design, 115*, 960-966. http://10.1115/1.2919294

Shih, A. J. (1993). Kinematics of the cycloidal internal combustion engine mechanism. *ASME J. Mech. Design, 115*, 953-959. http://10.1115/1.2919293

9

Cableless In-piping Magnetic Actuator for Transportation over Long-range

Tomohiro Izumikawa[1], Ryuichi Watanabe[1] & Hiroyuki Yaguchi[1]

[1] Faculty of Engineering, Tohoku Gakuin University, Miyagi, Japan

Correspondence: Hiroyuki Yaguchi, Faculty of Engineering, Tohoku Gakuin University, 1-13-1 Chuo, Tagajo, Miyagi, Japan. E-mail: yaguchi@mail.tohoku-gakuin.ac.jp

Abstract

This paper proposes a cableless In-piping magnetic actuator capable of locomotion over long range of 1,000 m within a pipe having an inner diameter of 10 mm. The cableless magnetic actuator is moved by using resonance energy of a mass-spring system excited by using an electromagnetic force. The proposed actuator incorporates a new type of an electrical inverter that directly transforms DC from button batteries into AC. The electrical DC-AC inverter incorporates a mass-spring system, two reed switches and two curved permanent magnets that switch under an electromagnetic force. The conventional DC-AC inverter and the newly proposed inverter were compared, and the effect of the inverter on the motive properties of the cableless magnetic actuator was examined. The influence of the consumption current of the battery on the range of the actuator was examined. Experimental result demonstrates that the cableless magnetic actuator was able to move horizontally at 1,588 m, and horizontal speed at 176.5 mm/s when two reed switches were used.

Keywords: cableless actuator, electrical DC-AC inverter, pipe inside mover, new motion principle, locomotion

1. Introduction

The demand of in-piping robot to transport a medicine and bacteria from a laboratory to a laboratory in the same building or another building is increasing. Safety and shortening of the transportation time increases by using a pipe transportation method. By this approach, the inside air of the pipe is completely intercepted with outside air. If an accident occurs in the pipe, we can evade a situation scattering hazardous materials such as virus outside a pipe. However, the realization for in-piping cableless robots capable of locomotion over long range inside the pipe is required. A many studies have investigated the mechanisms of an actuator having an attached electrical cable to provide locomotion in the pipe using devices such as piezoelectric elements (Saito et al., 2000; Fujita et al., 2004), air cylinders (Ohno et al., 2006), shape memory alloys (Reynaerts et al., 1996; Kwon et al., 2005), and electromagnetic motors (Choi et al., 2007; Miyagawa et al., 2008; Bocko et al., 2009). There are several potential problems in the design, however, with regard to extending the range and avoiding tangling of the electrical cable, and thus it is clearly desirable to adopt a cable-free system. However, little research has been conducted on cableless robots (Fukuda et al., 1991; Tsuruta et al., 2002) and electromagnetic vibration type (Yaguchi et al., 2009; Yaguchi et al., 2011; Izumikawa et al., 2012) proposed by authors. Experiment results were compared to analytical results and the validity of the novel motion principle was completely verified (Yaguchi et al., 2013). In addition, the additional devices of a CCD camera, two LED lights, a transmitter with a wire antenna, and various types of batteries were attached to the proposed cableless magnetic actuator system. The inspection range of this actuator system was a horizontal distance of 402 m and a vertical motion of 261 m for over 50 minutes (Yaguchi et al., 2014) proposed by authors.

The actuator is moved due to mechanical resonance energy of the mass-spring system produced by an electromagnetic force. In-piping cableless actuators are not capable of movement over long range of 1,000 m.

This paper proposes a novel cableless magnetic actuator system with a new type of electrical DC-AC inverter that is coupled to two reed switches and two permanent magnets. The waveform of the interchange is generated by a DC battery, and the technique used to drive the cableless actuator is demonstrated. We compare the conventional DC-AC inverter and the newly proposed inverter, and examine the effect of the inverter on the motive properties of the cableless actuator. A driving test of the cableless actuator capable of movement inside a pipe having an inner diameter of 10 mm was carried out. Additionally, the influence of the consumption current of the battery on the range of the actuator was examined. Experimental results demonstrated that the total range of the cableless actuator was more than 1,000 m. The results hold great promise for the creation of highly mobile actuators capable of moving in pipes with diameters of under 10 mm.

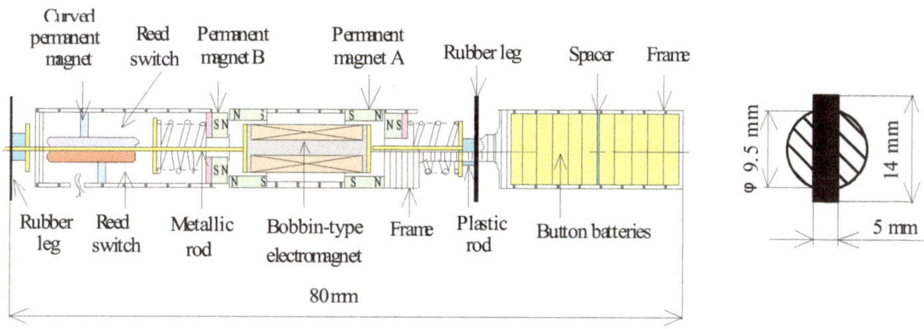

Figure 1. Structure of the cableless magnetic actuator

2. Structure of the Cableless Magnetic Actuator

Figure 1 shows a diagram of the proposed cableless magnetic actuator, which is capable of moving within a pipe having an inner diameter of 10 mm. The cableless actuator consists of a propulsion module, an electrical DC-AC inverter composed of two reed switches and two curved permanent magnets, a frame, and a ten button batteries. The propulsion module is composed of two identical ring-type permanent magnets, labeled A and B, two translational springs, two metallic rods with plastic circular plates, and a bobbin-type electromagnet. Permanent magnets A and B are NdFeB magnets and are magnetized in the axial direction. Permanent magnet A has an outer diameter of 7.8 mm, an inner diameter of 5.8 mm, and a thickness of 2.5 mm. Permanent magnet B has an outer diameter of 6.5 mm, an inner diameter of 2 mm, and a thickness of 1.5 mm. The two identical translational springs are stainless steel compression-coil-type springs and have an outer diameter of 5.5 mm, a free length of 6.5 mm, and a spring constant of k = 2,400 N/m. The bobbin-type electromagnet consists of an iron core with 3,200 turns of 0.07-mm-diameter copper wire. The electrical resistance of the electromagnet is 225 Ω. The legs used to support the actuator are constructed of natural rubber and have a total length of 10 mm, a thickness of 0.5 mm, and width of 5 mm. The actuator which contains ten batteries has a length of 80 mm and a total mass M_a of 12 g. The magnetic actuator is able to move by the difference in frictional force between forward and backward supporting force of the natural rubber Yaguchi et al. (2013).

3. Principle of the DC-AC Inverter

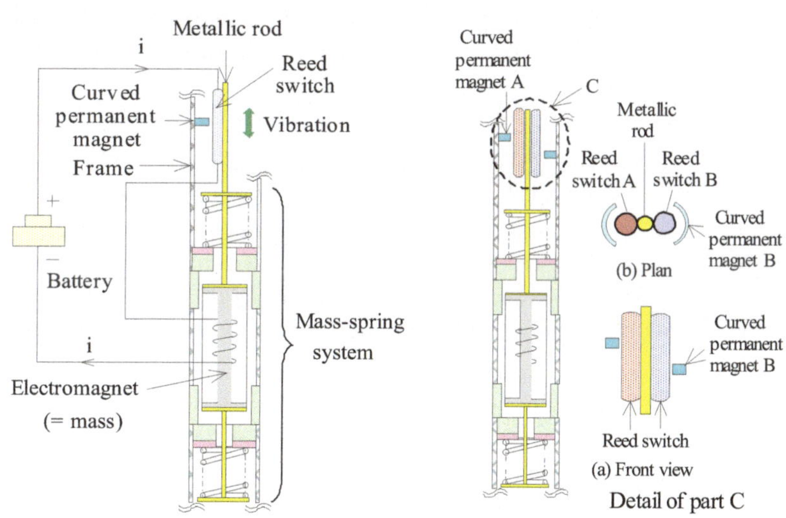

(a) One reed switch (b) Two reed switches

Figure 2. Principle of the electrical DC-AC inverter

Figure 2(a) illustrates the principle of the electrical DC-AC inverter proposed in the previous paper (Yaguchi et al., 2011; Izumikawa et al., 2012; Yaguchi et al., 2013), which was developed in order to realize a cableless magnetic actuator. Figure 2(b) illustrates the principle of the proposed electrical DC-AC inverter. The proposed inverter is composed of two reed switches and two curved permanent magnets. Each reed switch has a diameter of 1.8 mm and a length of 10 mm. The two curved permanent magnets, A and B, are arc type NdFeB magnets having a width of 1 mm and a thickness of 1 mm and are magnetized in the axial direction. In the experiment, ten button batteries were divided into two packs by five packages.

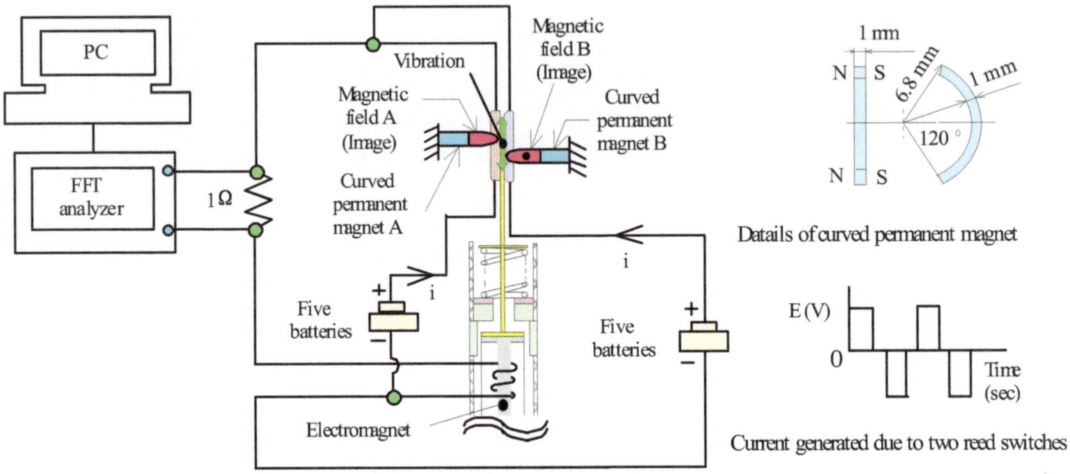

Figure 3. Concept of the AC-DC inverter

Figure 3 shows the connection of an electrical wire and the concept of the electrical DC-AC inverter using two reed switches. Such a connection generates an alternating current waveform from the DC battery. Reed switch A and B attached to the electromagnet vibrates around two curved permanent magnets I and II, respectively. The two curved permanent magnets, I and II, are arc type NdFeB magnets having a width of 1 mm and a thickness of 1 mm and are magnetized in the axial direction. The magnetic field between the reed switch and the curved permanent magnet is changed. The two cantilever beams inserted into each reed switch make and break contact as the switch is cycled on and off. As a result, the DC voltage is converted into a square alternating current waveform, and the resulting magnetic force acts on the mass-spring system. The on-areas of reed switch A and B, and the off-areas of switch A and B indicated in previous study (Yaguchi et al., 2014). The DC source for the experiment was two packs of five button batteries. When the electromagnet was connected to two battery packs, the direct current into the electromagnet was 0.05 A. The current waveform generated during switching of the reed switch was stored in a personal computer (PC) via a fast Fourier transform (FFT) analyzer as shown in Figure 3.

4. Locomotion Characteristics

Figure 4 shows photograph of this cableless actuator and an experimental apparatus. The cableless actuator was inserted in an acrylic pipe of 10mm inside diameter. Two coil A and B were placed outside of an acrylic pipe. The movement speed of the cableless actuator was measured by the time lag between two coils by using FFT analyzer. Maxell SR920W silver-oxide button batteries were used to drive the propulsion module of the cableless magnetic actuator. Each battery is 9.5 mm in diameter and has a mass of 0.6 g and a capacity of 39 mAh at a nominal output of 1.55 V. The supporting force of the actuator in the pipe was set to 0.12 N.

The solid line in Figure 5 shows the relationship between the number of button batteries and the vertical upward speed of the actuator when two reed switches were loaded in the cableless actuator. For the cases of ten batteries and six batteries, the vertical upward speeds were 121 mm/s and 40 mm/s, respectively. In this figure, the broken line indicates the results for one reed switch. When powered by ten batteries, the maximum vertical speed was 225 mm/s as shown in previous study (Yaguchi et al., 2014).

The solid line in Figure 6 shows the relationship between the tilt angle α and the speed of the actuator using ten button batteries when two reed switches were loaded. Ten button batteries were divided into two packs as shown in Figure 3. The tilt angle was varied from α = -90° (straight downward) to α = 90° (straight upward). The broken line shows the results for one reed switch when ten batteries were used.

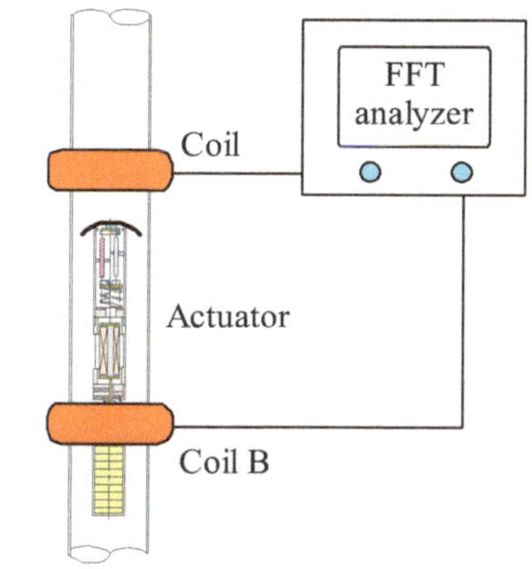

(a) Experimental apparatus (b) Photograph of the cableless actuator

Figure 4. Experimental apparatus and cableless actuator

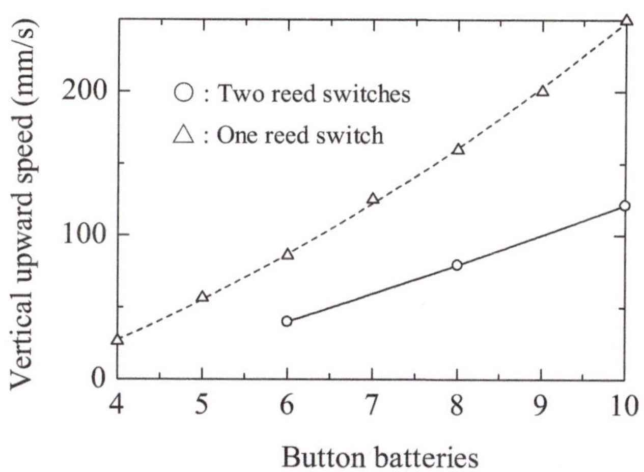

Figure 5. Relationship between number of batteries and vertical upward speed

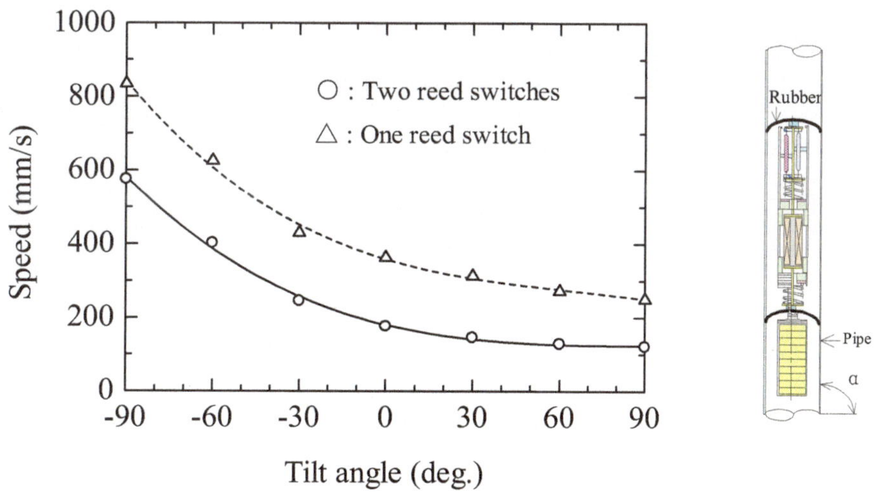

Figure 6. Relationship between tilt angle α and speed.

The solid line in Figure 7 shows the relationship between the load mass and the vertical upward speed of the actuator for ten button batteries. The figure indicates that the actuator can climb upward at 43.5 mm/s when pulling a load mass of 10 g, which is a moderately high performance. In this figure, the broken line shows the results for one reed switch as shown in previous paper (Yaguchi et al., 2014). The average of the consumption current in the case of one reed switch was 18 mA. However, that of two reed switches was 13 mA. Accordingly, the average of the consumption current by using five button batteries becomes 6.5 mA. This cableless actuator can move by a little consumption current.

Figure 8 shows the relationship between the load mass and self-propellant efficiency of the actuator for ten batteries. The self-propellant efficiency is expressed as follows:

$$\eta\,[\%] = \frac{(M_a + M_L)\, v_{up} \times G}{V \times I} \times 100 \tag{1}$$

where M_a is the total mass of the cableless actuator, M_L is the load mass, v_{up} is the vertical speed, G is the acceleration due to gravity, V is the input voltage and I is average of the consumption current. The maximum efficiency is 34 % for ten batteries when two reed switches were loaded. The maximum efficiency is thought to be high for an electromagnetic actuator of this size. The broken line in Figure 8 shows the results when one reed switch was loaded.

Figure 9 shows the relationship between the consumption current and the life of the button battery. If there are many consumption currents, the time of battery life drops sharply. It is approximated by the fifth-order equation of battery life time t minutes when I mA is input current into the actuator.

$$t(\text{min}) = 517.51 - 99.71 \times I + 8.95 \times I^2 - 0.411 \times I^3 + 9.31 \times 10^{-3} \times I^4 - 8.31 \times 10^{-5} \times I^5 \tag{2}$$

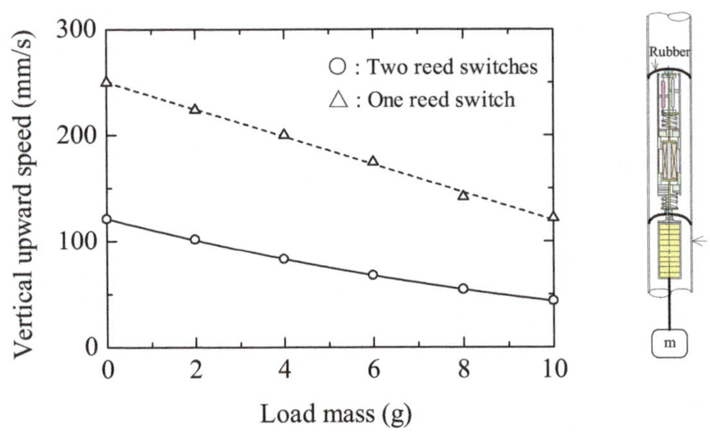

Figure 7. Relationship between load mass and vertical upward speed

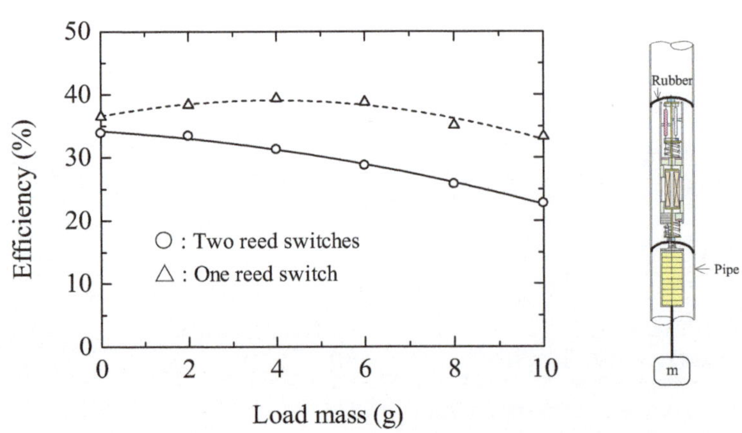

Figure 8. Relationship between load mass and efficiency.

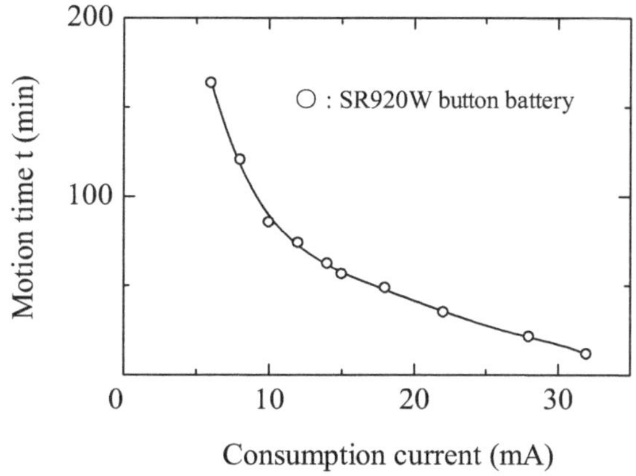

Figure 9. Relationship between consumption current and continuous motion time.

Table 1. Continuous motion time and range (The case of one reed switch)

Number of batteries	Consumption current (mA)	Vertical speed (mm/s)	Time (min)	Range (m)
4	8	26.3	117	184
5	10	56.6	89	302
6	12	86.2	72	372
7	14	125	61	457
8	15	159.8	57	546
9	16.5	200.2	52	624
10	18	250	47	705

Table 2. Continuous motion time and range (The case of two reed switches)

Number of batteries	Consumption current (mA)	Vertical speed (mm/s)	Time (min)	Range (m)
6	4	40	237	568
8	5.5	79.5	178	849
10	6.5	121	150	1089

Table 3. Continuous motion time and range (The case of two reed switches)

Number of batteries	Consumption current (mA)	Horizontal speed (mm/s)	Time (min)	Range (m)
10	6.5	176.5	150	1588

Table 1 shows the continuous motion time and range when one reed switch was loaded in the actuator. The maximum range of this cableless actuator was 765 m for vertical motion over 51 minutes when powered by 10 button batteries with no load mass.

Table 2 and Table 3 show the continuous motion time and range when two reed switches were loaded in the cableless actuator with no load mass. For vertical direction, this cableless actuator can move range of 1,089 m. On the other hand, that at horizontal direction was 1,588 m. The moving speed is lower than one reed switch type, however two reed switches type is capable of movement over long range of 1,000 m.

Figure 10 shows the relationship between the consumption current and total range at vertical upward direction of the cableless actuator. When the consumption current is small, the battery life is prolonged drastically.

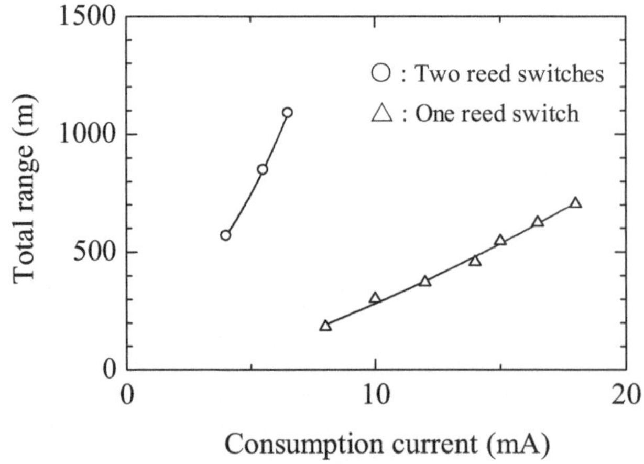

Figure 10. Relationship between total range and consumption current

5. Conclusion

A cableless magnetic actuator with a new electrical DC-AC inverter has been proposed. When powered by two battery packs of button batteries using two reed switches, the cableless actuator displayed speeds of 575.9 mm/s moving vertical downward, 176.5 mm/s moving horizontally, and 121 mm/s moving vertical upward. Since it moves downward much more quickly than it climbs, the actuator must be lightened to eliminate this large disparity. Experimental results demonstrated that the cableless actuator was found to be fully applicable to locomotion in a thin pipe.

The total operating range of this cableless magnetic actuator is 1,588 m in the horizontal direction or 1,089 m in the vertical direction over an operating time of 150 min when powered by two battery packs. The moving speed is lower than one reed switch model, however the actuator of two reed switches type is capable of movement over long range of 1,000 m. On the otherhand, the maximum efficiency is 34 % when two reed switches were loaded. The maximum efficiency is thought to be high for an electromagnetic actuator of this size.

If the power requirement of the propulsion module can be lowered, it appears possible to extend the range of the actuator to five kilometer. In the future, the magnetic circuit of the proplusion module for the actuator and the connection of the actuator body will be improved. Future research will be directed toward these goals.

References

Saito, H., Sato, K., Kudo, K., & Sato, K. (2000). Fundamental Study of Mover Travel inside a Small Diameter Pipe. *Trans. Japan Soc. Mec. Eng.*, 66(641), 346-353.

Fujita, S., Shiraogawa, Y., Yamamoto, S., & Kato, T. (2004). Study of mobile mechanism with elastic fibers by using vibration. *Trans. Japan Soc. Mec. Eng., 70*(698), 22-26.

Ohno, M., Hamano, T., & Kato, S. (2006). Modeling and fabrication of a mobile inspection microrobot driven by a pneumatic bellows actuator for long pipes. *Journal of Robotics and Mechatoronics*, *18*(1), 11-17.

Kwon, J., Park, S., Kim, B., & Park, J. O. (2005). Bio-material property measurement system for locomotive mechanism in gastro-intestinal tract. In *Robotics and Automation, 2005. ICRA 2005. Proceedings of the 2005 IEEE International Conference on* (pp. 1303-1308). IEEE.

Reynaerts, D., Peirs, J., & Van Brussel, H. (1996). Design of a shape memory actuated gastrointestinal intervention system. *Proc. Actuator, 96.*

Miyagawa, T., & Iwatsuki, N. (2008). Moving characteristics in bent pipes of in-pipe mobile robot with wheel drive mechanism using planetary gear drive. *Journal of the Japan Society for Precision Engineering*, *74*(12), 1346-1350.

Bocko, J., Kelemen, M., Kelemenova, T., & Jezný, J. (2009). Wheeled locomotion inside pipe. *Bulletin of applied mechanics, 5*(18), 34-36.

Choi, H. R., & Roh, S. G. (2007). *In-pipe robot with active steering capability for moving inside of pipelines. Bioinspiration and Robotics Walking and Climbing Robots, 23*, 375-400.

Fukuda, T., Hosoi, H., & Ohyama, H. (1991). A Study on In-Pipe Inspection Robots. *Trans. Japan Soc. Mec. Eng., 57*(537), 243-7248.

Tsuruta, K., Shibata, T., Mitsumoto, N., Sasaya, T., & Kawahara, M. (2002). Compact and low power consumed control circuit for wireless micromachine. *Transactions of the Institute of Electric Engineers of Japan, 122*(2), 67-72.

Yaguchi, H., Ishikawa, K., & Zanma, T. (2009). A novel cableless magnetic actuator capable of reversible motion in a thin pipe. *Magnetics, IEEE Transactions on, 45*(10), 4530-4533.

Yaguchi, H., & Izumikawa, T. (2011). Cableless magnetic actuator capable of locomotion in a thin pipe by means of a new motion principle. *Magnetics, IEEE Transactions on, 47*(10), 4290-4293.

Izumikawa, T., & Yaguchi, H. (2012). Movement of a cableless in-piping magnetic actuator with a new propulsion module. *Magnetics, IEEE Transactions on, 48*(11), 4196-4199.

Yaguchi, H., & Izumikawa, T. (2013). Wireless in-piping actuator capable of high-speed locomotion by a new motion principle. *Mechatronics, IEEE/ASME Transactions on, 18*(4), 1367-1376.

Yaguchi, H., & Izumikawa, T. (2014). Cableless In-piping Magnetic Actuator System Capable of Inspection Over Long Distances. *Magnetics, IEEE Transactions on, 50*(11), 1-4.

10

Derivation of Generalized Young's Equation for Wetting of Cylindrical Droplets on Rough Solid Surface

Xiao-Song Wang[1]

[1] Institute of Mechanical and Power Engineering, Henan Polytechnic University, No. 2001, Century Avenue, Jiaozuo, Henan 454003, China

Correspondence: Xiao-Song Wang, Institute of Mechanical and Power Engineering, Henan Polytechnic University, No. 2001, Century Avenue, Jiaozuo, Henan 454003, China. E-mail: wangxs2016@163.com

Abstract

The surface tension depends on the radius of curvature of the liquid-vapor interface. For nano-scale wetting phenomena of cylindrical droplets, we should consider the curvature effects of the surface tension and the line tension. However, previous works have not analyzed the influence of the curvature effects of the surface tension. In this paper, we discuss the influence of the curvature effects of the surface tension on the contact angles based the Kim-Lee-Han-Park equation. The hydrophilic wetting of cylindrical droplets on rough and chemically homogeneous non-deformable substrates were studied by methods of thermodynamics. A generalized Young's equation for wetting of cylindrical droplets on chemically homogeneous and rough non-deformable substrates was derived based on the thermodynamic equilibrium conditions. This equation reduces to the Wenzel equation if we ignore the influence of line tension. For contact angles of cylindrical droplets with sufficiently large radii, a generalized Young's equations were derived considering the curvature effects of the surface tension.

Keywords: contact angle, wetting, Young's equation, Wenzel equation, surface tension, curvature effect, cylindrical droplet

1. Introduction

Wetting phenomena are common in solid-liquid-gas systems, for instance, wetting of liquid droplets on solid surfaces, adhesives, lubricants and capillary penetration in to porous media (Adamson, 1990; Gennes, Brochard-Wyart, & Quere, 2004). Wetting abilities are important in many industrial applications, for example, the wetting abilities of electrolytes on electrodes plays a key role in improving the specific energy density of supercapacitors (Kim, Koo, Lee, & Braun, 2014) and lithium-ion batteries (Pfleging & Proella, 2014).

In 1805, Thomas Young argued that the contact angle θ_Y for the wetting of spherical droplets on rough and chemically homogeneous substrates is determined by the following equation (Young, 1805)

$$\cos\theta_Y = \frac{\sigma_{SG} - \sigma_{SL}}{\sigma_{LG}} \tag{1.1}$$

where θ_Y is the contact angle, σ_{LG} is the surface tension of the liquid - vapor interface corresponds to the choice of the surface of tension as a dividing surface, σ_{SG} is the surface free energy per unit area of the solid - vapor interface, σ_{SL} is the surface free energy per unit area of the solid - liquid interface.

Now, Equation (1.1) is called the Young's equation. The Young's equation Equation (1.1) is widely applied to macroscopic capillary phenomena (Pfleging & Proella, 2014; Xiao-Song et al., 2014).

In 1878, Gibbs for the first time gave a theoretical derivation of the Young's equation Equation (1.1) based on the theory of thermodynamics (Gibbs, 1928). Since then, many theoretical research works have been carried out (Pfleging & Proella, 2014).

The surface tension depends on the radius of curvature of the liquid-vapor interface. For nano-scale wetting phenomena of cylindrical droplets, we should consider the curvature effects of the surface tension and the line tension. However, previous works have not analyzed the influence of the curvature effects of the surface tension.

The purpose of this paper is to present a theoretical study the contact angles of cylindrical droplets on hydrophilic substrates and discuss the influence of the curvature effects of the surface tension on the contact angles.

2. The Helmholtz Free Energy for Wetting of a Cylindrical Droplet on Rough Substrate

Consider a single-component cylindrical liquid droplet in contact with chemically homogeneous and rough substrates.

If the contact angle less than 90°, then we say that the solid surface is hydrophilic. If the contact angle larger than 90°, then we say that the solid surface is hydrophobic. We only consider the wetting of a cylindrical droplet on hydrophilic solid surface. An illustration of hydrophilic wetting is shown in Figure 1.

Introducing Gibbs's concept of dividing surface and the concept of dividing line (Gibbs, 1928; Ono & Kondo, 1960; Rowlinson & Widom, 1982), the above solid-liquid-vapor system can be divided into six subsystems, i.e. liquid phase, vapor phase, the liquid-vapor interface, the solid-liquid interface, the solid-vapor interface and the three-phase contact line.

Therefore, the total Helmholtz free energy F of the system is the sum of the Helmholtz free energies of these seven parts. Thus, we have

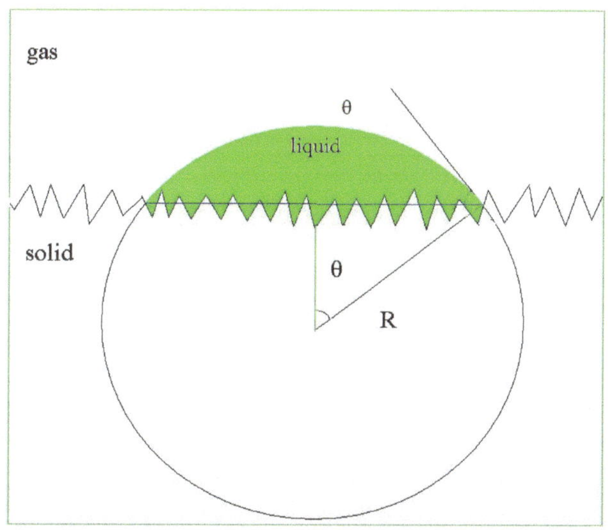

Figure 1 an illustration of hydrophilic wetting of a cylindrical droplet on a rough substrate.

$$F = F_L + F_G + F_{LG} + F_{SL} + F_{SG} + F_{SLG} \tag{2.1}$$

where F is the total Helmholtz free energy, F_L, F_G, F_{LG}, F_{SL}, F_{SG} and F_{SLG} are the Helmholtz free energies of the seven parts respectively.

The Helmholtz free energies of these seven parts can be written as

$$F_L = -p_L V_L + \mu_L N_L \tag{2.2}$$

$$F_G = -p_G V_G + \mu_G N_G \tag{2.3}$$

$$F_{LG} = \sigma_{LG} A_{LG} + \mu_{LG} N_{LG} \tag{2.4}$$

$$F_{SL} = \sigma_{SL} A_{SL} + \mu_{SL} N_{SL} \tag{2.5}$$

$$F_{SG} = \sigma_{SG} A_{SG} + \mu_{SG} N_{SG} \tag{2.6}$$

$$F_{SLG} = \kappa L_{SLG} + \mu_{SLG} N_{SLG} \tag{2.7}$$

where p_L and p_G are the pressures of the liquid phase and the vapor phase respectively, V_L and V_G are the volumes of the liquid phase and the vapor phase respectively, μ is the chemical potential of the six subsystems, $N_L, N_G, N_{LG}, N_{SL}, N_{SG}$ and N_{SLG} are the mole numbers of molecules of the liquid phase, the vapor phase, the liquid - vapor interface, the solid - liquid interface, the solid - vapor interface and the three - phase contact line respectively, $A_{LG,}$ A_{SL} and A_{SG} are the surface area of the liquid - vapor interface, the solid - liquid interface, the solid - vapor interface respectively, σ_{LG}, σ_{SL} and σ_{SG} are the surface tensions of the liquid - vapor interface, the solid - liquid interface, the solid - vapor interface respectively, L_{SLG} is the value of the length of the three - phase contact line, k is the line tension.

In order to calculate the geometrical quantities in the above equations, we may introduce the following assumption:

Assumption 1: Suppose the equilibrium shape of a droplet on a rough and homogeneous solid substrate is a part of a cylinder which was cut by a plane parallel to the axis of the cylinder.

Based on Assumption 1, the total Helmholtz free energy F of the system is

$$\begin{aligned} F= & -(p_L - p_G)(\theta - \sin\theta\cos\theta)R^2 L \\ & - p_G V_t + \mu N_L + \mu N_G \\ & + \sigma_{LG} 2\theta RL + \mu N_{SL} + \mu N_{LG} \\ & + r_s(\sigma_{SL} - \sigma_{SG}) 2R\sin\theta L \\ & + r_s \sigma_{SG} A_t + \mu N_{SG} \\ & + 2 r_L L\kappa + \mu N_{SLG}, \end{aligned} \tag{2.8}$$

where R is the radius of the cylindrical liquid droplet, θ is the contact angle, L is the length of the cylindrical liquid droplet, A_t is the total surface area, r_s is the surface roughness factor, r_L is the line roughness factor.

3. Generalized Young's Equations based on Thermodynamic Equilibrium Condition

The purpose of this section is to derive a generalized Young's equations for cylindrical droplets on rough and homogeneous solids by methods of thermodynamics.

According to Gibbs's concept of dividing surface (Ono & Kondo, 1960), we can choose an arbitrary conformal surface as a dividing surface. Now, we suppose that the radius R of the dividing surface has already been chosen according to some fixed conditions. The contact angle θ now becomes variable. The thermodynamic equilibrium condition at a fixed temperature T of an open system is (Nijmeijer, Bruin, Woerkom, & Bakker, 1992)

$$\left(\frac{\partial F}{\partial \theta}\right)_{T,\mu} = 0, \tag{3.}$$

where the subscript T and μ stands for fixed temperature T and fixed chemical potential μ.

It is convenient to introduce the concept of grand potential to treat an open system. The definition of the grand potential Ω of a system is (Rowlinson & Widom, 1982)

$$\Omega = \sum_{i=1}^{\tau} (F_i - \mu_i N_i), \tag{3.2}$$

where τ is the number of subsystems of the system, F_i is the Helmholtz free energy of the i - th subsystem, μ_i is the chemical potential of the the i-th subsystem, N_i is the mole numbers of molecule of the the i-th subsystem.

Putting Equation (3.2) into Equation (3.1), the thermodynamic equilibrium condition becomes (Nijmeijer, Bruin, Woerkom, & Bakker, 1992)

$$\left(\frac{\partial \Omega}{\partial \theta}\right)_{T,\mu} = 0. \tag{3.3}$$

Putting Equation (2.8) into Equation (3.2), the total grand potential Ω of the above system is

$$\begin{aligned} \Omega= & -(p_L - p_G)(\theta - \sin\theta\cos\theta)R^2 L \\ & - p_G V_t + \sigma_{LG} 2\theta RL \\ & + r_s(\sigma_{SL} - \sigma_{SG}) 2R\sin\theta L \\ & + r_s \sigma_{SG} A_t + 2 r_L L\kappa. \end{aligned} \tag{3.4}$$

Putting Equation (3.4) into Equation (3.3), we have

$$
\begin{aligned}
&-(p_L - p_G)\left(\frac{\partial f_1}{\partial \theta}\right)_{T,\mu} - f_1\left(\frac{\partial (p_L - p_G)}{\partial \theta}\right)_{T,\mu} \\
&+\sigma_{LG}\left(\frac{\partial f_2}{\partial \theta}\right)_{T,\mu} + f_2\left(\frac{\partial \sigma_{LG}}{\partial \theta}\right)_{T,\mu} \\
&+r_s(\sigma_{SG} - \sigma_{SL})\left(\frac{\partial f_3}{\partial \theta}\right)_{T,\mu} + f_3 r_s\left(\frac{\partial (\sigma_{SG} - \sigma_{SL})}{\partial \theta}\right)_{T,\mu} \\
&+r_s\sigma_{SG}\left(\frac{\partial f_4}{\partial \theta}\right)_{T,\mu} + f_4 r_s\left(\frac{\partial \sigma_{SG}}{\partial \theta}\right)_{T,\mu} \\
&+r_L\kappa\left(\frac{\partial f_5}{\partial \theta}\right)_{T,\mu} + f_5 r_L\left(\frac{\partial \kappa}{\partial \theta}\right)_{T,\mu} = 0,
\end{aligned}
\tag{3.5}
$$

where

$$f_1 = (\theta - \sin\theta\cos\theta)R^2L, \tag{3.6}$$

$$f_2 = 2\theta RL, \tag{3.7}$$

$$f_3 = 2LR\sin\theta, \tag{3.8}$$

$$f_4 = A_t, \tag{3.9}$$

$$f_5 = 2L. \tag{3.10}$$

According to Gibbs's concept of dividing surface (Gibbs, 1961; Rowlinson & Widom, 1982), we can choose an arbitrary conformal surface as a dividing surface. Now, the radius R of the dividing surface has already been chosen according to some fixed conditions. At the fixed temperature T and fixed chemical potential, the pressure p_L, p_G and surface tension will not influence the contact angle θ. Thus, in order to simplify Equation (3.5), we introduce the following assumption.

Assumption 2: Suppose the following equations are valid for the wetting of cylindrical droplets on rough and chemically homogeneous non - deformable substrates

$$\left(\frac{\partial (p_L - p_G)}{\partial \theta}\right)_{T,\mu} = 0, \tag{3.11}$$

$$\left(\frac{\partial \sigma_{SL}}{\partial \theta}\right)_{T,\mu} = 0, \tag{3.12}$$

$$\left(\frac{\partial \sigma_{SG}}{\partial \theta}\right)_{T,\mu} = 0, \tag{3.13}$$

$$\left(\frac{\partial \sigma_{LG}}{\partial \theta}\right)_{T,\mu} = 0. \tag{3.14}$$

We have the following results

$$\left(\frac{\partial f_1}{\partial \theta}\right)_{T,\mu} = 2R^2L\sin^2\theta, \tag{3.15}$$

$$\left(\frac{\partial f_2}{\partial \theta}\right)_{T,\mu} = 2RL, \tag{3.16}$$

$$\left(\frac{\partial f_3}{\partial \theta}\right)_{T,\mu} = 2RL\cos\theta, \tag{3.17}$$

$$\left(\frac{\partial f_4}{\partial \theta}\right)_{T,\mu} = 0, \tag{3.18}$$

$$\left(\frac{\partial f_5}{\partial \theta}\right)_{T,\mu} = 0, \tag{3.19}$$

Putting Equations (3.11-3.19) into Equation (3.5), we obtain

$$-(p_L - p_G)\ R\sin^2\theta + \sigma_{LG} + r_s(\sigma_{SL} - \sigma_{SG})\cos\theta + \frac{r_L}{R}\left(\frac{\partial\kappa}{\partial\theta}\right)_{T,\mu i} = 0. \tag{3.20}$$

It is known that a generalized Laplace's equation of a free cylindrical droplet in vapor can be written as (Wenzel, 1936)

$$p_L - p_G = \frac{\sigma_{LG}}{R} + \left[\frac{d\sigma_{LG}}{dR}\right], \tag{3.21}$$

Where the differential in square bracket[ ]denotes the change resulted from a mathematical variation of the position of this dividing surface by the amount dR in the same physical system under the same fixed physical state.

Applying Equation (3.21), Equation (3.20) becomes

$$\cos\theta = r_s\frac{\sigma_{SG} - \sigma_{SL}}{\sigma_{LG}} + \frac{R\sin\theta\tan\theta}{\sigma_{LG}}\left[\frac{d\sigma_{LG}}{dR}\right] - \frac{r_L}{\sigma_{LG}\,R\cos\theta}\left(\frac{\partial\kappa}{\partial\theta}\right)_{T,\mu i}. \tag{3.22}$$

Equation (3.22) is a generalized Young's equation for wetting of a cylindrical droplet on chemically homogeneous and rough non - deformable substrates. Equation (43) is the main results of this work.

Following Gibbs (Kim, Lee, Han, & Park, 2006; Nijmeijer, Bruin, Woerkom, & Bakker, 1992), we introduce the concept of surface of tension M_s as follows

$$\left[\frac{d\sigma_{LG}}{dR}\right]\Bigg|_{R=R_s} = 0, \tag{3.23}$$

Where R_s is the radius of the surface of tension M_s

If we choose the surface of tension M_s as the dividing surface, then Equation (43) becomes

$$\cos\theta = r_s\frac{\sigma_{SG} - \sigma_{SL}}{\sigma_{LG}} - \frac{r_L}{\sigma_{LG}\,R\cos\theta}\left(\frac{\partial\kappa}{\partial\theta}\right)_{T,\mu i}. \tag{3.24}$$

Equation (3.24) is a useful generalization of the Young's equations Equation (1.1) for wetting of cylindrical droplets.

If we neglect the second term on the right side of Equation (3.24), we have

$$\cos\theta = r_s\frac{\sigma_{SG} - \sigma_{SL}}{\sigma_{LG}}. \tag{3.25}$$

Equation (3.25) is the Wenzel equation (Wenzel, 1936).

4. Brief Review of Curvature Effects of Surface Tension of Cylindrical Droplet

In 2006, Kim, Lee, Han, and Park (2006) obtained the following equation for the curvature effects of surface tension of cylindrical droplets

$$\frac{1}{\sigma}\frac{d\sigma_{LG}(R_s)}{dR_s} = \frac{\dfrac{\delta_G}{R_s^2}\left(1 + \dfrac{\delta_G}{2R_s}\right)}{1 + \dfrac{\delta_G}{R_s} + \dfrac{\delta_G^2}{2R_s^2}}, \tag{4.1}$$

where $\sigma_{LG}(R_s)$ is the surface tension of a cylindrical droplet with radius R_s, R_s is the radius of Gibbs' surface of tension, $\delta_G \equiv R_e - R_s$, R_e is the radius corresponding to the equimolar dividing surface.

For sufficiently large droplets, we neglect δ_G^3 / R_s^3 and δ_G^2 / R_s^2, and treat δ_G as the constant δ_T, Equation (4.1) can be solved. The result is

$$\frac{\sigma_{LG}(R_s)}{\sigma_\infty} = \exp\left(-\frac{\delta_T}{R_s}\right), \tag{4.2}$$

where σ_∞ is the surface tension of plane surface with radius $R_s = \infty$, $\delta_T \equiv \lim_{R_s \to \infty} \delta_G$.
δ_T is usually called the Tolman length in the literatures (Rowlinson & Widom, 1982).
We have the following Taylor's expansion

$$\exp(-x) = 1 - x + \frac{x^2}{2!} - \frac{x^3}{3!} + \ldots \tag{4.3}$$

Applying Equation (4.3), Equation (4.2) can be expanded as

$$\frac{\sigma_{LG}(R_s)}{\sigma_\infty} = 1 - \frac{\delta_T}{R_s} + \frac{\delta_T^2}{R_s^2} + \ldots \tag{4.4}$$

5. Generalized Young's Equation Considering the Curvature Effects of the Surface Tension

Using Equation (4.2), Equation (3.24) can be written as

$$\sigma_\infty \exp\left(-\frac{\delta_T}{R_s}\right)\cos\theta = r_s\left(\sigma_{SG} - \sigma_{SL}\right) - \frac{r_L}{\mathrm{R}\cos\theta}\left(\frac{\partial \kappa}{\partial \theta}\right)_{T,\mu}. \tag{5.1}$$

Equation (5.1) is a generalized Young's equation for spherical droplets on rough but chemically heterogeneous non-deformable substrates considering the curvature effects of the surface tension
σ_{LG}. Equation (5.1) is one of the main results of this work.

Applying Equation (4.3), Equation (5.1) becomes

$$\sigma_\infty \left(1 - \frac{\delta_T}{R_s} + \frac{\delta_T^2}{R_s^2} + \ldots\right)\cos\theta = r_s\left(\sigma_{SG} - \sigma_{SL}\right) - \frac{r_L}{\mathrm{R}\cos\theta}\left(\frac{\partial \kappa}{\partial \theta}\right)_{T,\mu}. \tag{5.2}$$

According to experiments (Adamson, 1990), the curvature effects of the surface tension σ_{LG} of macroscopic cylindrical droplets can be neglected. However, for nano-scale cylindrical droplets, the curvature effects of the surface tension σ_{LG} are important (Wenzel, 1936; Rowlinson & Widom, 1982; Adamson, 1990; Nijmeijer, Bruin, Woerkom, & Bakker, 1992; Gennes, Brochard-Wyart, & Quere, 2004;). For droplets with sufficiently large radii, if we neglect $\frac{\delta_T^2}{R_s^2}$ and higher terms in Equation (5.2), then we have

$$\sigma_\infty \left(1 - \frac{\delta_T}{R_s}\right)\cos\theta = r_s\left(\sigma_{SG} - \sigma_{SL}\right) - \frac{r_L}{\mathrm{R}\cos\theta}\left(\frac{\partial \kappa}{\partial \theta}\right)_{T,\mu}. \tag{5.3}$$

6. Conclusion

The hydrophilic wetting of cylindrical droplets on rough and chemically homogeneous non-deformable substrates were studied by methods of thermodynamics. A generalized Young's equation for wetting of cylindrical droplets on chemically homogeneous and rough non-deformable substrates was derived based on the thermodynamic equilibrium conditions. This equation reduces to the Wenzel equation if we ignore the influence of line tension. It is known that the surface tension depend on the radius of curvature of the liquid-vapor interface. For nano-scale wetting phenomena, we should consider the curvature effects of the surface tension and the line tension. However, previous works have not analyzed the influence of the curvature effects of the surface tension. Based the Kim-Lee-Han-Park equation, we discuss the influence of the curvature effects of the surface tension on the contact angles. For contact angles of cylindrical droplets with sufficiently large radii, a generalized Young's equations were derived considering the curvature effects of the surface tension.

Acknowledgments

This work was supported by the Key Project of Science of the Education Bureau of Henan Province (Grant No. 15A130001) and the Doctor Research Foundation of Henan Polytechnic University (Grant No. 72515 - 466).

References

Adamson, A. W. (1990). *Physical Chemistry of Surfaces*. New York: John Wiley Sons.

Gennes, P. G. d., Brochard-Wyart, F., & Quere, D. (2004). *Capillarity and wetting phenomena: drops, bubbles, pearls waves*. New York: Springer-Verlag.

Gibbs, J. W. (1928). *The Collected Works of J. Willard Gibbs* (vol. 1, Thermodynamics). New Haven: Yale Univ. Press.

Gibbs, J. W. (1961). *The Scientific Papers of J. W. Gibbs* (Vol. 1). New York: Dover.

Kim, B. G., Lee, J. S., Han, M., & Park, S. (2006). A molecular dynamics study on stability and thermophysical properties of nanoscale liquid threads. *Nanoscale and Microscale Thermophysical Engineering 10*, 283. http://dx.doi.org/10.1080/15567260600902061

Kim, S. K., Koo, H. J., Lee, A., & Braun, P. V. (2014). Selective Wetting-Induced Micro-Electrode Patterning for Flexible Micro-Supercapacitors. *Advanced Materials, 26*, 5108. http://dx.doi.org/10.1002/adma.201401525

Nijmeijer, M. J. P., Bruin, C., Woerkom, A. B. V., & Bakker, A. F. (1992). Molecular dynamics of the surface tension of a drop. *J. Phys. Chem., 96*, 565. http://dx.doi.org/10.1063/1.462495

Ono, S., & Kondo, S. (1960). Molecular Theory of Surface Tension in Liquids. In S. Flugge (Ed.), *Encyclopedia of Physics* (volume 10). Berlin: Springer-Verlag.

Pfleging, W., & Proella, J. (2014). A new approach for rapid electrolyte wetting in tape cast electrodes for lithium-ion batteries. *Journal of Materials Chemistry A, 2*, 14918. http://dx.doi.org/10.1039/C4TA02353F

Rowlinson, J. S., & Widom, B. (1982). *Molecular Theory of Capillarity*. Oxford: Clarendon Press.

Wenzel, R. N. (1936). Resistance of solid surfaces to wetting by water. *Industrial & Engineering Chemistry, 28*(8), 988-994.

Xiao-Song, W., Shu-Wen, C., Long, Z., Sheng-Hua, X., Zhi-Wei, S., & Ru-Zeng, Z. (2014). A Generalized Young's Equation for Contact Angles of Droplets on Homogeneous and Rough Substrates. *J. Adhesion Sci. Tech., 28*, 161. http://dx.doi.org/10.1080/01694243.2013.833401

Young, T. (1805). An essay on the cohesion of fluids. *Philos. Trans. Roy. Soc. London, 65*.

11

An Overview of Used-Products Remanufacturing

Magdalene Andrew-Munot[1], Raafat N Ibrahim[2] & Ervina Junaidi[1]

[1] Department of Mechanical & Manufacturing Engineering, Faculty of Engineering, Universiti Malaysia Sarawak, Sarawak, Malaysia

[2] Department of Mechanical and Aerospace Engineering, Faculty of Engineering, Monash University, Australia

Correspondence: Magdalene Andrew-Munot, Department of Mechanical & Manufacturing Engineering, Faculty of Engineering, Universiti Malaysia Sarawak, 94300 Kota Samarahan Sarawak, Malaysia.
E-mail: ammagdal@feng.unimas.my

Abstract

Remanufacturing is an important production activity, which is mainly driven by the increasingly strict environmental regulations imposed by governmental bodies, growing customers' awareness of green environment and potential economic benefits of remanufacturing. In this paper, the key motivating factors for companies to engage in remanufacturing program are examined. The major sources for acquiring used-products and the subsequent markets for selling remanufactured products are also examined. More importantly, the generic-key remanufacturing processes are examined and the corresponding unique characteristics are highlighted. Four examples of remanufacturing process of different products are presented to demonstrate that exact number and sequence of remanufacturing process largely depended on the type of product being remanufactured. In sum, this paper presents an overview of used-products remanufacturing, which in many ways would benefit the general public, environment, manufacturers/remanufacturers, as well as researchers.

Keywords: used-products, remanufacturing, product recovery

1. Introduction

Recently, remanufacture of used/worn-out/broken products has become a vital part of production activity in many companies (Ferrer & Swaminathan, 2009; Schulz & Ferretti, 2011). The automotive sector particularly has a strong and long history of remanufacturing (Seitz, 2007; Golinska, & Kawa, 2011) where the alternator has the highest remanufacturing rate (Schau, Traverso, & Finkbeiner, 2012). Remanufacturing is also gaining scientific significance in a variety of industries that include single use-devices for hospitals, such as wheelchairs and hearing aids (Srivastava, 2004) cellular phones (Guide Jr, Jayaraman, & Linton, 2003b) and truck tyres (Lebreton & Tuma, 2006). The diverse nature of remanufactured products together with different degrees of remanufacturing efforts creates a variety of applications (Zikopoulos, 2012). At one end is the direct re-use of slightly used commercial used-products, to the high-touch remanufacture of expensive modules, which are reassembled in complex new products, at the other end.

Remanufacture of used/worn-out/broken products (remanufacturing), refers to an industrial process in which used/worn-out/broken products (henceforth called used-products) are transformed into "new products" (Lund, 1984b). Hereafter, these "new products" are referred to as remanufactured products to distinguish them from a completely new product. Generally, any manufactured products or devices can be remanufactured, provided that these are discarded used-products with lower materials and components reprocessing costs compared to the market value of remanufactured items (Lund, 1984b). Also as outlined by Hauser and Lund (2003), used-products must be; (i) durable products, (ii) products that fails functionally, (iii) standardized products with interchangeable parts, (iv) products with high remaining value-added, (v) products with low acquisition cost, (vi) products with stable technology, and (vii) customer awareness of the remanufactured version.

Remanufacturing process could be sometimes confused with other similar activities such as repairing, refurbishing and reconditioning. Repaired, refurbished and reconditioned products are very close to remanufactured ones and these terms can regularly be considered synonymous with each other (Ijomah, McMahon, Hammond, & Newman, 2007b). Though these terms have similarities, however in term of functionality and performance, the weightage and meanings are different. For example, repairing is described as corrections of specified faults in a product;

when repaired products have warranties, the warranty does not cover the whole product but only on the component that has been replaced. In refurbishing, a used-product is cleaned up, tested, repackaged and made available for resale. A refurbished product has some nonworking parts that are replaced to restore the product to working condition, with warranty less than that of a new product. For reconditioning, a used-product is cleaned up and tested extensively with possible repair before repackaged for resale at a satisfactory working condition, with warranty less than that of a newly manufactured equivalent, in which the warranty applies to all major wearing parts. Finally, the term recycling describes the processing of used material into new products through discarding used materials into production of different or same product.

This paper aims is to provide the general public, potential remanufacturers, as well as researchers with valuable insights into remanufacturing program by examining remanufacturing activities and identifying major opportunities and concerns. The remainder of this paper begins in Section 2, which examines the three major motivating factors for remanufacturing. The next section (Section 3), examines the sources for acquiring used-products and markets for selling remanufactured products. Section 4 examines the different types of remanufacturing companies. Section 5 presents the generic remanufacturing process and four examples of real-industry remanufacturing process for different products. Finally in Section 6, the concluding remarks on the overview of used-products remanufacture are presented.

2. Motivations and Benefits of Remanufacturing

The three major factors that are responsible for motivating companies to engage in remanufacturing are product take-back legislations, customers' awareness of green environment and economic benefit.

Product Take-Back Legislations - These emerging legislations imposed by governmental bodies force companies to extend their manufacturers' responsibility to include products that have been used and retired. This in turn, forces companies to design products and production process that would make it possible to recover used-products through remanufacturing activity. In Europe, for example, the directive on waste electrical and electronic equipment (Directive/2002/96/EC) states that "Where appropriate, priority should be given to the reuse of waste electrical and electronic equipment (WEEE) and its components, subassemblies and consumables. Where reuse is not preferable, all WEEE collected separately should be sent for recovery, in the course of which a high level of recycling and recovery should be achieved. In addition, producers should be encouraged to integrate recycled material in new equipment". Similarly, the directive on legislation of end-of-life vehicle (European Council Directive 2000/53/EC 2000) has forced the European automobile manufacturers to assume full responsibility for disposal of cars manufactured from 2002 onwards with no cost to the final owners. The directive also states that "the requirements for dismantling, reuse and recycling of end-of life vehicles and their components should be integrated in the design and production of new vehicles". These strict environmental regulations results in greenhouse gas emission from remanufacturing that is about 90% less than the manufacturing of new product; a study done on remanufacture of compressors (Gamage, Ijomah, & Windmill, 2013).

Customers' awareness of green environment - Customers' are becoming aware of the implication of disposing used-products on the environment. For instance, more consumers are procuring products from manufacturers who take back their products after the end of the products' useful life (Gungor & Gupta, 1999). In another report by Ginsberg and Bloom (2004) the percentage of consumers who are interested in some form of green products are between 15% and 46%. For the manufacturers, taking back and reprocessing their used-products in an environmentally-friendly manner creates the opportunity to boost the value of their products. For example, Fuji Xerox Australia (2007a) has taken the initiative to be one of the pioneering companies that implements remanufacturing program for printers and copier parts. In 2006, for example, such remanufacturing practice has saved the company approximately $13 million and created revenue of $5.4m in exports. Similarly, Hewlett-Packard, HP (2014) also embraces the green environment culture through recycling programs, which help to reduce the environmental impact of the companies' activities. In a comparable effort in 2012, IBM (2014) worldwide product end-of-life management operations processed 36,100 metric tons of end-of-life products for reuse or recycling, sending only 0.3 percent of the total to landfills or to incineration facilities for treatment.

Economic Benefit - Remanufacture of-used products incurred cost that is typically 40% - 65% less that those incurred in the manufacture of new products (Dowlatshahi, 2000; Giuntini & Gaudette, 2003) and remanufactured products are typically sold off at 45% - 65% of the prices of new products (Lund & Hausner, 2003). This results from the availability of raw materials (recovered components), which are cheaper than those newly manufactured components; recovered components do not have to be redesigned and ordered from suppliers. As reported in a publication by the All-Party Parliamentary Sustainable Resource Group (2014) and shown in Figure 1, the economic advantage of remanufacturing is vast. For producing remanufactured products (sold at 20% discount), the input costs is significantly lower than the input costs for manufacturing new products from virgin resources; the same case applies for the remanufacture of lease products (e.g. leased Ricoh photocopiers). Also, as shown in the figure 1, remanufacture of leased products yielded the highest gross profit when compared to the gross profit of manufacture of products from virgin resources. From the remanufacturers' perspective, they benefited economically from the significant discounts for the remanufactured products with the same warranty and quality (Lu, Goh, Garg & Souza, 2014). It is also well recognized that reuse of components in the production of subsequent product generations, results in cost savings (Bhattacharya, Guide Jr., & Van Wassenhove, 2006). In terms of energy and materials consumption, remanufacture of used-products contribute to less usage than that is needed to manufacture new products (Kim, Skerlos, Severengiz, & Seliger, 2009; Boustani, Sahni, Graves & Gutowski, 2010; Warsen, Laumer, & Momberg, 2011; Sundin & Lee, 2012). To give an example, as reported in Kim et. al., (2009), remanufacture of automotive engine can reduced metal consumption by 7,650 tonnes, conserved energy as much as 16 million kW h and decreased emission of about 11,300 to 15,300 tonnes CO2 equivalent (CO2eq).

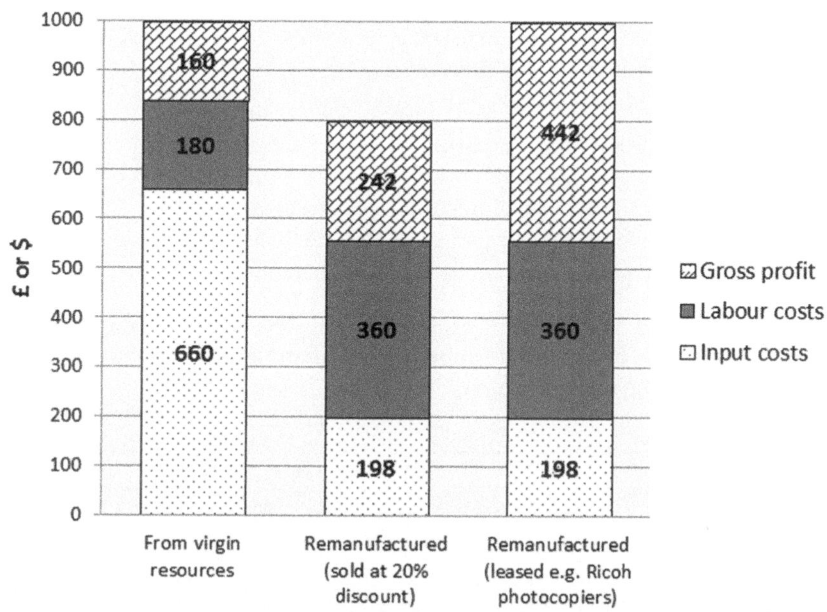

Figure 1. Potential Economic Benefits of Remanufacturing (reproduced from a report retrieved from http://www.policyconnect.org.uk/apsrg/sites/site_apsrg/files/apsrg_-_remanufacturing_report.pdf)

Until recently, companies have considered environmental regulations and consumers' awareness as additional operational cost, rather than opportunities to generate alternative profit and boost corporate image. However, with the increasing pressure to become environmentally responsible and sourcing for inexpensive raw materials, more companies are considering remanufacturing as a strategy to generate profit, boost company image and maintain competitiveness.

3. Sources of Used-Products and Market for Remanufactured Products

The two major types of product remanufactured are used capital goods and used consumer goods (Dobos & Floriska, 2008). These used-products are acquired either from the waste-stream or market-stream (Guide & Wassenhove, 2001; Jayaraman, 2006). The waste-stream generates used-products that are no longer useful, which includes vehicles that are damaged during accidents and sent to wrecking facilities; or malfunctioned electrical home appliances that are sent to third-party products recovery facilities. As such, the waste-stream generates used-products with uncontrollable quality and quantity (Ghoreishi, Jakiela, & Nekouzadeh, 2011). The market-stream generates used-products that are still useful but are no longer needed by the owners. For example, old models of mobile phone are traded-in for new models with advanced and sophisticated features. Thus, the quality conditions and quantity of these used-products are controllable by giving the appropriate amount and type of incentives for returning the used-products. Also, as reported by Guide and Wassenhove (2001), the quality of used-products could be controlled by having a system to grade the used-products based on their condition and age and pays the financial incentives accordingly. Alternatively, as demonstrated in a study by Kumar, Liou, Balakrishnan & Kumar (2013) the use of radio-frequency identification (RFID) technology could effectively improve visibility of data on the quality conditions of used-products, which consequently improve inventory control and operational efficiency at collection, disassembly and refurbishing centres. Irrespective of their sources, the quantity of used-products available for remanufacturing is bound by the quantity of new products sold in the previous periods (Esenduran, Emahlıo˘glu-Ziya, & Swaminathan, 2012).

From the suppliers perspective, there are five different ways to acquire used-products from the suppliers, namely Direct-Order, Reman-Contract, Deposit-Based, Credit-Based and Buy-Back (Sundin & Dunbäck, 2013). In direct-order, the customers are the suppliers themselves, and they send their used-products to the remanufacturer. The remanufactured products would be sent back to the customers/suppliers. This is typical for complex products such as engines and other end user items. Similarly, for reman-contract, the customers are the suppliers themselves, who gives an order to remanufacture the used-products. In this case, there is a closer collaboration between the remanufacturer and customers/suppliers, it usually invoves greater quantities of used-products and common towards original equipment manufacturers (OEM). The deposit-based used-products are common in the automotive industry; this requires the customers to send used-products (e.g. brake calipers) that are similar to new parts that they needed. Comparable to the Direct-Order and Contract-Reman, in the credit-based, the customers are also the suppliers. In this situation, for customers who send their used-products, credits are given to them, which could be utilized as a discount for buying remanufactured products. Finally, the Buy-Back used-products are often obtained from the used-product brokers, scrap yards or end-users.

In the world, the U.S. market for remanufactured goods is the largest, where the production of remanufactured goods increased from $37.3 billion (in 2009) to $43.5 billion in 2011 (United States International Trade Commission, 2012). The types of these remanufactured goods are from production, from import and for export (Table 1). Specifically, the top three industry that contribute to these remanufactured goods are Aerospace, heavy-duty and off-road (HDOR) equipment and motor vehicle parts (Table 2).

Table 1. Remanufactured goods: U.S. market (apparent consumption) for remanufactured goods, 2009 – 2011 2011 (reproduced from (United States International Trade Commission, 2012))

Type	2009	2010	2011	% change, 2009 - 2011
		Thousands $		
Production	37,280,784	39,882,004	43,000,144	15.3
Imports	6,256,042	9,508,060	10,263,056	64.1
Exports	7,502,991	8,805,067	11,735,665	56.4
U.S. apparent consumption	36,033,835	40,584,997	41,527,535	15.2

Note: Totals may not sum due to rounding.

Table 2. Remanufactured goods: production of remanufactured goods by sector, 2009 - 2011(reproduced from (United States International Trade Commission, 2012))

Sector	2009	2010	2011	% change, 2009 - 2011
		Thousands $		
Aerospace	11,691,316	12,677,916	13,045,513	11.6
HDOR equipment	5,152,938	6,244,302	7,770,586	50.8
Motor vehicle parts	7,018,303	6,969,823	6,211,838	-11.5
Machinery	4,059,570	4,774,291	5,795,105	42.8
IT products	2,709,170	2,592,831	2,681,603	-1.0
Medical devices	1,307,588	1,367,739	1,463,313	11.9
Retreaded tires	1,038,679	1,188,315	1,399,088	34.7
Consumer products	557,612	567,320	659,175	18.2
All other[a]	3,745,608	3,499,468	3,973,923	6.1
Total	37,280,784	39,882,004	43,000,144	15.3

Note: Totals may not sum due to rounding.

[a]The electrical apparatus, locomotive, office furniture, and restaurant equipment remanufacturing sectors.

Generally, remanufactured products are sold either in the primary-market or secondary-market. Within the primary product market, remanufactured products are perfect substitutes for new products (Souza & Ketzenberg, 2002). This refers to original equipment manufacturer (OEM) used-products, which are remanufactured to a quality standard that is as good as new at the OEM's remanufacturing facility. For example Katun Corporation (Katun Corporation, 2014), one of the world's leading providers of OEM-compatible imaging supplies, photoreceptors, and parts for copiers, printers, MFPs and other imaging equipment, remanufacturers cartridges for use in HP Color Printers/MFP's. For this company, ensuring customer satisfaction means offering the highest and most consistent remanufactured toner cartridge quality possible. The remanufactured products are sold in the same market as the new products, at a cheaper price than the new ones (Ferrer & Swaminathan, 2009). By selling remanufactured products together with new ones, companies area able to better segment their demand and seize sales to the "low-end" customers, who will not purchase the new product. This requires that the price and quantity of remanufactured products are properly selected (Ovchinikov, 2011). These remanufactured products could also be sold at the same price as newly manufactured products (Gallo, Romano, & Santillo, 2012). However, as reported in (Atasu, Guide, & Van Wassenhove, 2010) there are concerns that selling remanufactured products in the same market with new products might cannibalized the sales of new products.

For the secondary product market, remanufactured products are popular amongst customers who may have financial restrictions. For example, remanufactured multifunctional digital copiers or multi-function peripherals (MFPs) are sold to customers who are satisfied with the cheaper and lower-level functions MFPs (Chang, Bae, Yi, 2013). Furthermore, as reported in Osibanjo and Nnrom (2008), it is estimated that the demand for remanufactured mobile phones in Nigeria is still higher than their availability. For the secondary-market, remanufactured products are usually produced by third-party remanufacturers who harvest the economic benefits of remanufacturing. In this case, usually, the remanufactured products (e.g., computer systems, auto components, and office equipment) have a lower quality standard and price than newly manufactured products (Ayres, Ferrer, & van Leynseele, 1997; Ferrer, 1997). The boundaries of the primary and secondary markets are clearly separated in the case of remanufactured and new mobile phones. However, in the case of remanufactured and new automotive components, these two markets overlap (Gallo, et. al., 2012).

4. Types of remanufacturer

Remanufacturing could either be carried out by original equipment manufacturer (OEM) remanufacturers, contract remanufacturers or independent remanufacturers. For OEM remanufacturers, remanufacturing could co-exist with the normal manufacturing activity, in which resources could either be dedicated to each activity or shared amongst the two activities. For the OEM remanufacturers, remanufacturing activity is an alternative way to generate extra income (from the production of remanufactured spare-parts and products) and obtain low cost raw materials. Most importantly, by remanufacturing its own products, OEM manufacturers could avoid competition from independent remanufacturers (Gutowski, Sahni, Boustani, & Graves, 2011). Some examples of OEM remanufacturers are (i) Fuji Xerox for parts and sub-assemblies from photocopier (Fuji Xerox Australia, 2007a), (ii) Alpha1 Technologies for heads and cameras (Alpha1Technologies, 2014), (iii) Robert Bosch Tool for alternators and starters (Robert Bosch Tool Corporation, 2014), and (iv) Hewlett-Packard for inkjet printers (Hewlett-Packard, 2014).

For both contract and independent remanufacturers, remanufacturing activity is an alternative source for generating profit. Contract remanufacturers are companies that are contracted by OEM to remanufacture their used-products; for example Hewlett-Packard's used-printers are remanufactured in a network of hardware recovery centers (Kumar, Guide Jr, & Van Wassenhove, 2002). In this case Hewlett-Packard carefully protect their intellectual capital and brand name through a contractual agreements, as well as enforcing their quality standards. Independent remanufacturers are companies that realize the economic benefits of remanufacturing, and thus would remanufacture any type of used-products, as long as there are markets for them.

Even though remanufacturing market is dominated by OEMs and contract remanufacturers, the majority of the remanufacturers are independent small-scale operators (Matsumoto & Ijomah, 2013). However, the OEM remanufacturers with their available resources, would be able to effectively remanufacture their own used-products compared to the independent remanufacturers. The resources as reported in Boustani (2010) are (i) full knowledge of the durability, reliability, replacement parts, and proper disassembly methodologies of their own used-products, (ii) established network for the supply and distribution of the remanufactured products, (iii) established relationship with suppliers of parts and materials, (iv) knowledge of customers use patterns, psychology and potential reasons for using remanufactured products, (v) market power, established manufacturing platform, and available workforce to venture into remanufacturing practices.

5. Remanufacturing Process

In general, remanufacturing process consists of four generic-key process, where different activity takes place at each process (Figure 1). The exact number of process and sequence depend on the nature of used-products being remanufactured (Ilgin & Gupta, 2012). The four process can be further expanded to twelve processes, as reported in Boustani (2010) to include (i) warehouse of incoming used-products, parts, and outgoing remanufactured products, (ii) sorting of incoming used-products, (iii) cleaning of used-products, (iv) disassembly of used-products and subassemblies, (v) inspection of used-products, subassemblies, and parts, (vi) cleaning of specific parts and subassemblies, (vii) repair or renewal of parts, (viii) testing of parts and subassemblies, (ix) reassembly of parts, subassemblies and products, (x) testing of subassemblies and finished products, (xi) packing, and (xii) shipping.

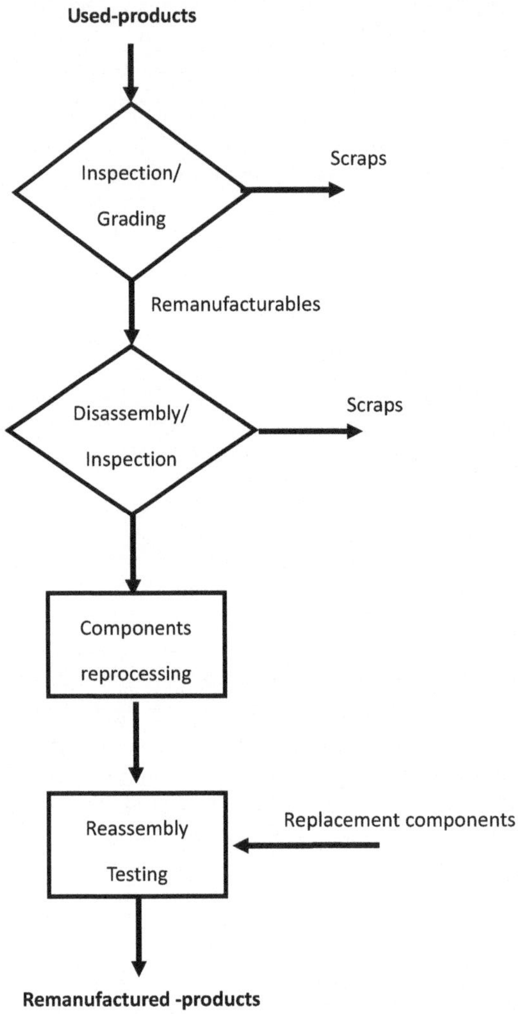

Figure 2. Generic Remanufacturing Process [constructed based on the description provided in Lund (1984a)]

Inspection/Grading process – During this process, used-products are inspected to assess their remanufacturability status, which could either be remanufacturables or scraps. Unlike, manufacturing, remanufacturing requires 100% inspection of used-products (Errington & Childe, 2013). This process usually involves complete visual inspection, where inspection times are identical for the same type of used-products that originate from the same source (e.g. used-cellular phones from the market stream). On the other hand, the waste-stream used-products with high variability of quality conditions require different inspection times for each unit. Scraps are either disposed off or sold to scrap brokers, whereas remanufacturables are sent to the disassembly/inspection process.

The inspection yield (fraction of used-products graded as remanufacturables) could vary from one batch to another due to the uncertain quality conditions of used-products. Remanufacturables could be further classified into several quality groups, in which the best quality group is given the highest priority to remanufacture (Aras, Boyaci,

& Verter, 2004; Aras, Verter, & Boyaci, 2006). The waste-stream used-products with high variability of quality conditions, would yield more quality groups than the market-stream.

Disassembly/inspection - Each unit of remanufacturable is disassembled into its corresponding modules, which are further disassembled into their constituent components. This process usually involves general purpose tools such as power drill, although in some cases, robot arms might be necessary for disassembly of complex products (Steinhilper, 1998) or hazardous parts (Zussman & Seliger, 1999). Despite the quality groups and source, disassembly time for the same type of remanufacturables would be identical. This time mainly depends on the product structure complexity, where remanufacturables with a simple product structure, would require a shorter disassembly time than remanufacturables with a more complex product structure.

Depending on the product structure and volume, disassembly of remanufacturables and inspection of the constituent components could either take place simultaneously or sequentially. For high volume of remanufacturables with simple product structures, disassembly and inspection process could take place simultaneously. Likewise, for low volume of remanufacturables with complex product structure, disassembly and inspection process could takes plane simultaneously. By contrast, for high volume of remanufacturables with complex product structure, disassembly and inspection process usually take place sequentially in a two-stage disassembly line (Steinhilper, 1998).

During inspection, each unit of constituent component could either be classified as remanufacturable or scrap. Scrap components are replaced with newly manufactured components, which are ordered either from the internal production lines (for OEMs remanufacturers) or external sources (for contract or independent remanufacturers). Replacement components are very critical, particurly for a remanufacture-to-order strategy, where customers send their used-products for remanufacture and get the same item back.

The fraction remanufacturable components (i.e. disassembly yield), might differs from one batch to another; this results from the uncertain quality conditions of remanufacturables. Remanufacturable components could be further classified into different quality groups, where the best component quality group should be given the highest priority to remanufacture. The waste-stream remanufacturables would result in more quality component groups than the market-stream remanufacturables.

Irrespective of the product structure, the product design would also affect the disassembly yield of constituent components (Ferrer, 2001). Products that are originally design for disassembly would generate a higher disassembly yields than those that are not originally designed for disassembly because of the damaged during disassembly. Thus, it is important to integrate product design in the product development process in order to reduce damaged components (Ijomah, McMahon, Hammond, & Newman, 2007a; Ijomah, 2009; Barquet, Rozenfeld & Forcellini, 2013).

Components reprocessing - This process typically involves cleaning, repairing (e.g. machining worn-out holes) and surface finishing (Sundin, & Bras, 2005). The actual number of reprocessesing steps and corresponding time depend on the component quality group. For example, the best component quality group would perhaps need a simple process (eg cleaning and surface finishing) and short reprocessing time. By contrast, the worst component quality group probably requires a more complex reprocessing steps (eg cleaning, repairing and surface finishing) and longer reprocessing time. For complex component design, multiple repair steps, such as welding and trimming might be necessary in order to restore the components to their original condition. Equally, for some components such as bulb, electrical wire or cellular phone casing, there is no repair process required because these components are merely replaced with new ones.

Reassembly/testing - This process typically involves general purpose tools to reassemble the reprocessed components into remanufactured products. In some cases, it is necessary to use robot arms to reassemble constituent components into remanufactured products with complex structure. Finally, testing ensures that the assembled products, i.e. remanufactured products meet their specified function.

As shown earlier in Figure 1, remanufacturing process includes inspection/grading, disassembly/inspection, component reprocessing and reassembly/testing. However, the exact number of process and sequence depend largely on the types of used-products being remanufactured. This can be observed in Figures 3 - 6, which illustrate examples of remanufacturing process of four different products. Figure 2 shows the remanufacturing process of photocopiers, figure 3 shows the remanufacturing process of automotive engines, figure 4 shows the remanufacturing process of ink cartridges and figure 5 shows the remanufacturing process of medical devices.

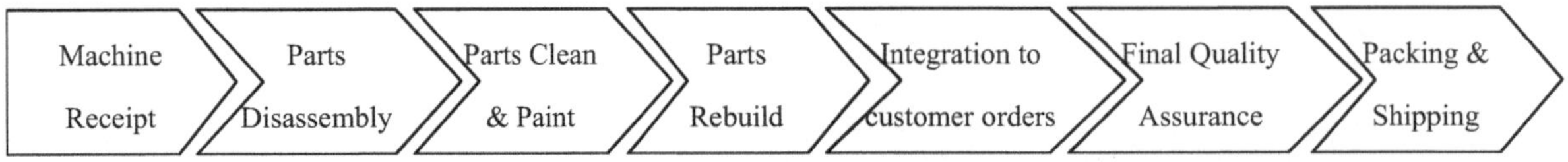

Figure 3. Remanufacturing process of photocopiers, Xerox UK (King, Miemczyk & Buffon, 2006)

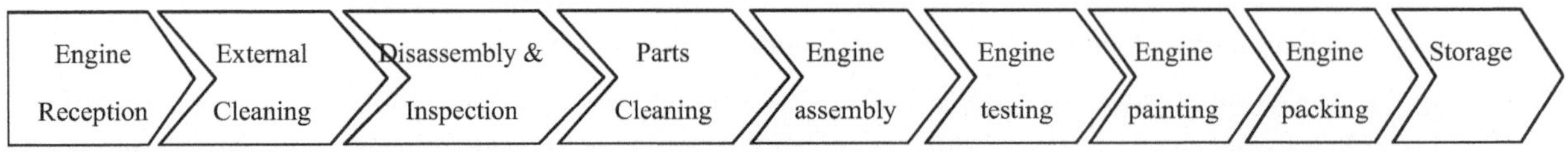

Figure 4. Remanufacturing process of automotive engines [constructed based on information obtain from mti (2014)

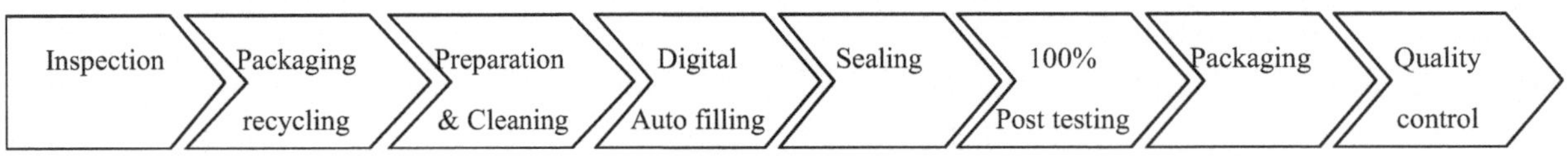

Figure 5. Remanufacturing process of ink cartridges (constructed based on information obtain from Clover Technologies Group (2014)

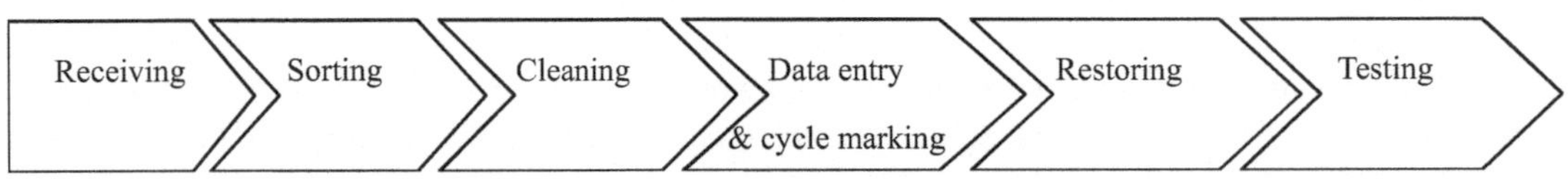

Figure 6. Remanufacturing process of medical devices (constructed based on information obtained from Stryker Sustainability Solutions (2014)

6. Concluding Remarks

Until recently, not many companies engaged in remanufacturing activities. Nonetheless, with the increasingly strict environmental regulations, growing consumers' awareness of green environment and potential economic benefits, more companies are involve in remanufacturing programs. This article makes contributes in the following ways. Firstly, this article highlights the key factors that motivate companies to engage in remanufacturing activities. Secondly, this article identifies the potential sources for acquiring used-products and the markets for selling remanufactured products. Thirdly, this article highlights the generic-key remanufacturing process to be inspection/grading, disassembly, components reprocessing and reassembly.. Fourthly, this article highlights four unique remanufacturing characteristics that are uncertain quality conditions of used-products, variable inspection-yields of used-products, variable disassembly-yield of constituent components and variable reprocessing efforts of constituent components. Finally, this article illustrates with example that the exact number of process & sequence depended on the type of used-product being remanufactured. In short, remanufacturing program, despite its fundamental requirements and unique characteristics, would in many ways benefit the general public, environment, manufacturers/remanufacturers and researchers.

Acknowledgment

The authors would like to acknowledge Universiti Malaysia Sarawak (UNIMAS) for giving financial support towards this project.

References

All-Party Parliamentary Sustainable Resource Group. (2014). *Remanufacturing: Towards a Resource Efficient Economy: report from the three-month inquiry into the need for UK government to improve UK's remanufacturing*. Retrieved from http://www.policyconnect.org.uk/apsrg/sites/site_apsrg/files/apsrg_-_remanufacturing_report.pdf

Alpha1Technolgies. (2014). Information retrieved from http://alpha1technologies.com/remanufactured-heads-and-cameras

Aras, N., Boyaci, T., & Verter, V. (2004). The effect of categorizing returned products in remanufacturing. *IIE Transactions, 36*(4), 319-331. http://dx.doi.org/10.1080/07408170490279561

Aras, N., Verter, V., & Boyaci, T. (2006). Coordination and priority decisions in hybrid manufacturing/remanufacturing systems. *Production and Operations Management, 15*(4), 528-543. http://dx.doi.org/10.1111/j.1937-5956.2006.tb00161.x

Atasu, A., V. Daniel R. Guide, Jr., & Van Wassenhove, L. N. (2010). So what if remanufacturing cannibalizes my new product sales?. *California Management Review, 52*(2).

Ayres, R., Ferrer, G., & Van Leyseele, T. (1997). Eco-efficiency, asset recovery and remanufacturing. *European Management Journal, 15*(5), 447-574. http://dx.doi.org10.1016/S0263-2373(97)00035-2

Barquet, A. P., Rozenfeld, H., & Forcellini, F. A. (2013). An integrated approach to remanufacturing: model of a remanufacturing system. *Journal of Remanufacturing, 3*(1). http://dx.doi.org/10.1186/2210-4690-3-1

Bhattacharya, S., Guide Jr., V. D. R., & Van Wassenhove, L. N. (2006). Optimal order quantities with remanufacturing across new products generations. *Production and Operations Management, 15*(3), 421-431. http://dx.doi.org/ 10.1111/j.1937-5956.2006.tb00255.x

Boustani, A. (2010). Remanufacturing and Energy Savings. Master of Science Thesis

Boustani, A., Sahni, S., Graves, S. C., & Gutowski, T. G. (2010). Appliance remanufacturing and life cycle energy and economic savings. In *Sustainable Systems and Technology (ISSST)* (pp. 1-6). 2010 IEEE International Symposium, Arlington, VA. http://dx.doi.org/10.1109/ISSST.2010.5507713

Chang, Y., Bae, J. H., & Yi, H. C. (2013). Ultrasonic Cleaning of Used Plastic Parts for Remanufacturing of Multifunctional Digital Copier. *International Journal Of Precision Engineering And Manufacturing, 14*(6), 951-956. http://dx.doi.org/10.1007/s12541-013-0125-8

Clover Technologies Group. (2014). *8-Step Ink Cartridge Remanufacturing Process*. Retrieved from http://www.clovertech.com/eight-step-ink-cartridge-remanufacturing-process

Directive 2000/53/EC of the European Parliament and of the Council of 18 September 2000 on end-of life vehicles (Directive 2000/53/EC - the "ELV) (2000). Retrieved from http://ec.europa.eu/environment/waste/elv_index.htm.

Directive 2002/96/EC of the European Parliament and of the Council of 27 January 2003 on waste electrical and electronic equipment (WEEE) (Directive/2002/96/EC). (2003). Retrieved from http://www.environ.ie/en/Legislation/Environment/Waste/WEEE/FileDownLoad,613,en.pdf.

Dobos, I., & Floriska, A. (2008). The efficiency of remanufacturing in a dynamic input-output model. *Central European Journal of Operations Research, 16*(3), 317-328. http://dx.doi.org/10.1007/s10100-008-0060-4.

Dowlatshahi, S. (2000). Developing a theory of reverse logistics. *Interfaces, 30*, 143-155. http://dx.doi.org/10.1287/inte.30.3.143.11670

Errington, M., & Childe, S. J. (2011). A business process model of inspection in remanufacturing. *Journal of Remanufacturing, 3*(7). http://dx.doi.org/ 10.1186/2210-4690-3-7.

Esenduran G., Emahlıo˘glu-Ziya, E., & Swaminathan, J. M. (2012). Product Take-Back Legislation and Its Impact on Recycling and Remanufacturing Industries in T. Boone et al. (Eds.), Sustainable Supply Chains, 129. *International Series in Operations Research & Management Science, 174*. http://dx.doi.org/10.1007/978-1-4419-6105-1

Ferrer, G. (1997). The Economics of Tire Remanufacturing. *Resources, Conservation and Recycling, 19*(4), 221-255. http://dx.doi.org/10.1016/S0921-3449(96)01181-0

Ferrer, G. (2001). On the Widget Remanufacturing Operation. *European Journal of Operational Research, 135*(2), 373-393. http://dx.doi.org/10.1016/S0377-2217(00)00318-0

Ferrer, G., & Swaminathan, J. M. (2009). Managing New and Differentiated Remanufactured Products. *European Journal of Operational Research*. http://dx.doi.org/10.1016/j.ejor.2009.08.007.

FujiXerox, Australia (2007a). Information. Retrieved from http://www.fujixerox.com.au/about/eco_manufacturing.jsp

Gallo, M., Romano, E., & Santillo, L. C. (2012). A Perspective on Remanufacturing Business: Issues and Opportunities. *International Trade from Economic and Policy Perspective*, 209. http://dx.doi.org/10.5772/2726.

Gamage, J. R., Ijomah, W. L., & Windmill, J. (2013). What makes cleaning a costly operation in remanufacturing? In G. Seliger (Ed.), Proceedings of the 11th Global Conference on Sustainable Manufacturing - Innovative Solutions.

Ghoreishi, N., Jakiela, M. J., & Nekouzadeh, A., (2011). A cost model for optimizing the take back phase of used product recovery. *Journal of Remanufacturing, 1*(1). http://dx.doi.org/ 10.1186/2210-4690-1-1.

Ginsberg, J. M., & Bloom, P. (2004). Choosing the Right Green Manufacturing Strategy. *MIT Sloan Management Review, 46*(1), 79 - 84.

Giuntini, R., & Gaudette, K. (2003). Remanufacturing: The next great opportunity for boosting US productivity. *Business Horizons,* 41-48. http://dx.doi.org/10.1016/S0007-6813(03)00087-9

Golinska, P., & Kawa, A. (2011). Remanufacturing in automotive industry: Challenges and limitations. *Journal of Industrial Engineering and Management, 4*(3), 453-466. http://dx.doi.org/10.3926/jiem.2011.v4n3.p453-466

Guide Jr., V. D. R., & Van Wassenhove, L. N. (2001). Managing product returns for remanufacturing. *Production and Operations Management, 10*(2), 142-155. http://dx.doi.org/10.1111/j.1937-5956.2001.tb00075.x.

Guide Jr., V. D. R., Jayaraman, V., & Linton, J.D. (2003b). Building contingency planning for closed-loop supply chains with product recovery. *Journal of Operations Management, 21*(3), 259-279. http://dx.doi.org/10.1016/S0272-6963(02)00110-9

Gungor, A., & Gupta, S. M. (1999). Issues in environmentally conscious manufacturing and product recovery: a survey. *Computers and Industrial Engineering, 36*(4), 811-853. http://dx.doi.org/10.1016/S0360-8352(99)00167-9.

Gutowski, T. G., Sahni, S., Boustani, A., & Graves, S. C. (2011). Remanufacturing and energy savings. *Environmental Science & Technology, 45*(10), 4540-4547. http://dx.doi.org/ 10.1021/es102598b.

Hauser, W., & Lund, R. T. (2003). Remanufacturing: An American resource. Teaching materials on remanufacturing. Retrieved from http://www.bu.edu/reman/RemanSlides.pdf

Hewlett Packard. (2014). Retrieved from http://www8.hp.com/us/en/hp-information/environment/recycling-reuse.html#.UyeltvmSx-5.

IBM. (2014). Information retrieved from http://www.ibm.com/ibm/environment/products/recycling.shtml

Ijomah, W. L. (2009). Addressing decision making for remanufacturing operations and design-for remanufacture. *International Journal of Sustainable Engineering, 2(*2), 91-102. http://dx.doi.org/10.1080/19397030902953080

Ijomah, W. L., McMahon, C. A., Hammond, G. P., & Newman, S. T. (2007a). Development of robust design-forremanufacturing guidelines to further the aims of sustainable development. *International Journal of Production Research, 45*(18-19), 4513-4536. http://dx.doi.org/ 10.1080/00207540701450138

Ijomah, W. L., McMahon, C., Hammond, G., & Newman, S. (2007b). Development of design for remanufacturing guidelines to Support sustainable manufacturing. *Robotics and Computer Integrated Manufacturing, 23*(6), 712-719. http://dx.doi.org/10.1016/j.rcim.2007.02.017.

Ilgin, M. A., & Gupta, S. M. (2012). Remanufacturing Modelling and Analysis. Boca Raton Florida, CRC Press.

Jayaraman, V. (2006). Production planning for closed-loop supply chains with product recovery and reuse: an analytical approach. *International Journal of Production Research, 44*(5), 981-998. http://dx.doi.org/10.1080/00207540500250507.

Katun Corporation. (2014). Information retrieved from http://www.katun.com/ap/wp-content/uploads/2011/05/5031_HP_Color_TNR_PL.pdf

Kim, H.-J., Skerlos, S., Severengiz, S., & Seliger, G. (2009). Characteristics of the automotive remanufacturing enterprise with an economic and environmental evaluation of alternator products. *International Journal of Sustainable Manufacturing, 1*(4), 437-449. http://dx.doi.org/10.1504/IJSM.2009.031363

King, A., Miemczyk, J., & Bufton, D. (2006). Photocopier remanufacturing at Xerox UK: A description of the process and consideration of future policy issues. In D. Brissaud et al. (Eds.), *Innovation in Life Cycle Engineering and Sustainable Development* (pp. 173-186). Springer. Printed in Netherlands

Kumar, N., Guide Jr, V. D. R., & Van Wassenhove, L. N. (2002). Managing product returns at Hewlett, Packard, teaching case 05/2002-4940, INSEAD.

Kumar, V. V., Liou, F. W., Balakrishnan, S. N., & Kumar, V. (2013). Economical impact of RFID implementation in remanufacturing: a Chaos-based Interactive Artificial Bee Colony approach. *Journal of Intelligent Manufacturing*. http://dx.doi.org/10.1007/s10845-013-0836-9.

Lebreton, B., & Tuma, A. (2006). A quantitative approach to assessing the profitability of car and truck tire remanufacturing. *International Journal of Production Economics, 104*(2), 639-652. http://dx.doi.org/10.1016/j.ijpe.2004.11.010.

Lu, Q., Goh, M., Garg, M. & Souza, R. D. (2014). Remanufacturing in Asia: location choice and outsourcing. *International Journal of Logistics Management, 25*(1), 20 – 34. http://dx.doi.org/10.1108/IJLM-10-2012-0125.

Lund, R. T. & Hauser, W. M. (2003). Remanufacturing - An American Perspective. Retrieved from http://www.reman.org/Papers/ICRM%20APRA%20version.pdf

Lund, R. T. (1984a). Remanufacturing: The experience of the United States and implications for developing countries. *World Bank Technical Paper, 31*.

Lund, R. T. (1984b). Remanufacturing. *Technology Review 87, 18*(8).

Matsumoto, M., & Ijomah, W. (2013). Remanufacturing. In J. Kauffman, & K. M. Lee (Eds.), *Handbook of Sustainable Engineering*. http://dx.doi.org/10.1007/978-1-4020-8939-8-93.

Mti. (2014). Retrieved from http://www.mti.net.my/remanufacturingworkflow

Osibanjo O., & Nnrom, I. C. (2008). Material Flows of Mobile Phones and Accessories in Nigeria: Environmental Implications and Sound End-of-Life Management Options. *Environmental Impact Assessment Review, 28,* 198-213. http://dx.doi.org/10.1016/j.eiar.2007.06.002.

Ovchininnikov, A., (2011). Revenue and Cost Management for Remanufactured Products. *Production and Operations Management, 20*(6), 824-840. http://dx.doi.org/ 10.1111/j.1937-5956.2010.01214.x.

Robert Bosch LLCBosch. (2014, June 23). Retrieved June 23, 2014 from http://www.boschautoparts.com/Pages/default.aspx

Schau, E. M., Traverso, M. & Finkbeiner, M. (2012). Life cycle approach to sustainability assessment: a case study of remanufactured alternators. *Journal of Remanufacturin, 2*(5). http://www.journalofremanufacturing.com/content/2/1/5.

Schulz, T., & Ferretti, I. (2011). On the alignment of lot sizing decisions in a remanufacturing system in the presence of random yield. *Journal of Remanufacturing, 1*(3). http://dx.doi.org/10.1186/2210-4690-1-3.

Seitz, M. A. (2007). A critical assessment of motives for product recovery: the case of engine remanufacturing. *Journal of Cleaner Production, 15*(11-12), 1147-115. http://dx.doi.org/ 10.1016/j.jclepro.2006.05.029

Souza, G. C., & Ketzenberg, M. E. (2002). Two-stage make-to-order remanufacturing with service-level constraints. *International Journal of Production Research, 40*(2), 477-493. http://dx.doi.org/10.1080/00207540110090920.

Srivastava, R. (2004). Reusing Single Use Devices in Hospitals: A Case Study. In G. M. Surendra (Eds), *Environmentally Conscious Manufacturing III, Proceedings of SPIE Vol. 5264* (1-5). SPIE, Bellingham, WA.

Steinhilper, R. (1998). *Remanufacturing: The ultimate form of recycling*. Stuttgart: Fraunhofer IRB Verlag, D-70569.

Stryker Sustainability Solutions. (2014). Retrieved from http://sustainability.stryker.com/how-it-works/device-reprocessing

Sundin, E., & Bras, B. (2005). Making functional sales environmentally and economically beneficial through product remanufacturing. *Journal of Cleaner Production, 13*(9), 913-925. http://dx.doi.org/10.1016/j.jclepro.2004.04.006

Sundin, E., & Dunbäck, O. (2013). Reverse logistics challenges in remanufacturing of automotive mechatronic devices. *Journal of Remanufacturing, 3*(2). http://dx.doi.org/10.1186/2210-4690-3-2

Sundin, E., & Lee, H. M. (2012). In what way is remanufacturing good for the environment? In *Design for Innovative Value Towards a Sustainable Society: Proceedings of EcoDesign 2012: 7th International Symposium on Environmentally Conscious Design and Inverse Manufacturing* (552-557). New York: Springer.

United States International Trade Commision. (2012). *Remanufactured Goods: An Overview of the U.S. and Global Industries, Markets and Trade*. Retrieved from http://www.usitc.gov/publications/332/pub4356.pdf

Warsen, J., Laumer, M., & Momberg, W. (2011). Comparative life cycle assessment of remanufacturing and new manufacturing of a manual transmission. In J. Hesselbach, & C. Herrmann (Eds.), *Glocalized Solutions for Sustainability in Manufacturing* (67-72). Berlin, Heidelberg: Springer.

Zikopoulos, C. (2012). Remanufacturing lot-sizing under alternative perceptions of returned units' quality. *Int. Journal of Business Science and Applied Management, 7*(3).

Zussman, E., & Seliger, G. (1999). Product Remanufacturing. In Shimon Y. Nof (Eds), *Handbook of Industrial Robotics* (2nd ed., pp. 1037-1043), New York: Wiley & Sons.

12

The Influence of Design Parameters on Straight Bevel Gear Dynamics

Chiu-Fan Hsieh[1] & You-Qing Zhu[1]

[1] Department of Mechanical and Computer-Aided Engineering, National Formosa University, Taiwan

Correspondence: Chiu-Fan Hsieh, Department of Mechanical and Computer-Aided Engineering, National Formosa University, 64 Wunhua Road, Huwei, Yunlin, Taiwan. E-mail: cfhsieh@nfu.edu.tw

Abstract

This study analyzes the influence of design parameters on the dynamics of straight bevel gears by constructing a model that allows variation in the shaft angle, pressure angle, and backlash. According to the statistical analysis, the order of influence of these parameters on weight is shaft angle > pressure angle > backlash. When the shaft angle is 90°, the statistical results show the drive is stable and the stress fluctuation level is low. The pressure angle, on the other hand, can affect the gear's dynamic property by influencing the driving component force on the gear and the component force on the shaft. The results for the shaft and pressure angles are used to determine the appropriate backlash. Overall, the analysis not only provides designers with an important reference but explains the dominance in the market of gear designs with a 90° shaft angle and a 20° pressure angle.

Keywords: bevel gears, shaft angle, pressure angle, backlash, dynamics, stress

1. Introduction

Bevel gears, primarily of spiral, hypoid, and straight types, are widely used on the power transmission of concurrent axes and alternating axes. In general, the shaft angle of a bevel gear is designed to be 90° with a maximum reduction ratio of 10, leading to high efficiency and high precision. In particular, the design must meet the spatial surface meshing condition, and the gear blank and tooth form designs must be able to vary with different cutting methods (Litvin et al., 1987). The calculation of bevel gears is thus more complicated than for other gear types (Fong & Tsay, 1991, 1992) and requires the development of special design software. The most common type of bevel gear is the straight bevel gear, which is applied to power transmissions between concurrent axes. This gear is characterized by a straight tooth whose depth decreases gradually from large to tip end. The theoretical tooth form is involute (Al-Daccak et al., 1994), but the tooth form designs vary with different manufacturing methods. Compared to other bevel gears, however (e.g. the spiral bevel gear), the configuration is simple and the gear easily made, forged, and shaped. As a result, the bevel differential mechanism is used extensively in the market.

As regards research on the straight bevel gear, most studies have focused on the theoretical mathematical model of gear (Tsai & Chin, 1987; Chang & Tsay, 2000; Figliolini & Angeles, 2004), teeth contact analysis (Jiao & Cao, 2011; Kolivand et al., 2015), finite element modeling and analysis (Nalluveettil & Muthuveerappan, 1993), teeth deflection analysis (Gagnon et al., 1997), manufacturing techniques and cutting methods (Özel et al., 2005; Shih & Fong, 2008; Shih, 2012; Fuentes et al., 2011; Tsuji et al., 2013; Kawasaki et al., 2015), measurement technique (Kawasaki & Shinma, 2009), lubrication regime (Ghahnavieh et al., 2014) and dynamic behavior of two-stage straight bevel gear (Yassine et al., 2014). To the best of our knowledge, however, no work exists on the influence of dynamic contact stress and gear design parameters on straight bevel gear dynamics. Hence, this study assesses the influence of a straight bevel gear's main design parameters (shaft angle, pressure angle, and backlash) on its dynamics by constructing a dynamics analysis model that weights the influence of the various parameters. The results provide a useful reference for straight bevel gear design.

2. Dynamics Analysis Model for a Straight Bevel Gear

2.1 Straight Bevel Gear Modeling

First, using Solidworks, a straight bevel gear CAD model is built in which the tooth form is created based on a mathematical model. Here, the modulus is 2, the number of teeth is 15, the pressure angle is 20°, the shaft angle is 90°, and the backlash is 0.2. The building of this model is described below and linked to the illustrations in Figure 1:

Step1: A blank is created with a shaft angle of 90° (a);

Step2: The involute profile is drawn on the gear plane (b);

Step3: A straight bevel gear tooth form is cut along the conical point of involute profile (c);

Step4: The required teeth are cut in a ring array to complete a straight bevel gear (d); and then

Step5: Another straight bevel gear (45 teeth) is built following the above steps, and the two straight bevel gears are combined (e).

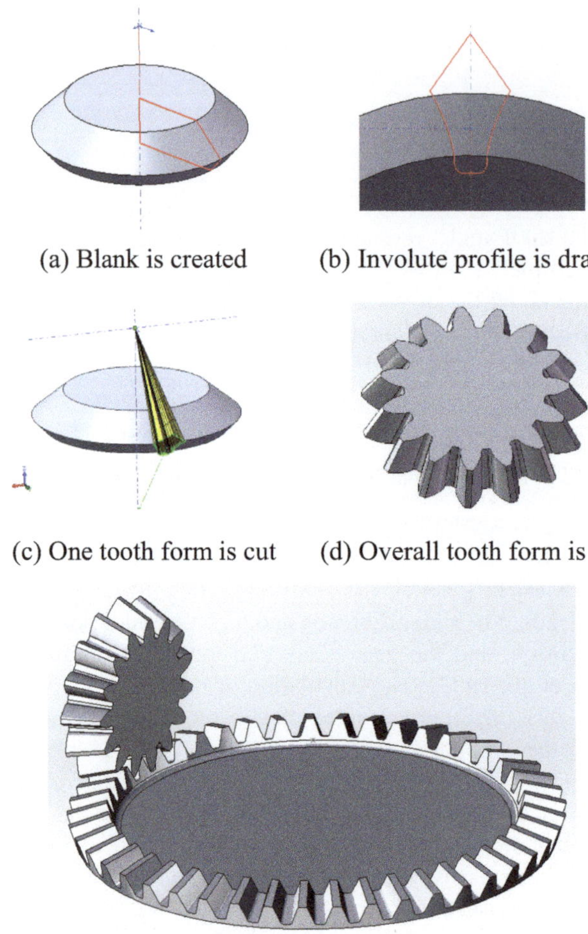

(a) Blank is created (b) Involute profile is drawn

(c) One tooth form is cut (d) Overall tooth form is cut

(e) The straight bevel gear is assembled

Figure 1. Construction of the straight bevel gear

2.2 Dynamics Analysis Model

To construct the dynamic analysis model, it employs a systematic dynamics calculation model for an experimentally validated cycloidal speed reducer (Hsieh, 2014), whose applications include testing traditional, improved, and novel designs and identifying their shortcomings and advantages (Hsieh, 2014). These modeling and calculation techniques (Hsieh, 2014) allow constructing a dynamic analysis model of the straight bevel gear. In this paper, the gear modulus is fixed at 2, the driving gear is set at 15 teeth, and the driven gear is at 45 teeth. The simulated conditions include total run duration of 1.5 sec and the driving gear rotation speed set at a constant 120 rpm (720 deg/s). The contact pattern is a 3D collision mode, and gravity acceleration is 9.81 m/sec^2. The material is alloy steel whose mechanical properties are given in Table 1.

In the stress analysis, stress can be calculated by the finite element method using SolidWorks (Hsieh, 2014). During bevel gear motion, the variations in gear contact stress can be measured over time. First, the triangular mesh is calculated. If the values obtained when calculating gear engagement are to be accurate, the grid density in the place without gear contact can be low, but the grid density in the place with gear contact must be high. Grid computation is shown in Figure 2(a) and stress calculation in Figure 2(b).

The material used in all models and the friction factor are the same. Assuming the loading at the driven gear is fixed as 5 N-m, the effect on dynamics was carried out by giving different driving torque, which includes 15 N-m, 30 N-m, 45 N-m, 60 N-m, 75 N-m and 90 N-m, respectively. Thus, different parameter design values are assigned to the straight bevel gear model that result in different motion and stress variations. The tooth profile design parameters are pressure angle, shaft angle, and backlash, whose parameter value settings are listed in Table 2. To assess the influence of parameter design on the straight bevel gear's dynamics, these parameter values are cross related, producing 27 groups as simulation subjects in each driving torque case.

Table 1. Mechanical properties of the material

Material	Mechanical properties N/mm^2			
	Elastic modulus	Shear modulus	Tensile strength	Yield strength
Alloy steel	210,000	79,000	723.1025	620.422

Table 2. Gear design parameter values in each driving torque case

Numerical	Parameter		
	Pressure angle	Backlash	Shaft angle
1	14.5°	0.1	45°
2	20°	0.3	60°
3	25°	0.5	90°

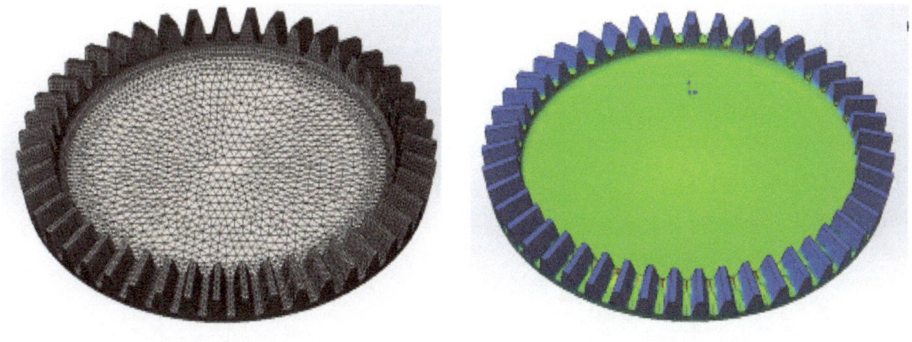

(a) Grid chart (b) Schematic diagram of stress

Figure 2. Stress analysis model

2.3 Dynamics Analysis

In the dynamics analysis, for example, the driving torque is 15 N-m, the fixed pressure angle is 20°, the backlash is 0.1 mm, and the variable shaft angle is 45°, 60°, and 90° respectively. The following results are obtained based on the data from one revolution of the driving gear. The time interval of calculation is 0.0013889 s. Figures 3 and 4 show the analytic results of the angular velocity and angular acceleration of the driven gear. The velocity fluctuation is regular in all three shaft angle designs, but the 60° shaft angle has the maximum fluctuation. The angular acceleration analysis further reveals that the 60° shaft angle has the maximum peak values. Although one explanation could be larger vibrations, when the shaft angle is 90°, the minimum angular velocity fluctuation and acceleration reach peak values, so the motion is relatively stable and the vibration slight.

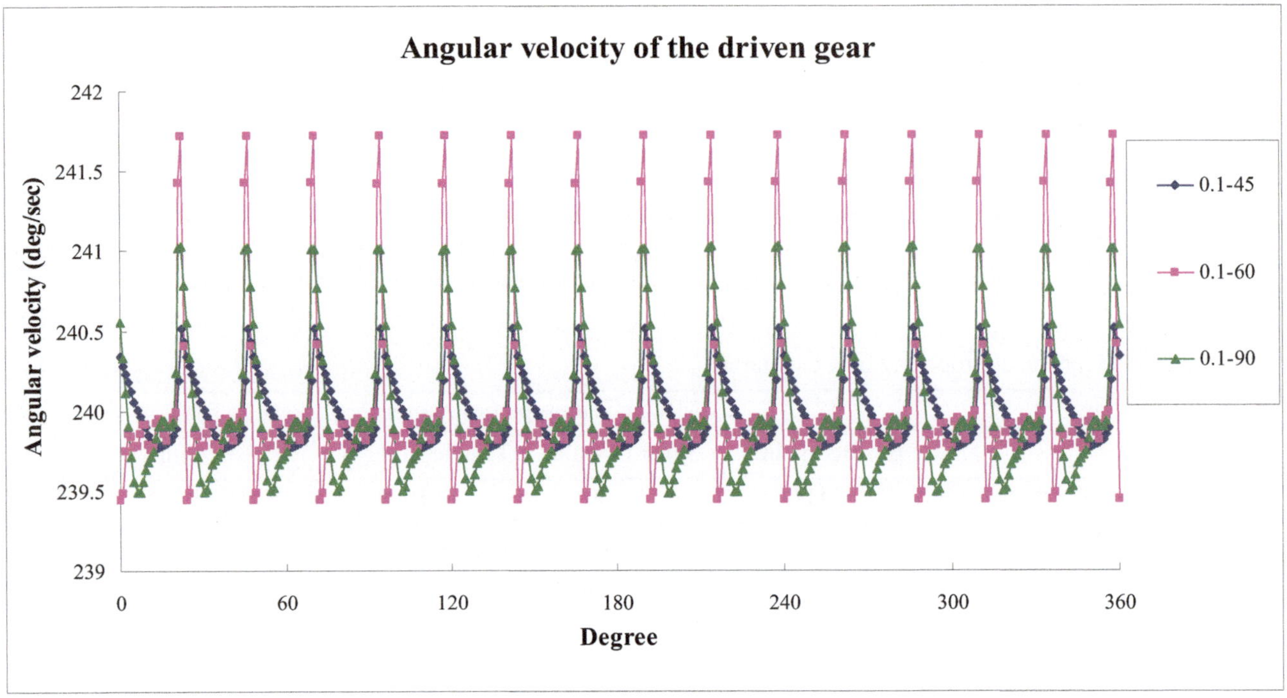

Figure 3. Angular velocity of the driven gear

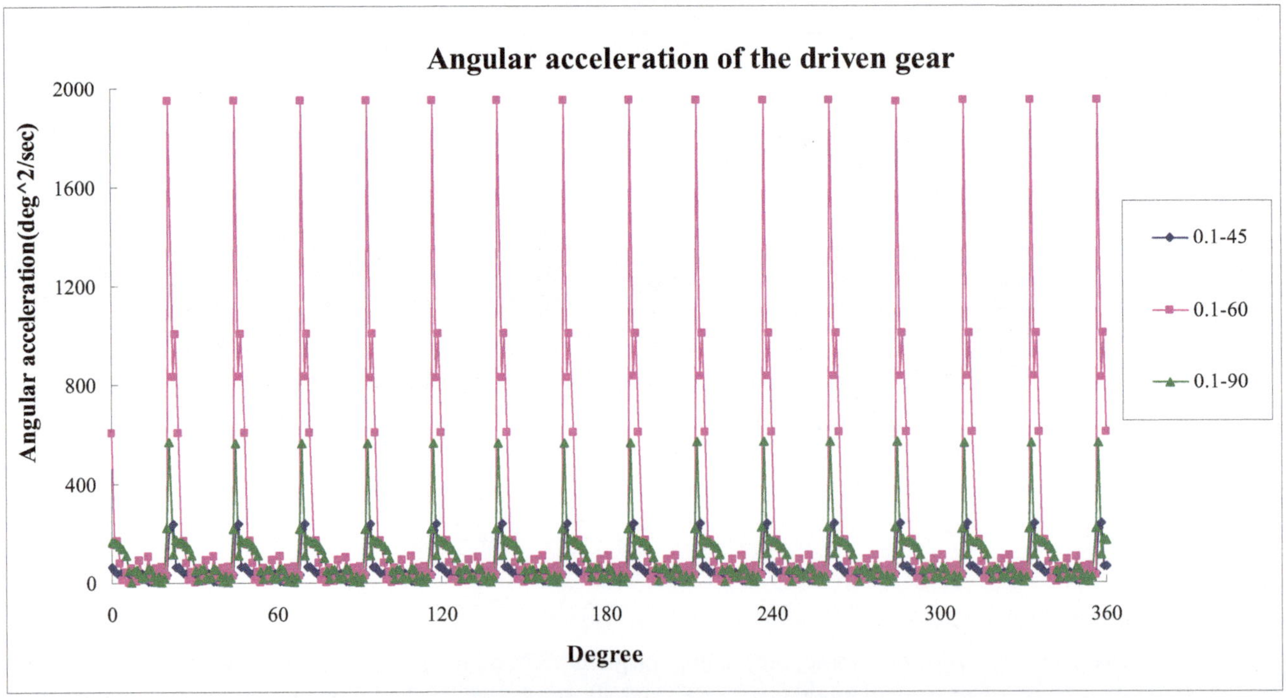

Figure 4. Angular acceleration of the driven gear

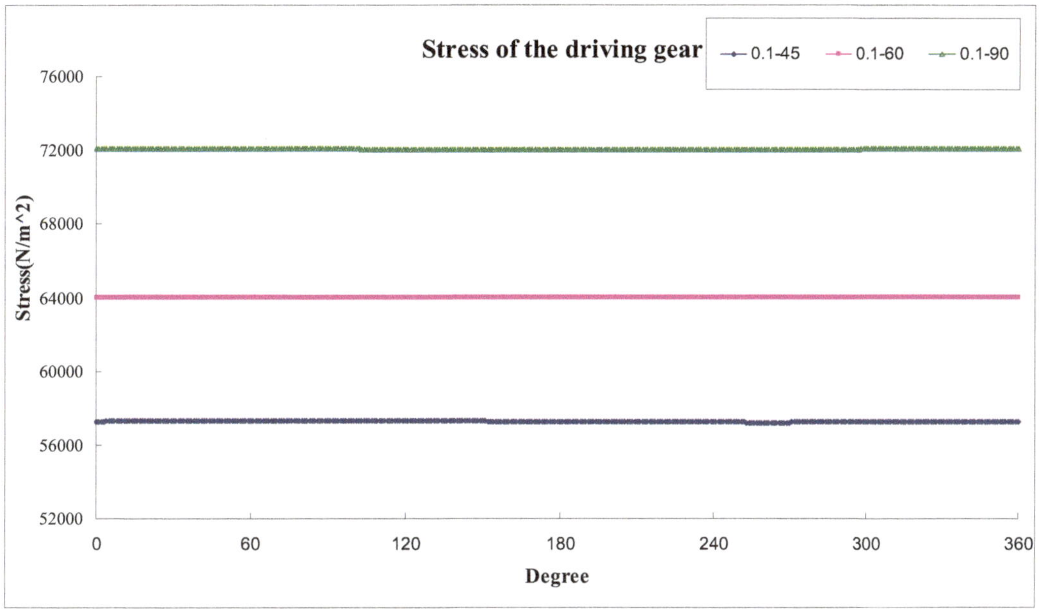

(a)

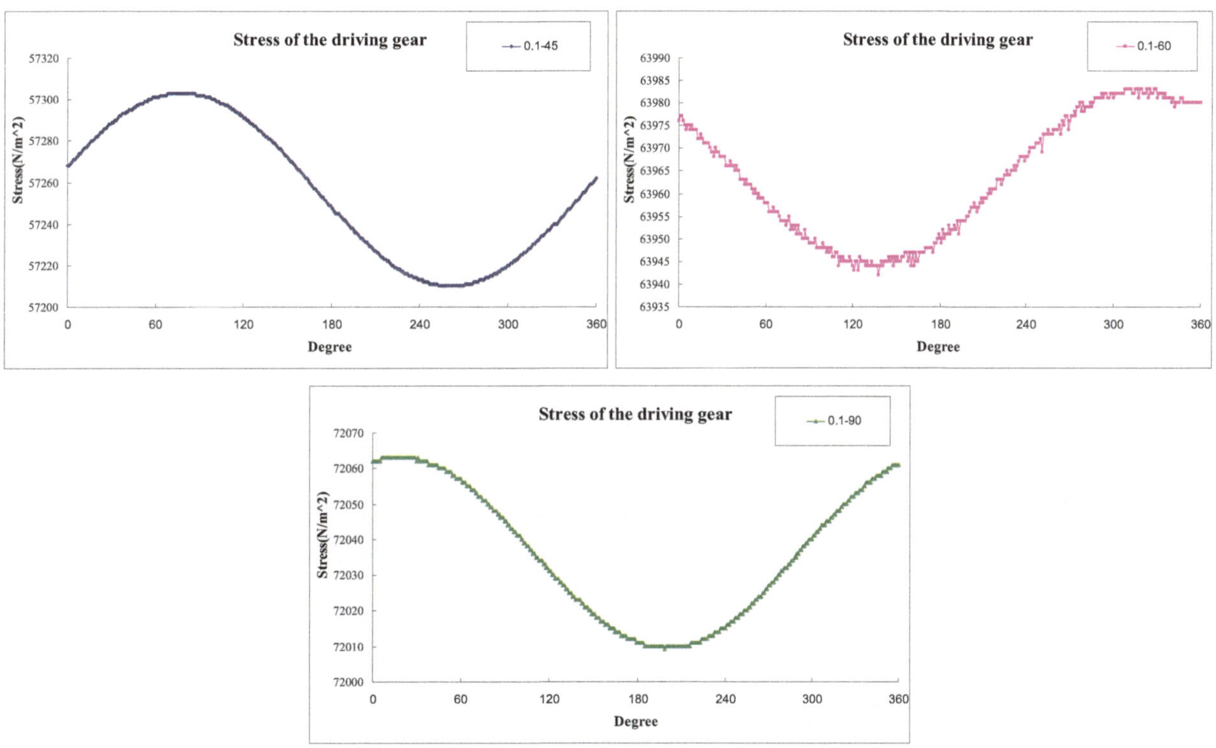

(b)

Figure 5. Stress variation of driving gear

According to the stress analysis, the axial component force increases with the shaft angle, while the stress of the driving gear increases gradually (Figure 5(a)). As Figure 5(b) shows, the three stress curves of the driving gear fluctuate regularly but slightly. When the shaft angle is 90°, the average stress is maximum about 72000 N/m^2; when the shaft angle is 60°, the average stress is about 64000 N/m^2; and when the shaft angle is 45°, the average stress is minimum about 57000 N/m^2.

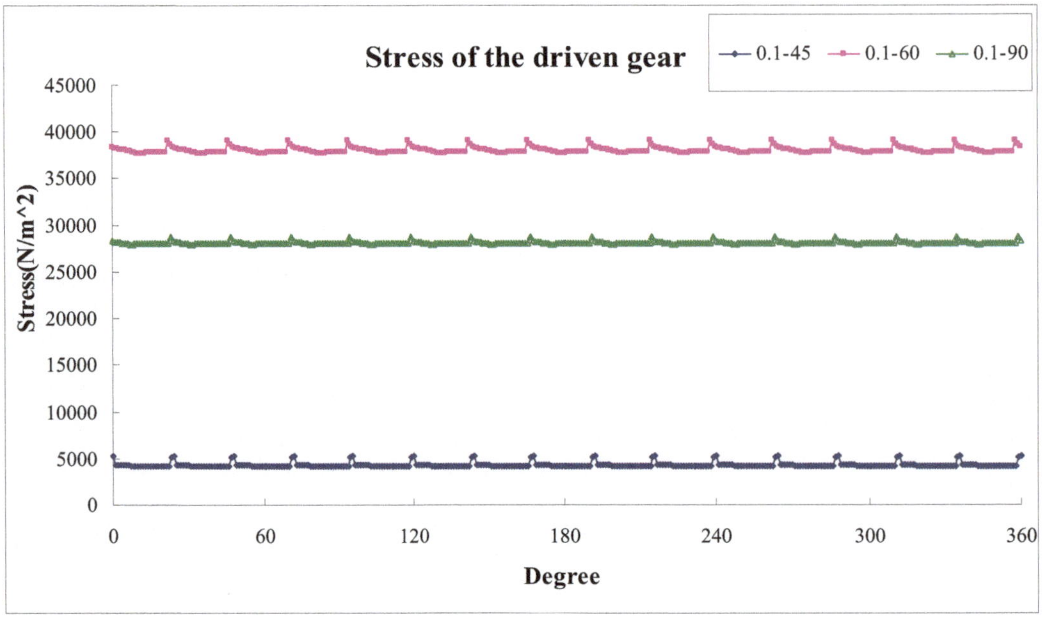

(a)

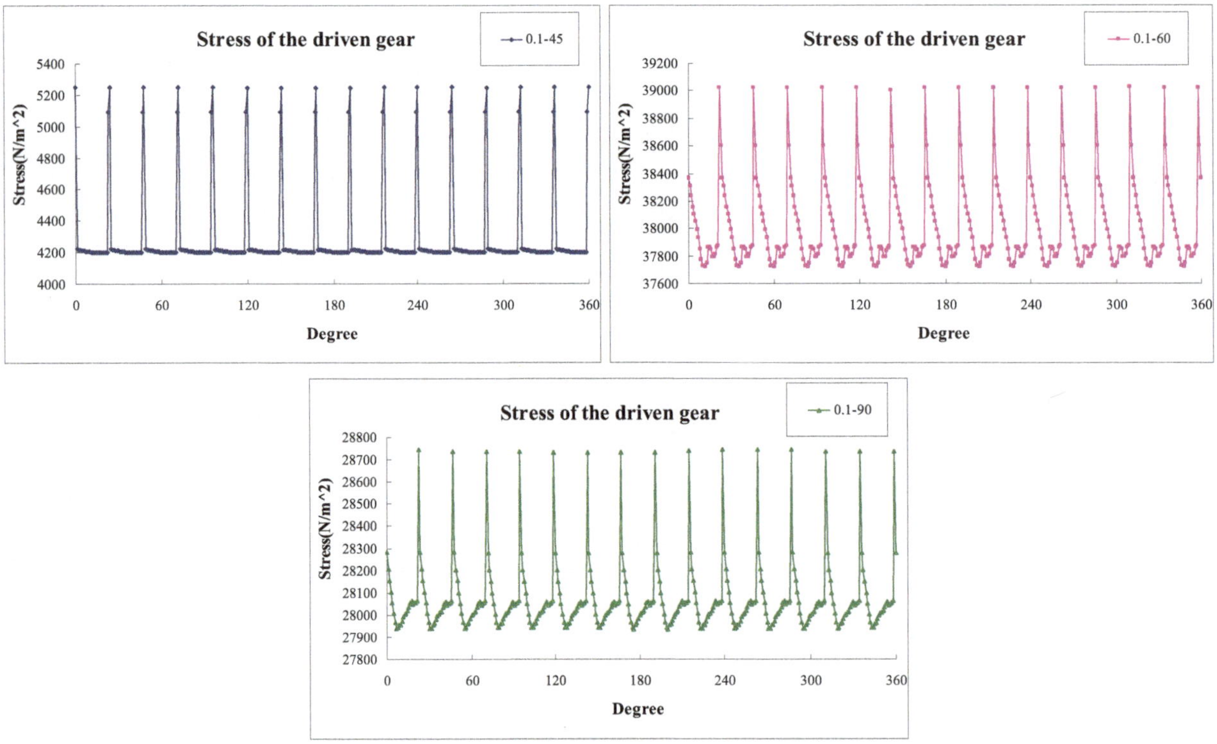

(b)

Figure 6. Stress variation of driven gear

The stress of the driven gear is as shown in Figure 6. The stress peaks are all regular (Figure 6(b)). In this case, the maximum average stress is the shaft angle 60° after that is the shaft angle 90°, and the minimum average stress is the shaft angle 45°. Since the stress ripple coefficient (Hsieh, 2014) can indicate the stress fluctuation level, the calculation result shows the hierarchy of the stress ripple level is as shaft angle 45°>shaft angle 60° >shaft angle 90°. This analysis thus suggests a shaft angle of 90° as an ideal design.

The stress analysis can indicate the contact stress condition, thus would affect the life of the components. For comparing, each driving torque case includes 27 sets of stress data.

Since the rotation speed setting is on the driving gear, only the stress of the driven gear is analyzed and discussed. The equation of standard deviation can be expressed as

$$S_d^{(i)} = \frac{1}{N}\sqrt{N\Sigma\sigma^2 - (\Sigma\sigma)^2}\ ,\ i = 1 \sim 27 \tag{1}$$

where N is the number of stress data for one revolution of the driving gear. σ is the stress data of the driven gear.

In order to compare the stress variation level, the coefficient of variation can be defined as

$$C_v^{(i)} = \frac{S_d}{\bar{\sigma}} \times 100,\ i = 1 \sim 27 \tag{2}$$

In which, $\bar{\sigma}$ means the average stress of the driven gear.

Here, the influences of the backlash, pressure angle and the shaft angle parameters, the equations are defined respectively as

$$B_m = \sum_{j=1}^{9} C_{v,m}^{(j)},\ m = 0.1,\ 0.3,\ 0.5 \tag{3}$$

$$P_n = \sum_{j=1}^{9} C_{v,n}^{(j)},\ n = 14.5,\ 20,\ 25 \tag{4}$$

$$S_o = \sum_{j=1}^{9} C_{v,o}^{(j)},\ o = 45,\ 60,\ 90 \tag{5}$$

In this paper, the tendency of a whole influence is calculated by sum of variation coefficient of each driving torque case as follows:

$$SB_m = B_m^{(15)} + B_m^{(30)} + B_m^{(45)} + B_m^{(60)} + B_m^{(75)} + B_m^{(90)},\ m = 0.1,\ 0.3,\ 0.5 \tag{6}$$

$$SP_n = P_n^{(15)} + P_n^{(30)} + P_n^{(45)} + P_n^{(60)} + P_n^{(75)} + P_n^{(90)},\ n = 14.5,\ 20,\ 25 \tag{7}$$

$$SS_o = S_o^{(15)} + S_o^{(30)} + S_o^{(45)} + S_o^{(60)} + S_o^{(75)} + S_o^{(90)},\ o = 45,\ 60,\ 90 \tag{8}$$

3. Statistical Results and Discussion

3.1 Influence on Dynamics of the Backlash Parameter

As shown in Figure 7 (a), when the backlash is small (0.1 mm), the gear collision frequency and gear vibration will increase and the gear contact travel will extend to increase the stress. Thus, the variation coefficients in most driving torque cases are the largest. The whole tendency shows that the stress ripple level would relieve with the increasing of backlash as shown in Figure 8(a). When the backlash is large, in contrast, the collision between gears can be reduced, although there will be energy loss in the power transmission, reducing transmission efficiency. The maximum variation coefficient in backlash parameter is the smallest (see Figure 8) among the three parameters, thus it has the minimum influence comparing to the other parameters. Accordingly, the main influence on stress and fluctuation occurs at the shaft angle and pressure angle, an appropriate backlash value should be determined based on their design and working condition.

3.2 Influence on Dynamics of the Pressure Angle Parameter

The variation coefficient results show the influence degree of the pressure angle parameter is between the backlash and shaft angle parameters in Figures 7 and 8. In the driven gear, the stress ripple level would be increased with the decreasing of the pressure angle especially at 14.5°, which load and fluctuation are the largest. Apparently, the gear driving component force and stress fluctuation increase as the pressure angle decreases. On the other hand, the larger the pressure angle (25°), the lower the gear driving component force, and the stress fluctuation decreases while the load on the shaft increases. Hence, according to the stress statistical result, a pressure angle of 20° is a better choice.

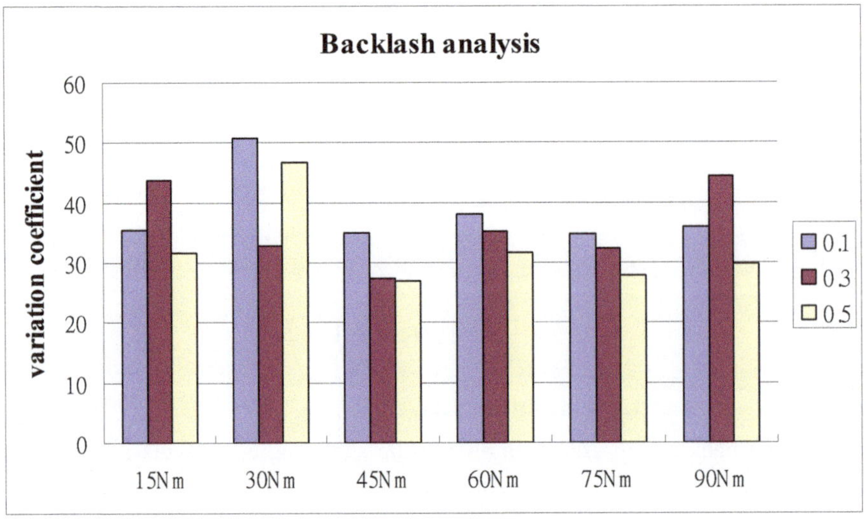

(a) Backlash

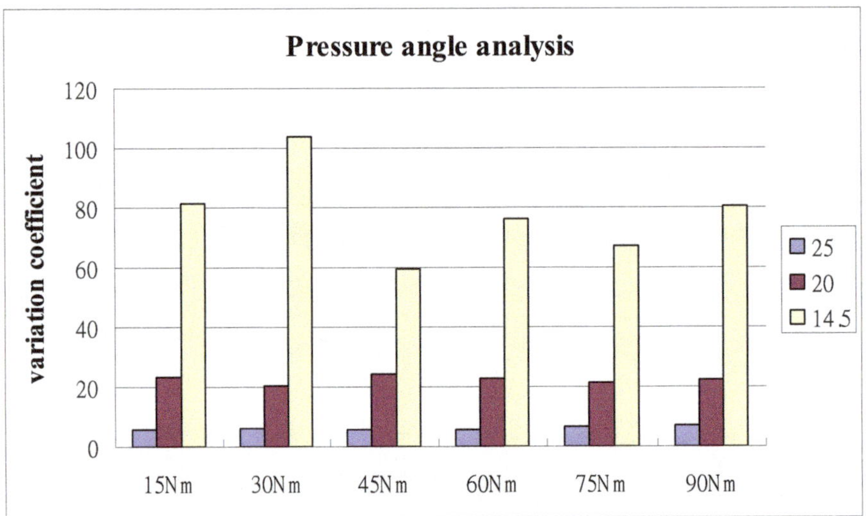

(b) Pressure angle

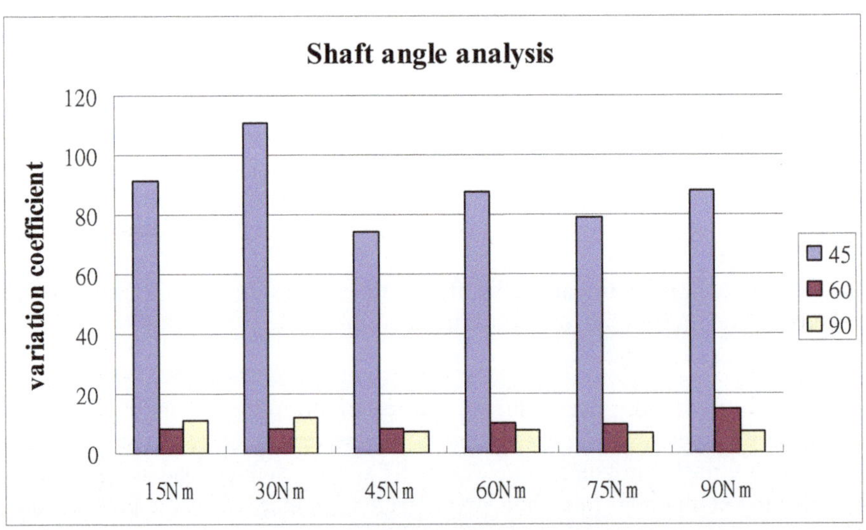

(c) Shaft angle

Figure 7. Statistical results in each driving torque case

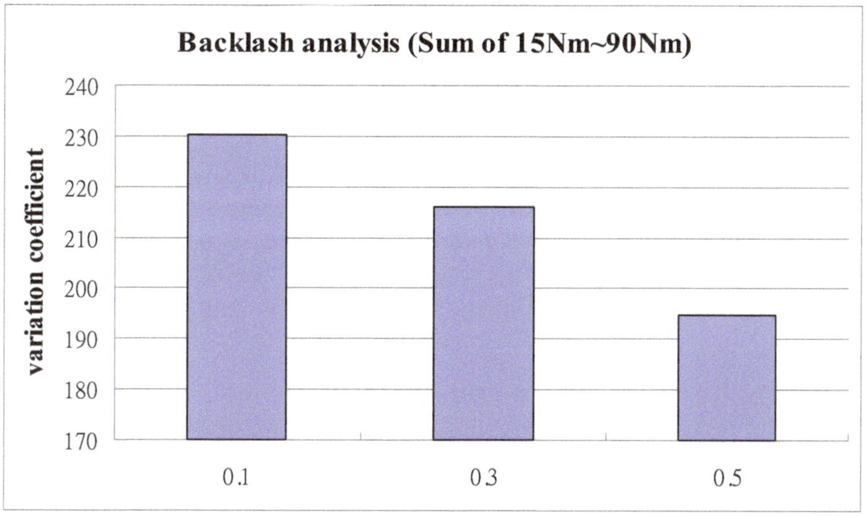

(a) Backlash

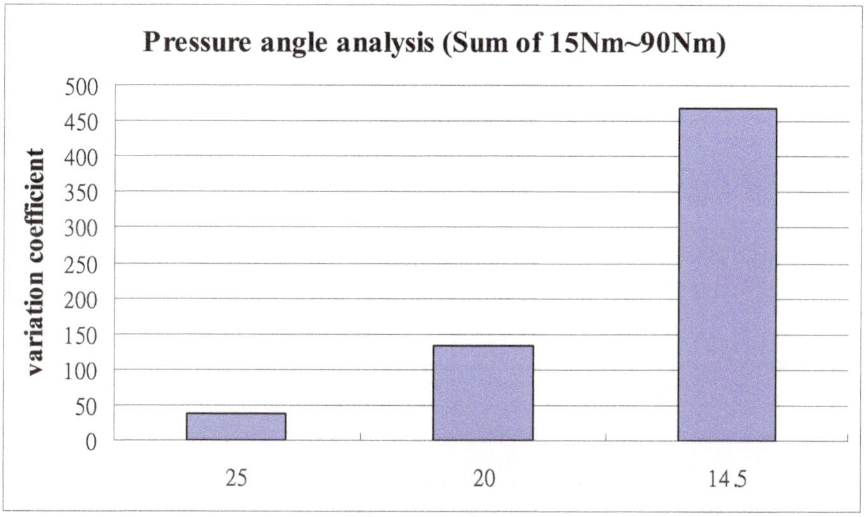

(b) Pressure angle

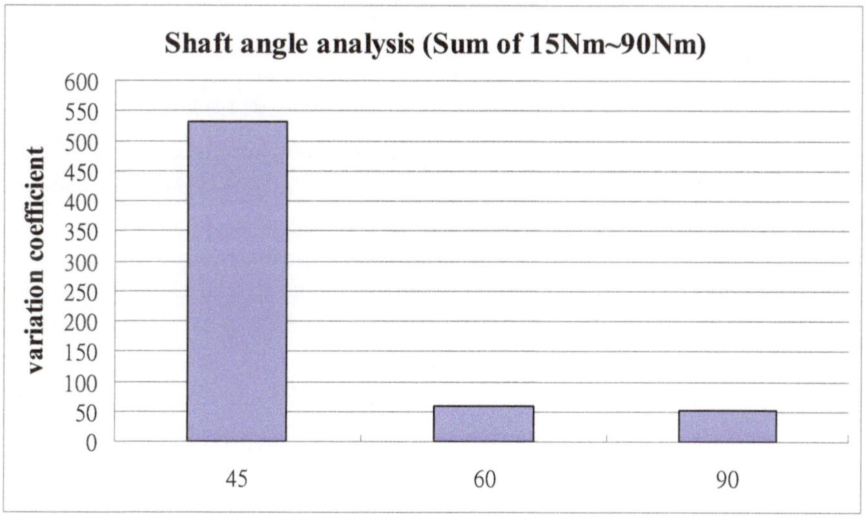

(c) Shaft angle

Figure 8. Statistical results by sum of variation coefficient of each driving torque case

3.3 Influence on Dynamics of the Shaft Angle Parameter

Since the maximum variation coefficient in shaft angle parameter is the largest as shown in Figures 7 and 8, it places the maximum influence on the shaft angle design. The variation coefficient of shaft angle at 45° is the largest in each driving torque case (see Figure 7(c)). When the shaft angle increases to 60°, the stress fluctuation frequency would obviously reduce. The shaft angle at 90° is the most stable one and thus has the least stress variation coefficient in most driving torque cases. Furthermore, the shaft angle at 90° can obtain more stable operation than the case at 60° under a larger driving torque. Besides, Figure 8 (c) represents the whole tendency in the smaller shaft angle (like shaft angle at 45°), however, its combination with other two parameters may not be a better condition. The shaft angle at 90° can get the lowest variation coefficient. Hence, unless there are special power requirements, the shaft angle at 90° is recommended as the design basis.

4. Conclusions

By constructing a dynamics analysis model for straight bevel gears, this analysis generates accurate shaft angle, pressure angle, and backlash design parameters and then measures the influence of each parameter on the gear's dynamics. From the statistical analysis, the influence of the backlash is not obvious because the variation coefficient is quite similar in different backlash size. However, the size of the backlash still can influence the gear's dynamic properties. The analysis also reveals that the pressure angle can be designed according to load. That is, when the pressure angle is large, the stress and stress fluctuation on the gear can be reduced, but the axial component force is increased, placing a burden on the bearing. The component having the maximum influence on dynamics is the shaft angle, and the shaft angle at 90° with other two design parameters would be recommended as a better design.

References

Al-Daccak, M. L., Angeles, J., & González-Palacios, M. A. (1994). The Modeling of Bevel Gears Using the Exact Spherical Involute. *ASME Journal of Mechanical Design, 116*, 364-368. http://dx.doi.org/10.1115/1.2919387

Chang, C. K., & Tsay, C. B. (2000). Mathematical Model of Straight Bevel Gears with Octoid Form. *Journal of the Chinese Society of Mechanical Engineers, 21*, 239–245.

Figliolini, G., & Angeles, J. (2004). Algorithms for involute and octoidal bevel-gear generation. *ASME Journal of Mechanical Design, 127*(4), 664-672. http://dx.doi.org/10.1115/1.1900147

Fong, Z. H., & Tsay, C. B. (1991). A Study on the Tooth Geometry and Cutting Machine Mechanisms of Spiral Bevel Gears. *ASME Journal of Mechanical Design, 113*, 174-181. http://dx.doi.org/10.1115/1.2912788

Fong, Z. H., & Tsay, C. B. (1992). Kinematical Optimization of Spiral Bevel Gears, *ASME Journal of Mechanical Design, 114*, 498-506. http://dx.doi.org/10.1115/1.2926578

Fong, Z. H., & Tsay, C.B. (1991). A Mathematical Model for the Tooth Geometry of Circular-Cut Spiral Bevel Gears. *ASME Journal of Mechanical Design, 113*, 346-351. http://dx.doi.org/10.1115/1.2912766

Fuentes, A., Iserte, J. L., Gonzalez-Perez, I., & Sanchez-Marin, F. T. (2011). Computerized Design of Advanced Straight, & Skew Bevel Gears Produced by Precision Forging. *Computer Methods in Applied Mechanics and Engineering, 200*, 2363-2377. http://dx.doi.org/10.1016/j.cma.2011.04.006

Gagnon, P., Gosselin, C., & Cloutier, L. (1997). Analysis of Spur and Straight Bevel Gear Teeth Deflection by the Finite Strip Method. *ASME Journal of Mechanical Design, 119*, 421-426. http://dx.doi.org/10.1115/1.2826385

Ghahnavieh, A. B., Akbarzadeh, S., & Mosaddegh P. (2014). A numerical study on the performance of straight bevel gears operating under mixed lubrication regime. *Mechanism and Machine Theory, 75*, 27-40. http://dx.doi.org/10.1016/j.mechmachtheory.2014.01.005

Hsieh C. F. (2014). Dynamics analysis of cycloidal speed reducers with pinwheel and nonpinwheel designs. *ASME Journal of Mechanical Design, 136*(9), 091008. http://dx.doi.org/10.1115/1.4027850

Hsieh C. F. (2014). The effect on dynamics of using a new transmission design for eccentric speed reducers. *Mech. Mach. Theory, 80*(1), 1-16. http://dx.doi.org/10.1016/j.mechmachtheory.2014.04.020

Hsieh C. F. (2014). Traditional versus improved designs for cycloidal speed reducers with a small tooth difference: The effect on dynamics. *Mech. Mach. Theory, 86*(1), 15-35. http://dx.doi.org/10.1016/j.mechmachtheory.2014.11.013

Jiao, J., & Cao, X. (2011). Generation and TCA of Straight Bevel Gear Drive with Modified Geometry. *Applied Mechanics and Materials, 86*, 403-406. http://dx.doi.org/10.4028/www.scientific.net/AMM.86.403

Kawasaki, K., & Shinma, K. (2009). Accuracy Measurement and Evaluation of Straight Bevel Gear Manufactured by End Mill Using CNC Milling Machine. *ASME Journal of Mechanical Design, 131*, 011001. http://dx.doi.org/10.1115/1.2988480

Kawasaki, K., Tsuji, I., Gunbara H., & Houjoh H. (2015). Method for remanufacturing large-sized skew bevel gears using CNC machining center. *Mechanism and Machine Theory, 92*, 213–229. http://dx.doi.org/10.1016/j.mechmachtheory.2015.05.013

Kolivand, M., Ligata, H., Steyer, G., Benedict, D. K., & Chen, J. (2015). Actual Tooth Contact Analysis of Straight Bevel Gears. *ASME Journal of Mechanical Design, 137*(9), 093302. http://dx.doi.org/10.1115/1.4031025

Litvin, F. L., Tsung, W. J., Coy, J. J., & Heine, C. (1987). Method for Generation of Spiral Bevel Gears with Conjugate Gear Tooth Surfaces. *ASME Journal of Mechanical Design, 109*, 163-170. http://dx.doi.org/10.1115/1.3267431

Nalluveettil, S. J., & Muthuveerappan, G. (1993). Finite element modeling and analysis of a straight bevel gear tooth. *Computers and Structures, 48*, 739-744. http://dx.doi.org/10.1016/0045-7949(93)90268-I

Özel,C., İnan, A., & Özler, L. (2005). An Investigation on Manufacturing of the Straight Bevel Gear Using End Mill by CNC Milling Machine. *ASME Journal of Manufacturing Science and Engineering, 127*, 503-511. http://dx.doi.org/10.1115/1.1863256

Shih, Y. P. (2012). Mathematical Model for Face-Hobbed Straight Bevel Gears. *ASME Journal of Mechanical Design, 134*, 091006. http://dx.doi.org/10.1115/1.4007151

Shih, Y. P., & Fong, Z. H. (2008). Flank Correction for Spiral Bevel and Hypoid Gears on a Six-Axis CNC Hypoid Generator. *ASME Journal of Mechanical Design, 130*, 062604. http://dx.doi.org/10.1115/1.2890112

Tsai, Y. C., & Chin, P. C. (1987). Surface Geometry of Straight and Spiral Bevel Gears. *ASME Journal of Mechanical Design,* 109, 443-449. http://dx.doi.org/10.1115/1.3258815

Tsuji, I., Kawasaki, K., Gunbara, H., Houjoh, H., & Matsumura, S. (2013). Tooth Contact Analysis and Manufacture on Multitasking Machine of Large-Sized Straight Bevel Gears with Equi-Depth Teeth. *ASME Journal of Mechanical Design, 135*, 034504. http://dx.doi.org/10.1115/1.4023324

Yassine, D., Ahmed, H., Lassaad, W., & Mohamed, H. (2014). Effects of gear mesh fluctuation and defaults on the dynamic behavior of two-stage straight bevel system. *Mechanism and Machine Theory*, 82, 71-86. http://dx.doi.org/10.1016/j.mechmachtheory.2014.07.013

13

Enhancement of the Capabilities of CNC Machines via the Addition of a New Counter boring Cycle with a Milling Cutter

Adel Taha Mohamed Abbas[1]

[1] Department of Mechanical Engineering, College of Engineering, King Saud University, Riyadh 11421, P.O. Box 800, Saudi Arabia

Correspondence: Adel Taha Mohamed Abbas, Department of Mechanical Engineering, College of Engineering, King Saud University, Riyadh 11421, P.O. Box 800, Saudi Arabia. E-mail: atabbas1954@yahoo.com

Abstract

The operations of the machine, such as milling or drilling are the processes of shape transformation in which the metal is removed from the material stock to generate a part. The contact amongst the work piece and the generating tool creates substantial force. The study intends to examine and analyze different techniques of the model-based procedure control. Moreover, this research also aims at assessing a feature based approach, i.e., CNC approach. The program of CNC consists of a combination of machine specific instructions and machine specific codes. After reviewing the manuals of all modern CNC units from the different manufacturers, no efficient cycle was found for boring and counter-boring holes using a milling cutter. This research encapsulates the an investigation for the development of general programming algorithm, which is applied for boring or counter-boring many holes of different diameters using only one standard milling cutter. This algorithm has been integrated to produce a user-defined cycle (G888) and a subroutine for boring or counter-boring holes to achieve specific counter-bore diameters with accurate tolerance and good surface finishing. This programming algorithm can be equipped with our user-defined cycle (G881) to enable the machining of many counter-bore holes lying in a straight line (row/column/diagonal) or holes lying in regular or inclined matrix form. The algorithm can also be equipped with the presented user-defined cycle (G890) to machine counter-bore holes that form a circular pattern.

Keywords: user-defined cycles, canned cycles, NC programming algorithms, CNC Machining

1. Introduction

Machine tools, which are automated for operations using encoded program commands, rather than using cams for mechanical automation or levers/ hand wheels for manual control, is referred as numerical control or NC. During the previous years, like 1940 and 1950, NC machines were first built by modifying the existing tools, which integrate motors and moving system controls over punched tape, following points fed. Servo mechanism of these machines was based on analog and digital computers for rapid argumentation; hence, creating the revolutionary process of manufacturing, computer numerical control (CNC). The modern CNC are highly automated in nature, as it possess contemporary programs (Anderberg & Kara, 2009). In addition to this, these systems also integrate CAD (computer aided design) and CAM (computer aided manufacturing), along with the end-to-end designs of the components.

It has been analyzed that in order to operate this type of machine, a computer file is generated with a program, which is used for the interpretation of the command via postprocessor; hence, leading CNC machine to load a file for the production of the command. A number of steps are found to be involved in the production of the command, which are highly automated and complex. More so, it is also observed that the entire process also matches with original design of CAD (Luggen, 1983; Groover & Zimmers, 1984).

Modern computer numerical control machines allow the programmer, among other things, to call up fixed manufacturing routines, which are usually known as cycles. These routines automatically perform pre-programmed tasks; hence, making the manufacturing process easier and more efficient. It is significant to bring into the notice that there are different cycles on machining centers, horizontal boring machines, milling machines, turning machines, drilling machines, and grinding machines, depending on their jobs. There are two types of cycles. The first type of cycle is known as canned cycles, which are pre-prepared by the CNC unit manufacturer. It has been established that there are a number of different canned cycles, which are available in drilling and milling machines for different purposes. These include drilling holes, deep drilling, boring, reaming, and tapping (Anderberg et al., 2012). In milling machines, there are many canned cycles for the purpose of face milling, contour milling, thread cutting, rectangular stud milling, and circular stud milling. In addition to this, they are also available for making a long hole on a circle, a slot on a circle, a circular slot, a rectangular pocket, and a circular pocket.

Different types of canned cycles are also available for a turning machine, such as a grooving cycle, an undercut cycle, a stock removal cycle, a thread cutting cycle, and chaining of thread. The use of canned cycles is found to be a good programming method, which can be adopted to drill one hole in a single block. In this method, the programmer has to calculate the center coordinate for each hole, while drilling many holes, which lie in a straight line (row/column/diagonal) or holes lie in a regular matrix form. It has been assessed that such calculations increase the risk of different programming errors and are also found to be time consuming in nature (Ernesto & Farouki, 2012). In most of the industrial cases, these holes do not only lie in a regular matrix form, but also may be lying in an inclined matrix form. In this case, the programmers need to have large amount of time, in order to calculate the coordinate centers for each hole; thus, the program will require a large amount of memory storage (Gidding & Lewis, 2000; Siemens, 2012; Gidding & Lewis, 2000; Siemens, 2012; Siemens, 2012)

This research intends to include and discuss some of the major techniques, which are being used for monitoring and controlling the phenomenon, which is usually occurred from the contact of the cutting tools and work piece. It has been established that modern CNC units are capable enough to create a second type of cycle that is similar to canned cycles, which is known as user-defined cycles, for diverse operations of machining. The user-defined cycles are prepared and developed by CNC researchers or professional programmers, in accordance with their experience and the requirements of their products. It is a notable aspect that the task of producing user-defined cycles is not easy. It is due to the fact that this process is highly complicated in nature and requires special experience. More so, it also needs to have significant efforts for validation and verification of the cycle, before its installation into the CNC unit and to ensure that the cycle is ready for the programmer to use (Conway et al., 2012). On the other hand, if all of these parameters are not fulfilled, the cycle may produce a machine error or machine crash.

In this study, the capabilities of the algorithm were tested by the help of two case studies. The first case study was developed for a Numeripth8000 -B- CNC unit, equipped with a horizontal Orion-500 machining center. However, the second case study was developed for a Sinumeric 840-D, equipped with a vertical EMCO machining center. During this activity, several work-pieces showed satisfactory outcomes, with an accuracy of up to 0.15 mm for the diameter and 0.80 micron for the surface finishing. The developed algorithm saves approximately 80% of the special boring tools, while saving 95% of the preparation time, which is usually needed for the part program and reduction in the errors of program. Abbas (1996) has discussed the preparation of a custom macro in modern CNC machines and its effect on time, which is usually required to prepare the CNC part programs and error reducing techniques, within the program.

Omirou et al. (2012) has presented an algorithm, in order to improve the turning operations' interpolation accuracies. The free-form profiles are resolved within the parts that are considered in the implementation of the Bezier formulation, turning in the region of a center linear axis. The production of parts through forging and casting plays a major role in meeting the demand to manufacture the geometric shapes. Furthermore, it has also been established that computer numerical control (CNC) is used as a method for turning machine, which works on the concept of locus tracing (Kadir et al., 2011). The potential algorithm applies a CNC interpolator on real-time, which supplies the turning machine with the highest possible capability and accuracy. This task of machining is programmed for a canned cycle, which tends to extend the existing capability of G70/G73 CNC for turning fixed cycles and feature-generation.

Sheth and Rathour (2014), applied macro parametric programming for developing a Bezier surface and canned cycle with hypo cycloid, which is merged with EP-hypo cycloid curve. It has also been established that the developed algorithm was useful in canned cycle and for a variety of applications including, surface milling, a lobe pump rotor, and a cycloidal speed reducer, which is exclusive of an arrangement of external CNC interpolator or CAD modeling.

Abbas (2003) has presented an approach that high tolerance and good surface finishing can be obtained by using CNC grinding machines. However, due to the high cost of these machines, manufacturers often consider interchanging boring bar or add a grinding spindle, for a turning lathe. Complex software is usually used to facilitate this addition as the programmer ought to have control over the complete grinding wheel surface and not only on the tip of the cutter or tool, in contrast to milling or turning (Lasemi et al., 2010). This paper aims at presenting simplification of the grinding program by using this general algorithm, which is applied by the programmer for generating a suitable machine subroutine. Afterwards, this subroutine can be applied to all the operations of grinding machine. In this algorithm, it is essential for the programmer to describe each work piece separately for the contour within a small program, as similar with the program of CNC lathe.

There experimentation generated satisfactory results, when this algorithm was applied to a numerical control, Sinumeric 840 C through a hollow spindle lathe as well as many gun barrels were turned and grounded with the suggested approach. Abbas (2004) observed a significant development within the technology of grinding complex geometrical parts. Nowadays, many factories consider additional grinding spindle within a turning lathe for avoiding the purchase of additional and expensive grinding machines.

There are two types of software, which are required for interchangeable use of grinding spindle, along with a turning boring bar. The first software to be used is a program that includes subroutine for controlling the movement of machine, through the work-piece profile of the grinding wheels. The first software to be used is a program, which generates a subroutine that are essential for creating profile of the wheels for new grinding with the respective work-piece of the grinding profiles and dressing them, after using the grinding wheels (Newman & Nassehi, 2009). A long-term programming is needed when the cut is very small in depth. Different software will be required for each of the grinding wheels, having different shape. The machine manuals and literature research has not provided that the detailed description of such problem, within the machine.

Thus, this paper aims at representing a description of a general algorithm for dressing programs and profile production, more simply for the subsequent wheel grinding. This algorithm will be implemented to create appropriate subroutine in the programming for the system of machine dressing. The algorithm can be integrated on the grinding machines as well as on the hollow spindle lathe of additional grinding spindle. The subsequent subroutine will be integrated for profiles of grinding wheels and dressing of the commonly shaped grinding wheels (Ridwan & Xu, 2013). The grinding wheel contour will only be needed by the programmer, in order to have clear and detailed description. Hollow spindle lathe has also undergone the implementation of this algorithm, which is applied on a numerical Sinumeric 840C control. Satisfactory results have been demonstrated with the implementation of this approach for many grinding wheels that were profiled and dressed through its application.

It has been indicated by Abbas (2004) that CNC programs involve complex task for grinding parts, involving different shapes. These are long programs that go through problematical calculations as the programmer need to have control over the grinding wheel's surface and not only on the tip of the cutter or the tool diameter like, milling or turning. Furthermore, there is additional software, which is required for grinding wheels' profiling and dressing. It observed that no detailed description has been provided in the literature and machine manuals, about the resolution of this issue. This paper represents a simple way for producing the programs for dressing and profile via general algorithm for grinding wheels with complex shape.

The programmer needs to apply this simple algorithm to produce a subroutine, which is applicable on CNC grinding machines and spindle on CNC lathes that are additional. When subroutine is stored in the CNC unit, than the programmer will describe the contour by filling in the grinding wheels parameters. The algorithm was implemented on an "add-on grinding spindle" along with a numerical control, Sinumeric 840C and a hollow spindle lathe. When the proposed approach was implemented it gave satisfactory results for many grinding wheels, having complicated shape that were profiled and dressed.

Abbas and Megahed (2005) described an examination of developing a programming algorithm, which is employed for drilling holes, which usually lie in a form of regular matrix. This paper explains the examination the process, which deals with developing a general programming algorithm to drill holes that lie within a general matrix or an inclined matrix form in G-code. The algorithm will be applied to develop cycles that are user-defined or subroutines to drill holes that are lying in an inclined matrix form, staggered matrix, regular matrix, column, sole row, and diagonal.

The programmed algorithm is helpful for machining drum and trammels screen, separators, and boiler-plates. To test the capabilities of the examined algorithm, there are many work-pieces, which have projected the satisfactory results. Therefore, some examples of these machined pieces are given. The study recognizes that the fundamental objective of the two sub-problems of finding tool path (CL) points and tool orientations is to completely and correctly machine the work piece, in order to obtain the design surface.

Additionally, due to economical and ecological constrains, the machining process must also be optimized for its performance and environmental footprint. Performance of a machining process is indicated by several parameters such as cycle time, cost, and tool life. However, the environmental footprint is characterized by its energy use, emissions, coolant use, etc. The common constraints include those, which are related to the gouging prevention, scallop height limit, and kinematic limits (Overby, 2010). The research has exploited these similarities in the sub-problem objectives and constraints by expressing them as those of a single optimization problem. Also, the optimization problem was defined in the combined parameter space, which is the full configuration space (C-space) of the cutter.

Abbas et al. (2011), explored the drilling machines for CNC with respect to optimum path planning, while considering products of a special class that include numerous holes, which are usually placed in a form of rectangular matrix. Some of these examples include different products, such as trammel screens, drum, and boiler plates, steel structures with connection flanges, and separators of food-processing, in addition to definite parts of printed circuit boards. CAD software packages that are mostly used for commercial purposes include modules that results in CNC code auto-generation.

The plan, which is generated for tool path using the commercial software CAD, is not found to be optimized, in terms of considering the distance for tool travel and allocated time for total machining. This process of optimization is usually incomplete, as the distance of tool travelling is minimized, which is known as travelling salesman problem (TSP). In the context of programming, TSP is found to be the most difficult problem for solution, without having any identified solution, which can be attained in polynomial time (Sarhan & Matsubara, 2014). A number of heuristic optimization algorithms are explained to solve the TSP, with diverse success factors. ACO (Ant colony optimization) is recognized as an integrated and one of the most successful algorithms, which assists in addressing the concerns in TSP; hence imitating the nature of ants' behavior.

This paper has applied the ACO algorithm to the CNC drilling tool path planning between rectangular matrixes of holes. Taking a benefit of the rectangular design of the holes, there were two variations, which were proposed for the basic ACO algorithm. Different case studies on simulation, point out the average path through modification of ACO algorithms, which displays a major decrease in the total distance, traveled by the tool.

A common algorithm has been developed, which is practical to allow the drilling of holes and is circularly organized within multiple patterns. This potential algorithm can be simply adapted by using program within any modern CNC unit, just by creating a cycle of user-defined subroutine (Yusof et al., 2009). In order to test the effectiveness of the potential algorithm, there were two cases that enhanced understanding of the algorithm. The first used the Nwneripth800-B CNC unit, which was equipped with a horizontal Orion-500 machining center. On the other hand, the second case study was developed by using the Sinumeric 840-D, which was equipped with a vertical EMCO machining center. During this process, the production of few samples was successful. This original algorithm exhibits potential competence in the part program development and proposes the benefit of diminishing the errors in encoding.

Several research have been performed, which were based on the optimization of the path-planning within a computer numerical control (CNC), where machining jobs takes place in machining center and involve many concentric holes for drilling, which are typically organized within circular patterns. It is assumed that this research may contribute to great benefits, while reducing the machining time for some components, which are utilized in the food separators, trammel screens, condensers, boilers, and heat exchangers (Conway et al., 2012; Omirou et al., 2012). The cost of machining and tool optimization are the two aspects that are typically overlooked when paths are generated for tool and CNC codes, with respect to the CAD software packages, which are commercially available (Ridwan & Xu, 2013). This problem of achieving minimum travelling distance for tool path can be represented as the travelling salesman problem (TSP).

A varied level of success has been achieved for this optimization problem of solving TSP, which is based on the literature. TSP problem is also addressed by using Ant Colony Optimization (ACO). ACO is found to be a prominent approach, which is established on the basis of imitating the natural behavior of the ant colonies. This research recommends a study by using hybrid ACO, which takes an advantage with biasing mechanic design of a local search and the hole, which is arranged in geometric pattern. The examples of simulation indicate that this approach comprises of better performance when compared with the classical approach of ACO, considering the use of commercial CAD software to generate simple spiral path as well as the approach of genetic algorithm (GA). This approach is then implemented within the food-industry for the planning of drilling path, which consists of a separator plate with two-thousand-hole.

(Majerik et al., 2015) the objective of the article was to suggest the possibility of creating CNC program production process components on a rotary chuck, or with minimal handling. This is achieved by shortening the time of the final machining while eliminating inaccuracies caused by manual switching of the workpiece. The results presented in this paper can be further exploited in the process of teaching courses and programming CNC machines and technical practice for upgrading older solutions and processes of the main and additional times, while maintaining the dimensional accuracy of machine parts. In our future research plans e.g. comparing to the existing workshop CNC machining centers, have the presented simulation methodology have a number of advantages. First, with the same interpolation steps, the contour accuracy is significantly higher. Second, the programming module is easier to use more simple G code lines. And last, machining cost is lower because of the reduced machining time.

Usually, CNC unit manufacturers may not provide satisfactory explanation for cycles in a user-defined CNC program. User-defined cycles are considered as the black box in CNC manufacturing unit manufacturers, i.e., CNC programmer is only aware of the input and the results. Thus, this research develops a simple way of boring or counter-boring with a general programming algorithm and only one standard milling cutter, comprising of numerous holes with various diameters. This algorithm is used for achieving specific counter-bore diameters to create a new cycle of subroutine or user-defined (G888), applied for the accuracy, tolerance, and good surface finish.

2. Algorithm of User-Defined Cycle

The G-code (G81-G89) is used within standard canned cycles and they are cancelled by (G80), while activation within user-defined cycles takes place by means of the G-code (G801-G899) and cancelled by (G800). This algorithm is used by the programmer, in order to create suitable machine subroutine or G888 (user-defined cycle) that fits to the machine. These cycles applied for boring or counter-boring holes. This cycle can be equipped with our user-defined cycle (G881) to bore or counter-bore holes that lie within a regular matrix, an inclined matrix, a single row, a staggered matrix, a column, and a diagonal. In addition, this cycle can be equipped with our user-defined cycle (G890) for boring or counter-boring holes, which lie within a circular form.

When the user-defined cycle (G888) is stored for the CNC unit, then the programmer just needs to call out for the particular cycle and filling within one block of parameters, in contrast to write several blocks to describe the route of milling cutter. The buffer page or the directory does not display the subroutine when the execution of the program takes place; as a result, the operator cannot adjust the contents of the subroutine. To pass the values within the global parameters of the user-defined cycle, letter addresses are used (indicated by A through Z); such as, the letter address A is a value which is stored within parameter (P201) and the letter address B is a value, which is stored in (P202). Therefore, the potential algorithm precisely and clearly describes method, which is used in the creation of statements for user-defined cycle (G888) for boring or counter-boring many holes with one standard milling cutter. In concurrent algorithm, parameters to be input are similar to the following:

A = Milling-cutter diameter or End-mill diameter (P201).

B= Initial hole diameter (P202).

D = Final counter-bore diameter (P204).

X = Coordinate center of the initial hole in the X-axis direction (P224).

Y= Coordinate center of the initial hole in the Y-axis direction (P225).

R= Level of Approach within the Z-axis direction (P218).

Z= Depth of the counter-bore diameter in the Z-axis direction (P226).

E= Level of Retraction within the Z-axis direction (P205).

This system will create three calculated points, i.e., P1, P2 and P3. Figure 1 displays all of the features of a general algorithm, which is appropriate to generate a subroutine or a user-defined cycle (G888) for its use on any modified CNC controls. The prime focus of the potential algorithm is to take all of the calculations and complicated formula as input, which are required for controlling the center of the milling cutter movements, according to the diameter of the milling cutter or the end-mill, the initial diameter of the hole, and the final diameter of the counter-bore hole. The subroutine will start to move the center of the milling cutter with rapid motion to the center of the hole, according to input data in parameters X = P224 and Y = P225. Next, the front of the milling cutter will move with rapid motion to approach the level in the Z direction where, R = P218. Subsequently, the algorithm will have three routes:

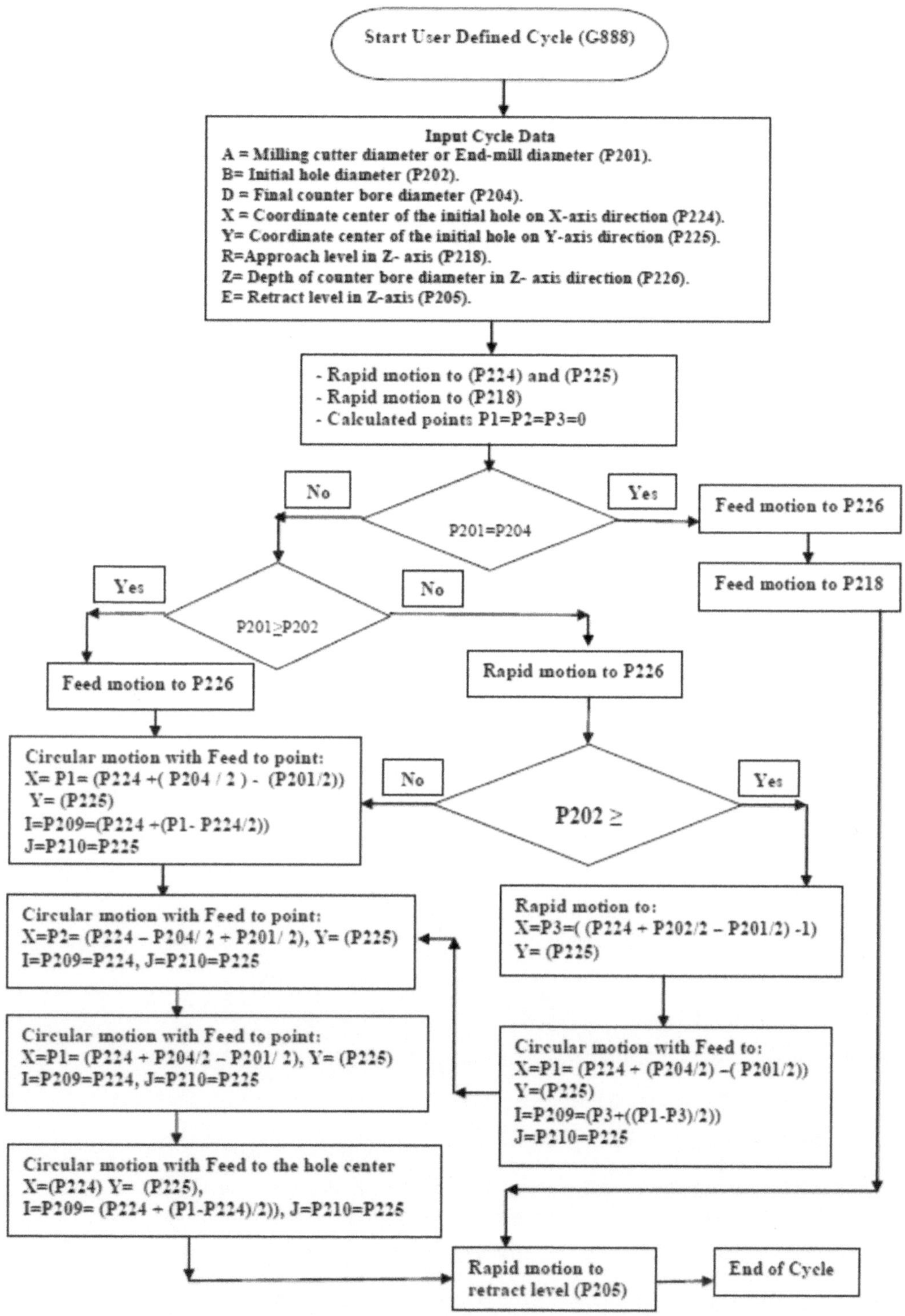

Figure 1. A General Algorithm for User-Defined Cycle (G888)

2.1 The First Route

If the size of the milling cutter diameter is equal to the final diameter of the hole, then the cutter will move with feed motion until, the boring depth in the (Z) direction where, (Z=P226). Next, the cutter will return to the approach level in the (Z) direction where, (R=P218), with feed to produce a good surface finish. Finally, the cutter will move with rapid motion in the (Z) direction to retract to the level where (E=P205) to save time. This route will be similar to the regular boring operation.

2.2 The Second Route

If the diameter of the cutter is not equal to the final diameter of the bore and is larger than the initial hole diameter, then the cutter will proceed with feed motion until the counter-bore depth in the (Z) direction where, (Z= P226), to avoid the tool crashing with the work-piece and to enlarge the hole diameter; subsequently, the center of the cutter will move with a circular motion to point (P1) to position the outside surface of the cutter to be tangent with the final diameter, followed by another circular movement to point (P2), and then a circular movement to point (P1) again. Next, the center of the cutter will return back to the hole center, as shown in figure 2a . Finally, the cutter will rapidly move to the retracting level in the (Z) direction where, (E= P205).

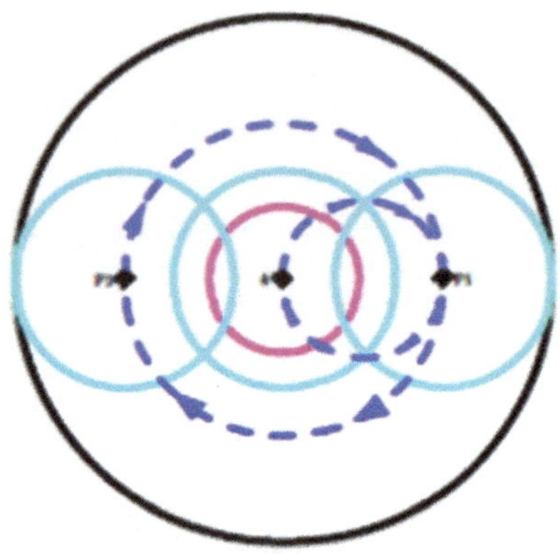

Figure 2a. Milling Tool Path for the Second Route

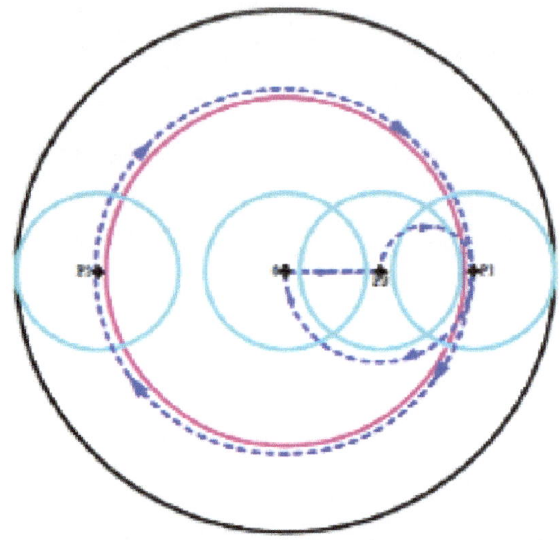

Figure 2b. Milling Tool Path for the Third Route (First Sub Route)

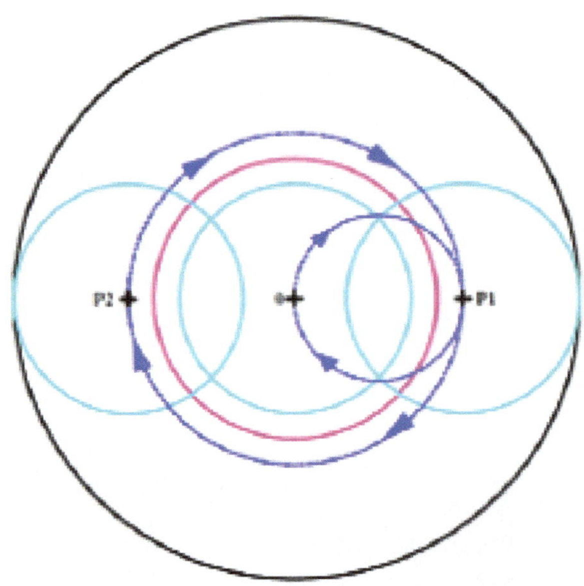

Figure 2c. Milling Tool Path for the Third Route (Second Sub Route)

2.3 The Third Route

If the diameter of the cutter is less than the initial diameter of the hole, then the cutter will move with rapid motion in the (Z) direction until, reaching the counter-bore depth where, (Z= P226) to save time. Next, the cutter has two sub-routes of movements.

2.3.1 The First Sub-Route

If the initial diameter is greater than twice the cutter diameter, then the cutter will be moved to the side with rapid linear motion in the (X) direction to the calculated point (P3), in order to save the time. Next, the cutter will move with feed by circular movement to the calculated point (P1) to be a tangent with the final diameter of the counter-bore, followed by circular movement to point (P2), circular movement to point (P1), and then returning to the hole center, as shown in figure 2b.

2.3.2 The Second Sub-Route

If the initial diameter is smaller than the twice of the cutter diameter, then the cutter will move directly with feed by circular movement to the calculated point (P1), followed by circular movement to point (P2), by circular movement to point (P1), and then returning to the hole center, as shown in figure 2c. The objective for all of the above circular movements is to achieve a smooth touch or smooth retract with the final surface to avoid the tool gauche.

3. Applications

The developed algorithm in the subsequent part of the research is applied for creating an appropriate user-defined cycle (G888) with a horizontal milling machine for Lewis machining center and grinding, which is equipped with an Vertical Machining Center, EMCO with Sinumeric 840-D and CNC unit Numeripath 800. Figures 3 and 4 provide the details of such a cycle (G888) for the respective control units. The modification of parameters can be obtained, with respect to a suitable Fanuc control unit, integrating a change in Parameter P1 to be #1 and Parameter P2 to be #2, etc. When the software to be utilized for the user-defined cycle (G888) is uploaded in the CNC unit, the CNC memory stores the software. The machine operator or programmer is unable to view the subroutine and they cannot change it.

```
%
N01 (PGM, G888, COUNTERBORE HOLES)
N02 ! A = Milling cutter diameter or End-mill diameter (P201).
N03 ! B= Initial hole diameter (P202).
N04 ! D = Final counter bore diameter (P204).
N05 ! X = Coordinate center of the initial hole on X-axis direction (P224).
N06 ! Y= Coordinate center of the initial hole on Y-axis direction (P225).
N07 ! R=Approach level in Z- axis (P218).
N08 !  Z=Depth of counter bore diameter in Z- axis direction (P226).
N09 !  E=Retract level in Z-axis (P205).
N10 ! P1, P2, P3= Calculated Points
N11 (LET, P1=0)
N12 (LET, P2=0)
N13 (LET, P3=0)
N14 G00 X=P224, Y=P225
N15 G00 Z=P218
N16 (IFS, P201 = P204) GO TO N32
N17 (IFS, P201 GE P202) GO TO (G01 Z=P226) (GTO, N20)
N18 G00 Z=P226
N19 (IFS, P202 GE 2*P201) (LET, P3= (P224 + P202/2 – P201/2) -1) (GTO,
N29)
N20 (LET, P1= (P224 + P204 / 2  -  P201/2))
N21 G02 (LET X=P1 , Y=P225, I=P209=(P224 +(P1- P224/2)) J=P210=P2)
N22 (LET, P2= (P224 – P204/ 2 + P201/ 2))
N23 G02 (LET, X=P2, Y=P225, I=P209=P224, J=P210=P225)
N24 (LET, P1= (P224 + P204/2 – P201/ 2) )
N25 G02 (LET, X=P1, Y225 I=P209=P224, J=P210=P225)
N26 G02 (LET, X=224, Y225, I=P209= (P224 + (P1-P224)/2)), J=P210=P225)
N27 G00 (LET, Z=P205)
N28GTO N100
N29 G00 (LET, X=P3, Y=P225)
N30 G02 (LET, X=P1, Y=P225,  I=P209=(P3+((P1-P3)/2))), J=P210=P225
N31 GTO N23
```

Figure 3. User defined cycle (G888) Numeripath 8000-B – Giddings & Lewis Control

```
Program CYCLE888.SPF
PROC CYCLE888(REAL _X,REAL _Y,REAL _Z,INT _E,INT _R,INT _A,INT
_B,INT _D), SAVE SBLOF
;Input cycle data
N0  ;_X = Coordinate center of the initial hole on X-axis direction
N01 ;_Y= Coordinate center of the initial hole on Y-axis direction
N02 ;_Z= Depth of counter bore diameter in Z- axis direction
N03 ;_E = Retract level in Z-axis
N04 ;_R = Approach level in Z- axis
N05 ; A = Milling cutter diameter or End-mill diameter
N06 ; B= Initial hole diameter
N07 ; D = Final counter bore diameter;_
N08 ;_P1= Calculated point
N09 ; P2= Calculated point
N10; P3= Calculated point
N11 P1=0
N12 P2=0
N13 P3=0
N14 G00 X=_X  Y=_Y
N15 G00 Z=_R
N16 IF (_A=_D) GOTO N32
N17 IF (_A≥_B ) GOTO (G01 Z=_Z) (GOTO, N20)
N18 G00 Z=_Z
N19 IF (_B ≥ 2*_A) (P3= (X + B/2 – A/2) -1) (GOTO, N29)
N20 P1= (X + D / 2  -  A/2)
N21 G02 X=P1  Y=_Y CR= (P1-_X) /2
N22 P2= (X – D/ 2 + A/ 2)
N23 G02 X=P2  Y=_Y  CR= (_X-P2)/2
N24 P1= X + _D/2 – _A/ 2)
N25 G02 X=P1  Y=_Y CR=(P1-P2) / 2
N26 G02 X=_X   Y=_Y CR= (P1-_X)/ 2
N27 G00 Z=_E
N28GOTO N100
N29 G00  X=P3  Y=_Y
N30 G02  X=P1 Y=_Y  CR= (P1-P3)/2
N31 GOTO N23
N32 G01 Z=_Z
N33 G01 Z= _R
N34 G00 Z=_E
N100 M17
```

Figure 4. User defined cycle (G888) for boring or counter boring with milling cutter suitable for Sinumeric 840-D-Siemens

This software comprises of all defined parameters and potential calculations, which are needed in machine motions. Then, the programmer can call the cycle by filling in the addresses, which are similar to the canned cycle. In order to test the validation and verification of the cycle (G888), many parts from the work-piece model shown in figure 5 with specifications in table 1, were machined with accuracies of up to 0.15 mm in diameter and 0.80 microns in surface finish, and satisfactory results were obtained. The CNC part programs required for counter-boring this part with one standard milling cutter are shown in figure 6 and figure 7 for Numeripath8000-B and Sinumeric 840-D, respectively.

Table 1. Work piece Specifications for Testing Cycle (G888) Capabilities

Hole No.	Milling Cutter Diameter = A = P201=45 mm						
	Hole Center		Counter bore Hole depth	Approach level	Retract level	Initial hole diameter	Final hole diameter
	X=P224	Y=P225	Z = P226	R=P218	E=P205	B=P202	D=P204
1	60	60	-20	5	100	40	$45^{+0.15}$
2	180	80	-20	5	100	45	$60^{+0.15}$
3	320	120	-20	5	100	60	$75^{+0.15}$
4	185	205	-20	5	100	90	$130^{+0.20}$
5	80	310	-20	5	100	100	$140^{+0.20}$
6	310	310	-20	5	100	100	$150^{+0.20}$

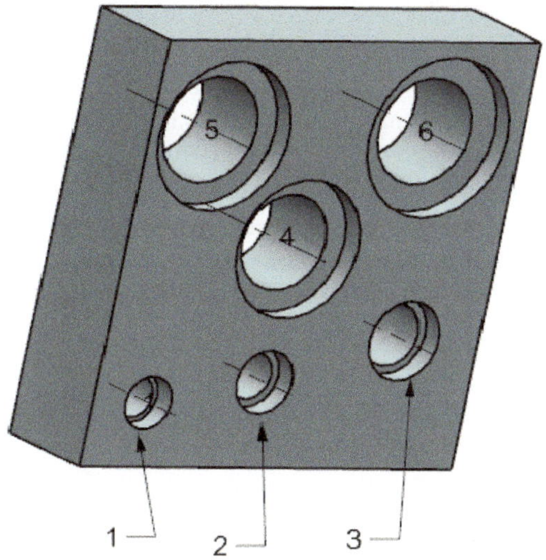

Figure 5. Work piece drawing for testing the user defined cycle capabilities

```
%
N10 (PGM, Counter Boring, Boring and Counter Boring Cycle-G888)
N20 T01 M06 S500 M08 F0.25
N30 G00 G90 G95 Z200 M03
N40 G888 X60 Y60 Z-20 R5 E100 A45 B40 D45  (Hole-1)
N50   X180 Y80 B45 D60  (Hole-2)
N60  X320 Y120 B60 D75 (Hole-3)
N70   X185 Y205 B90 D130 (Hole-4)
N80   X80 Y310 B100 D140 (Hole-5)
N90  X310 Y310 B100 D150 (Hole-6)
N100 G800 Z500 M05 M09
N70 M02
```

Figure 6. Part program for boring or counter boring with milling cutter for Numeripath 8000-B – Giddings & Lewis Control

```
%
;Program main.mpf
N10 G54
N20 M06 T1 D1
S500 M3 F100
N40 CYCLE888(60,60,-20,100,5,45,40,45)
N50 CYCLE888(180,80,-20,100,5,45,45,60)
N60 CYCLE888(320,120,-20,100,5,45,60,75)
N70 CYCLE888(185,205,-20,100,5,45,90,130)
N80 CYCLE888(80,310,-20,100,5,45,100,140)
N90 CYCLE888(310,310,-
20,100,5,45,100,150)
N100 G00 Z500 M05 M09
N110 M30
```

Figure 7. Part program for boring or counter boring with milling cutter for Sinumeric 840-D- Siemens

4. Feature Based Procedure Planning and Automatic Numerical control Part Programming

It is assumed that a well-integrated manufacturing and design system will provide a rich set of features; therefore a designer can manufacture the critical parts with utmost flexibility. Many systems of CAM/ CAD incorporate very few and limited set of design. In the research, the feature based-approach with respect to the integration of manufacturing and design is used. One of major objective behind this activity is to inculcate the sculptured surface as the features of higher order, which includes both computer-aided procedures and parametric design paradigm techniques (Sarhan & Matsubara, 2014). The advantage of the decomposition of tool-driven is the possibility of deploying multiple tools for the optimization of the milling operations. The figure 8 shows the decomposition of the pocket of hand-shaped into di roughing operations.

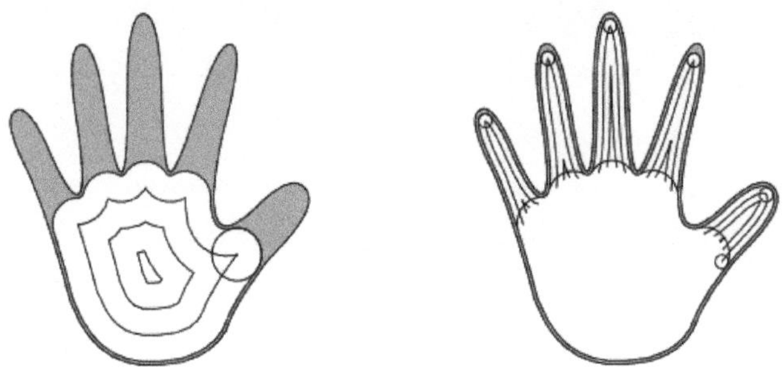

Figure 8. Feature Based Procedure Planning and Automatic Numerical control Part Programming

The left side indicates that the big tool is first deployed to machine, and away the area of the palm, it can be easily fixed. The shaded area represents the uncut region by the gigantic tool. However, the right figure represents the small tool that is deployed to cut the enduring figure area. The approach adopts several procedures so as to find the combination of tool diameters for the machine a milling feature, within a short span of time (Vichare et al., 2009). The tool, however, depends on the overhead time and the features. Therefore, a simplified linear overhead per-tool overhead is espoused so as to roughly approximate the component.

5. Process Control

The term "process control" can be understood as the operation of the process variables such as speed, depth-of-cut, and feed for the regulation of the processes. The operations of the machine tools, which are perform on-line and off-line, are governed by the adjustments of the speed and feeds to destroy chatter, while initiating an emergency termination in response against an event breakage of tool. It also performs the regeneration of the program so as to increase the depth-of-cut, in order to suppress burr development. Computer aided process of planning is found to be more cultured procedure, which utilizes process models off-line set in various different models for the selection of the variable (Tobon-Mejia et al., 2012). However, advancement in the modeling of the cutting mechanics is obligatory. Recently, models are resolute and empirically contain terms of non-linear that account the effects of unmolded. However, the biggest hindrances in its implementation include limited applicability and low reliability.

6. Discussion

This paper illustrates that the capabilities of CNC Machines using an additional counter boring cycle can be improved, which is currently saving 95% time for preparation and 80% cost for the special tools via milling cutter. The results of the current study are practical and consistent with the outcomes, which were established by (Abbas, 2012). Sona et al (2009) signifies the use of Numeripth800-B CNC unit within the machining center (Orion-500) through a case study that demonstrated the algorithm's capability, which has high academic value. The current study analyzes the capabilities of the CNC machine by using the user-defined cycle (G888) and attained results have shown great accuracy up to 0.15 mm for surface finishing.

Sona et al. (2009) presented a hybrid design for 5 axis CNC milling machine that can avoid substantial pitfalls and has several benefits for both mechanisms (parallel and serial). A further extension of the present research study involves the implication of same method to other kinds of five-axis CNC milling centers, while table tilting comprises of two rotations on each of the spindle tilting or tables (Yusof et al., 2009). Also, it can work with one rotation for each of the table and spindle types. Moreover, the same technique can be applied to develop a post processor for industrial robots. The compensation vectors can also be determined for same robot using principle of inverse kinematic transformation.

The current study is consistent with Sona et al. (2009), considering the high tolerance and capability advantage of the milling cutter. Another study by Anderberg (2012), demonstrated the method to improve CNC machining by improving its process planning. In this paper, we address same issue and improve efficiency of the CNC milling machines utilizing the milling cutter. To improve efficiency of the CNC machine, we focused on a new cycle in the boring for the milling. This study has identified and illustrated the inherent differences in between the current study and by the research of Anderberg (2012), as it was intended to achieve reduction in energy consumption behaviors using an efficient process. On the other hand, Sona et al. (2009) examined the 5-axis CNC machines; thus, treat them differently, to analyze, model, and suggest ways to improve efficiency of the machine consumption behavior.

The knowledge, techniques, and software tools developed by Abuthakeer et al. (2011) are used in manufacturing industry at various stages, in order to save energy (electricity). CNC machining encompasses a variety of manufacturing processes including cutting, milling, turning, and grinding. However, in this study we focused on CNC point-milling for analysis and modeling of the method, when new cycle is added to improve its capability. This research has a significant academic value, as it can be readily extended to other CNC manufacturing processes to improve their performance and energy usage. Sona et al. (2009) proposed a strategy to formulate the full 5-axis tool path planning problem by transforming it to a multi-objective non-linear optimization problem in the C-space of the cutter. The researcher has identified that an optimal solution to the 5-axis tool path planning problem necessitates simultaneous solutions of involving sub-problems, but none of the existing approaches address the inherent complexities.

The formulation method used in the study considers all the cutters simultaneously, without making any simplifying assumptions. A 5-axis tool path planning system based on our formulation may be able to find the optimal tool path solution in the full C-space of feasible solutions. It is impossible with the existing approaches to completely provide the solution and address all the complexities of the current algorithms and methods. However, the proposed approach can provide solution to the two sub-problems of cutter locations and cutter orientations simultaneously. Reducing energy consumption in CNC machining has been the ultimate achievement of the research, which was conducted by (Sona et al., 2009).

On the other hand, in this research has proposed and implemented few ideas, related to the CNC process planning. The proposed planning has demonstrated the need for considering the use of additional counter boring cycle during designing and comparing of manufacturing operations, along with the conventionally considered metrics of process productivity and cycle time. Sona et al. (2009) and Anderberg (2012) reported in their work a novel foundation over which, advanced 5-axis tool path planning systems may be built to obtain tool paths that minimize several, often conflicting objectives, simultaneously.

These studies highlighted the various other dimensions for the future research work in the field. Similarly, using the current research study further researches can be carried out in future and several other methods can be implemented, which enhances the theoretical value of the current study. With standards, requiring tangible progress toward saving energy, equipment cost, and offering substantial advantages for time effectiveness this study has a practical value for the industrial operations, as well as academic value for the further research work in the current field of study. Zhu et al. (2010) conducted the study using the CNC machine tools, specifically to gain capability advantages within the industrial product-service system.

The comparison with previous studies indicates that previous studies did not use the traditional milling cutter to transform the workability of the CNC machine. Therefore, this research has been added to the existing knowledge, which is available on this topic. Furthermore, the addition of a new counter boring cycle with a milling cutter has lead to significant improvements in the capabilities of CNC machines in comparison to the previous work done by researchers in this area. Future researchers can build on this study and research in other areas. The knowledge and the tools, developed in this research can be readily used, especially because they do not require any changes to the existing practices or infrastructure. As the research work extends for the sustainability, it is expected that the present work will contribute in saving energy, specifically in manufacturing sector. The future researches need to offer more advantage regarding time safety in the process of preparation and greater efficiencies regarding the use of equipment and cost advantages.

7. Conclusion

This paper describes an easy method to create subroutine or user-defined cycles (G888) for boring and counter-boring holes with different diameters using one standard milling cutter. These holes may be in random positions or lying in a row, a column, a rectangular matrix, or in circular patterns. This cycle is suitable for all modern CNC units or may require very slight modifications, depending on the names of the parameters, which are used in the CNC unit. This cycle can be used to counter-bore parts with an accuracy of up to 0.15 mm in diameter and 0.80 microns in surface finish. The cycle saves approximately 80% of the special boring tools, as well as saving 95% of the preparation time needed for the part program, additionally for reduction in the errors of the programming.

This research study has a significant practical and academic relevance, as the methods, procedures, and tools used to conduct the research work are practical and can be implemented easily. More so, the current study can also be considered as an imperative addition to the literature and further researches, which intends to understand CNC machining and ways to improve its capabilities, new implications of tool, methods, or procedures to enhance its efficiency, cost reduction within the process implementation, and time or energy consumption for carrying out the procedure. The current study offers a significant enhancement in performance of CNC machine, specifically when they are implemented with an additional counter boring cycle.

References

Abbas, A. T. (1996, April). Custom macro in CNC part programs. In *7th, App. Mechanics & Mech. Engineering Conference, Technical Military College* (Vol. 9, No. 11).

Abbas, A. T. (2003). Enhanced CNC lathe capability by addition of a grinding spindle. *International journal of production research, 41*(12), 2699-2709.

Abbas, A. T. (2004). A general algorithm for profiling and dressing grinding wheels when using a grinding spindle on a CNC lathe. *International journal of production research, 42*(18), 3995-4008.

Abbas, A. T. (2004). A general algorithm for profiling and dressing of complicated shape grinding wheels. *Robotics and Computer-Integrated Manufacturing, 20*(4), 313-327.

Abbas, A. T. M. (2012). Enhanced CNC machines capabilities by adding circular patterns cycle. *International Journal of Precision Engineering and Manufacturing, 13*(10), 1753-1758.

Abbas, A. T., & Megahed, S. M. (2005). A general algorithm for drilling holes lying in a matrix. *Robotics and Computer-Integrated Manufacturing, 21*(3), 235-239.

Abbas, A. T., Aly, M. F., & Hamza, K. (2011). Optimum drilling path planning for a rectangular matrix of holes using ant colony optimisation. *International Journal of Production Research, 49*(19), 5877-5891.

Abbas, A. T., Hamza, K., & Aly, M. F. (2014). CNC Machining Path Planning Optimization for Circular Hole Patterns via a Hybrid Ant Colony Optimization Approach. *Mechanical Engineering Research, 4*(2), p16.

Abuthakeer, S. S., Mohanram, P. V., & Kumar, G. M. (2011). Structural redesigning of a cnc lathe bed to improve its static and dynamic characteristics. *Annals of the Faculty of Engineering Hunedoara-International Journal of Engineering, 9*(3).

Anderberg, S. (2012). *Methods for improving performance of process planning for CNC machining-An approach based on surveys and analytical models*. Chalmers University of Technology.

Anderberg, S., & Kara, S. (2009). Energy and cost efficiency in CNC machining.

Anderberg, S., Beno, T., & Pejryd, L. (2012). Energy and cost efficiency in CNC machining from a process planning perspective. In *Sustainable Manufacturing* (pp. 393-398). Springer Berlin Heidelberg.

Conway, J. R., Ernesto, C. A., Farouki, R. T., & Zhang, M. (2012). Performance analysis of cross-coupled controllers for CNC machines based upon precise real-time contour error measurement. *International Journal of Machine Tools and Manufacture, 52*(1), 30-39.

Ernesto, C. A., & Farouki, R. T. (2012). High-speed cornering by CNC machines under prescribed bounds on axis accelerations and toolpath contour error. *The International Journal of Advanced Manufacturing Technology, 58*(1-4), 327-338.

Gidding and Lewis. (2000). Operator Instruction Manual for Numeripath, USA.

Gidding and Lewis. (2000). Programming Instruction Manual for Numeripath, USA.

Groover, M., & Zimmers, E. W. J. R. (1983). *CAD/CAM: computer-aided design and manufacturing*. Pearson Education.

Kadir, A. A., Xu, X., & Hämmerle, E. (2011). Virtual machine tools and virtual machining—a technological review. *Robotics and computer-integrated manufacturing, 27*(3), 494-508.

Lasemi, A., Xue, D., & Gu, P. (2010). Recent development in CNC machining of freeform surfaces: A state-of-the-art review. *Computer-Aided Design*, *42*(7), 641-654.

Luggen, W. W. (1983). Fundamental of Numerical Control. USA: Delmar Publishers.

Majerik, J., & Jambor, J. (2015). Computer Aided Design and Manufacturing Evaluation of Milling Cutter when High Speed Machining of Hardened Steels. *Procedia Engineering*, *100*, 450-459.

Newman, S. T., & Nassehi, A. (2009). Machine tool capability profile for intelligent process planning. *CIRP Annals-Manufacturing Technology*, *58*(1), 421-424.

Omirou, S. L., Rossides, S., & Lontos, A. (2012). A new CNC turning canned cycle for revolved parts with free-form profile. *The International Journal of Advanced Manufacturing Technology*, *60*(1-4), 201-209.

Overby, A. (2010). *CNC machining handbook: building, programming, and implementation*. McGraw-Hill, Inc..

Ridwan, F., & Xu, X. (2013). Advanced CNC system with in-process feed-rate optimisation. *Robotics and Computer-Integrated Manufacturing*, *29*(3), 12-20.

Sarhan, A. A., & Matsubara, A. (2015). Investigation about the characterization of machine tool spindle stiffness for intelligent CNC end milling. *Robotics and Computer-Integrated Manufacturing*, *34*, 133-139.

Sheth, S. N., & Rathour, A. N. Development of Canned Cycle for CNC Milling Machine. *nternational Journal of Engineering Research and General Science*.

Siemens. (2012). Cycle, Programming Manual for Sinumeric 840-D, Germany.

Siemens. (2012). Operator Manual for Sinumeric 840-D, Germany.

Siemens. (2012). Programming Manual for Sinumeric 840-D, Germany.

Son, S., Kim, T., Sarma, S. E., & Slocum, A. (2009). A hybrid 5-axis CNC milling machine. *Precision Engineering*, *33*(4), 430-446.

Tobon-Mejia, D. A., Medjaher, K., & Zerhouni, N. (2012). CNC machine tool's wear diagnostic and prognostic by using dynamic Bayesian networks. *Mechanical Systems and Signal Processing*, *28*, 167-182.

Vichare, P., Nassehi, A., Kumar, S., & Newman, S. T. (2009). A Unified Manufacturing Resource Model for representing CNC machining systems. *Robotics and Computer-Integrated Manufacturing*, *25*(6), 999-1007.

Yusof, Y., Newman, S., Nassehi, A., & Case, K. (2009). Interoperable CNC system for turning operations. In *Proceedings of world academy of science, engineering and technology* (Vol. 34, pp. 941-947). University of Bath.

Zhu, Q. Q., Jiang, P. Y., Huang, G. Q., & Qu, T. (2011). Implementing an industrial product-service system for CNC machine tool. *The International Journal of Advanced Manufacturing Technology*, *52*(9-12), 1133-1147.

14

Thermodynamic Analysis of Different Configurations of Combined Cycle Power Plants

Munzer S. Y. Ebaid[1] & Qusai Z. Al-hamdan[2]

[1] Department of Mechanical engineering, Philadelphia University, Amman, Jordan

[2] Perth College, University of the Highlands and Islands, Scotland, UK

Correspondence: Munzer S. Y. Ebaid, Department of Mechanical engineering, Philadelphia University, P. O. Box 1, Amman, Jordan. E-mail: mebaid2@philadelphia.edu.jo

Abstract

Several modifications have been made to the simple gas turbine cycle in order to increase its thermal efficiency but within the thermal and mechanical stress constrain, the efficiency still ranges between 38 and 42%. The concept of using combined cycle power or CPP plant would be more attractive in hot countries than the combined heat and power or CHP plant. The current work deals with the performance of different configurations of the gas turbine engine operating as a part of the combined cycle power plant. The results showed that the maximum CPP cycle efficiency would be at a point for which the gas turbine cycle would have neither its maximum efficiency nor its maximum specific work output. It has been shown that supplementary heating or gas turbine reheating would decrease the CPP cycle efficiency; hence, it could only be justified at low gas turbine inlet temperatures. Also it has been shown that although gas turbine intercooling would enhance the performance of the gas turbine cycle, it would have only a slight effect on the CPP cycle performance.

Keywords: gas turbines, steam turbines, CPP power plants, thermal efficiency

1. Introduction

1.1 Energy Scenario

Energy is one of the primary needs of human societies for their survival. It is needed for growing food, providing comfort and catering for a host of other application in all fields of activity such as agriculture, industry, transportation. The main sources of energy are fossil fuels, solar radiation fall out, winds, tidal, and geothermal. The conversion, distribution and utilisation of energy are the domain of engineering. The demand for energy throughout the world is increasing sharply because of growing world population, rising living standards and emphasis on developing energy intensive industries in almost all newly emerging countries to boost their economies in order to combat poverty and hardship.

Fossil fuels, coal, oil and gas, currently, provide more than 95% of the world's energy need. However, the reserves of fossil fuels on planet earth are finite and they are depleting rapidly. It should be noted also that the use of fossil fuels through combustion pollutes the environmental with toxic gases and contributes to global warming. Hence the continuing use of fossil fuels is undesirable both for energy conservation as well as for environmental protection.

Although sources of renewable energies appear to offer a promising alternative, their contribution to the world's total energy demand is still less than 10% and it is unlikely to change substantially in the near future. Hence, in order to conserve fossil fuels, increasing the efficiency of the current power generation systems is of paramount importance. This may be achieved either by modifying the thermal plant configuration or by using advanced thermodynamic cycles for power generation.

1.2 Review of Previous Work

Combined cycle researches dates back to the early part of the 20th century, however, research and development work on the combined gas turbine and steam turbine power generating plants started only in the late 1960s. Kehlhofer (1991) studied gas turbine power plant as a part of the combined cycle power plant and concluded that raising the gas turbine efficiency would not necessarily produce the best overall efficiency of the combined power plant. El-Masri (1985); El-Masri (1987) used the second law of thermodynamics and exergy analysis to locate and quantify the irreversibilities that cause loss of work output and of thermal efficiency of the gas turbine operating as a part of the combined power plant. He concluded that compressor inter-cooling will lead to an increase in specific work output, but this will happen with some reduction in thermal efficiency. Also, the dominant influence on cycle efficiency as turbine inlet temperature is raised will be the trade off between decreased combustion exergy losses and increased turbine blade cooling losses. Kail [5] analysed and evaluated different trends in combined cycle gas turbine power plant configurations. The various configurations have been compared with the simple cycle combined gas turbine/steam turbine power plant. The studied configurations were a combined reheat gas turbine/simple steam turbine power plant, a combined inter-cooled gas turbine/steam turbine power plant, a combined steam-cooled turbine blades gas turbine/steam turbine power plant, and a combined gas turbine/simple steam turbine power plant where the gas turbine has a closed-loop combustion chamber cooling system. In comparison with the 'simple' gas turbine, Kail concluded the followings:

i. Reheat of the gas turbine cannot transform its efficiency and output advantages into a lower cost of electrical power. The additional investments and higher maintenance costs overwhelm the thermodynamic advantages.
ii. Inter-cooling improves the efficiency and power output of the combined power plant.
iii. The concept of steam-cooled turbine blades places very stringent requirements on the blade materials, on the quality of the cooling steam and on the design of the closed cooling system.
iv. The gas turbine with a closed combustion chamber cooling system is less problematic than the gas turbine with a closed blade cooling system.
v. The simple combined gas turbine/steam turbine power plant achieves the lowest costs of electrical power and is therefore the best plant from an economic point of view.

Cerri (1987) studied the CPP plant and proposed thermodynamic parameters or indices to quantify the plant performance. He varied the maximum gas turbine inlet temperature between 800° C and 1400° C, at the same time gas turbine pressure ratio was varied between 2 to 24. Afterburning was also taken into consideration. His calculations produced both the CPP plant thermal efficiency and the specific work output. Cerri summarised his findings in the form of the following conclusions.

i. The thermal efficiency of the CPP plant is independent of the gas turbine pressure ratio but it would be influenced slightly by the steam pressure if it was sufficiently high.
ii. The thermal efficiency of the CPP plant would be positively influenced by adding an afterburner only if the turbine inlet temperature was significantly low.

Rufli (1987) analysed the CPP plant also by using the basic thermodynamic calculations for both the gas turbine and the steam turbine cycles. The maximum gas turbine inlet temperature was varied between 900 °C to 1350 °C, at the same time the gas turbine pressure ratio was varied from 8 to 22. Afterburning was also taken into consideration. Rufli's calculations produced values of the CPP plant thermal efficiency and of the total heat transfer area of the heat recovery steam generator. These calculations were simple and straightforward. Rufli presented a simple method for selecting the optimum parameters for the steam operating in a combined power and power plant cycle at any given gas turbine operating conditions. Bhinder and Mango (1995) used thermodynamics analysis to study the CPP plant performance. They concluded that the combined plant efficiency would be significantly higher than either the gas turbine efficiency or the steam turbine efficiency. The overall efficiency value of 60% for the CPP plant was shown to be achievable. In addition the thermal load on the environment was reduced to 59% of the gas turbine load working alone. The cycle calculations were simple and many of the losses were not included in the calculations. It would be difficult to achieve 60% thermal efficiency if all the losses were included in the calculations.

Horloock (1995, 1997) carried out an extensive study of combined power plants. The early history and recent developments and future prospects for the combined gas turbine/steam turbine plant were described. He adopted a graphical method of predicting the performance of the gas turbine cycle, developed by Hawthorne and Davis (1956) to determine the optimum pressure ratio of the gas turbine that would give maximum overall efficiency of the combined power and power plant. Bannister, Cheruvu, Little, & McQuiggan (1995) considered the techniques required to achieve energy conversion efficiencies greater than 60%. Their recommendations were improvements in operating process parameters for both gas turbine power plant and steam power plant by raising the gas turbine inlet temperature to 1427° C, introducing advanced cooling techniques in the gas turbine, utilisation of both cycles heat losses through greater integration between the two plants, and improving component efficiencies. Sarabchi and Polley (1994) examined the effect of key operating variables like compressor pressure ratio, turbine inlet temperature and heat recovery boiler pressure on the performance parameters of a simple combined cycle and comparison was made to a simple gas turbine cycle. Both thermal efficiency and specific net work output were examined as compressor pressure ratio and recovery boiler pressure were varied for each turbine inlet temperature. They concluded the followings:

For any given gas turbine inlet temperature, combined cycle maximum efficiency occurred at pressure ratios which were considerably less than those suitable for corresponding simple gas turbine maximum efficiency.

i. The combined cycle optimum pressure ratio is almost equal to the simple gas turbine optimum pressure ratio for maximum work output.

ii. The values of optimum pressure ratio and heat recovery boiler pressure for a combined cycle increased by increasing the gas turbine inlet temperature.

Shi, Agnew, and She (2011) presented a liquefied natural gas (LNG) gasification and power generation system integrated with a combined cycle power plant. A parametric analysis has been performed for the proposed combined system and they claimed that the net electrical efficiency and the work output of the proposed combined cycle can be increased by 3.8 per cent and 15.6 MW above those of the conventional combined cycle. Da Cunha Alves, De Franca Mendes Carneiro, Barbosa, Travieso, Pilidis, and Ramsden (2001) presented the concepts of intercooling and reheat for gas turbines, in a systematic way, using a model that includes the losses arised from turbomachinery and blades cooling, in order to evaluate their effects on the engine performance. They concluded that Intercooling promises large improvements in efficiency over the simple cycle, especially at high pressure ratios. Reheat on the other hand is much more suited to combined cycles.

Chodkiewicz, Krysinski, and Porochnicki (2002) presented two examples of applications of a recuperated gas turbine incorporating external heat sources in the combined gas-turbine cycle and have been analyzed from the economic and ecological viewpoint. Andreades, Dempsey, and Peterson (2014) discussed the necessary modifications and issues for coupling an external heat source to reheat air combined cycles (RACC). With the open-air configuration used in RACC power conversion, the ability to also inject natural gas or other fuel to boost power at times of high demand provides the electric grid with contingency and flexible capacity while also increasing revenues for the operator. This combination provided several distinct benefits over conventional stand-alone natural gas combined cycle and peaking plants.

Korakianitis, Grantstrom, Wassingbo, and Massardo (2005) used a computer program to evaluate the performance of various combined power plants using standard inputs for component efficiencies, and the design point for these plants is computed. It was found that the performance of the simple cycle gas turbine dominates the overall plant performance (plant efficiency and power). Furthermore, optimum parameters for the power plant based on design point power, hot water demand, and efficiency were shown. Gülen and Joseph (2011) described a simple, physics-based calculation method to estimate the off-design performance of a combined cycle power plant. A second law based approach to off-design performance estimation is found to be a highly viable tool for plant engineers and operators in cases where calculation speed with a small sacrifice in fidelity is of prime importance. Bassily (2015) studied the impacts of varying ambient temperature and relative humidity on the performance of the commercial combined cycle with the different gas turbine inlet-cooling techniques. The results indicated that the introduced innovative techniques for gas turbine inlet-cooling were the most suitable for hot and humid climates with improvements of up to 1.2 percentage point in the combined cycle efficiency and 1.4% in the power output when compared with the refrigeration and absorption gas turbine inlet-cooling systems.

Chiesa and Macchi (2005) investigated three different approaches to break 60% efficiency in combined cycle power plants. These are the conventional open-loop air cooling; the closed-loop steam cooling for stator vanes and blades; and the use of two independent closed-loop circuits (steam for stator vanes and air for rotor blades). Thermodynamic analysis showed that efficiency higher than 61% can be achieved in the frame of current available technologies. Bianci, Melino, and Peretto (2006) presented a study on the effect of both inlet evaporative and overspray fogging on a wide range of combined power plants utilizing gas turbines. Results showed that high pressure inlet fogging influences performance of a combined cycle power plant. Palestra, Barigozzi, and Perdichizzi (2008) studied the effect of air inlet cooling systems based on cool thermal storage, applied to a combined power cycles. They investigated two systems, namely, ice harvester and stratified chilled water, and considering different plant location sites to investigate the influence of climatic conditions. Their results showed that both systems performed similarly in terms of gross extra production of energy. However, the ice harvester showed higher parasitic load due to chiller consumption, and warmer climates of the plant site resulted in a greater increase in the amount of operational hours than power output augmentation. Rahim (2012) carried out a performance analysis of a combined cycle gas turbine power plant with various inlet air-cooling systems, such as evaporative cooling, fogging, mechanical (electric) chiller cooling, and absorption cooling. The performance characteristics were determined for a set of actual operational parameters including ambient conditions (temperature and relative humidity), turbine inlet temperature, pressure ratio, etc. The results showed that power and efficiency improvements depend on ambient air temperature. In addition, by decreasing the ambient temperature and increasing the humidity of the air, the output power can be increased.

Gülen and Smith (2010) carried out a study on the Rankin bottoming cycle (RBC) efficiency of the combined cycle (CC). They developed a CC_RBC performance model based on the second law and exergy concept. This model can enable the engineers to accurately estimate the performance that can be expected from a RBC for a given gas turbine exhaust gas temperature. Further work by Gulen S. C. (2011) to investigate the effect of auxiliary power consumption on combined cycle power plant efficiency. The results showed that the two largest contributors of auxiliary systems that effects the thermal efficiency of the combined plant are the boiler feed pumps and the heat rejection system.

The different combined cycle parametric studies, reviewed in the literature, gave different conclusions about the optimum conditions for the gas turbine cycle as part of the combined power and power cycle. Therefore, the choice of optimum parameters for the gas turbine plant operating in the CPP plant environment appears to be a matter of personal preference. Also, the range of gas turbine power plant design parameters, particularly cycle pressure ratio, depends whether the plant is to be designed for maximum thermal efficiency or maximum specific work output. Therefore, the choice of optimum parameters between the maximum efficiency and the maximum specific work depends on the application.

The main aim of the work reported in this paper was to investigate the potential gains that might be made in the overall efficiency of electrical power generation by combining gas turbine and steam turbine driven plants. Generally, the power generating plants efficiencies can be increased either by the utilization of cogeneration cycle (power and heat (CHP) plants or by combined cycle power (CPP) plants. Although the former is of simple structure and of higher thermal efficiency than the latter, this study is concerned mainly with latter type of plants (CPP). This is mainly due to the need for electricity and the hot climate nature of the developing countries.

2. Thermodynamic Analysis

The gas and steam turbine plants (CCP) represent a complex system consisting of a number of rotational and stationary parts, each part is characterised by its own behaviour. The overall performance of these plants depends on the performance of its individual components and component matching. In a combined cycle power CPP plant, the gas turbine power plant produces electricity as well as exhaust heat that can be used to produce high pressure steam to operate a steam turbine plant and thus generate more electricity. Therefore, the design of a CPP plant will involve greater complexity especially because of the coupling between the two different types of power producing systems. Obviously the parametric study of the combined plant will be the first step in deciding the design criteria of both plants by understanding the influence of the main parameters on the CPP plant. Different configurations of CPP plant can be constructed as described hereafter:

i. Simple gas turbine cycle combined with simple steam cycle.
ii. Simple gas turbine cycle combined with dual pressure steam cycle.
iii. Simple gas turbine cycle combined with dual pressure steam cycle.
iv. Reheat gas turbine cycle combined with simple steam cycle.
v. Gas turbine Intercooling cycle combined with simple steam cycle.
vi. Gas turbine Intercooling cycle combined with dual pressure steam cycle.

In the current parametric study different configurations for the combined cycle were investigated. The combined simple gas turbine with single pressure steam cycle was thermodynamically analysed. The same analysis procedure may be applied to any of the configurations listed above. It should be noted here that the analysis presented in the following sections is for the completeness of the work and for comparison purposes.

2.1 Gas Turbine Power Plant Cycle Analysis

A simple gas turbine plant was depicted schematically in Figure 1. This operates on the Joule/Brayton cycle and represented on the temperature-entropy diagram as shown in Figure 2.

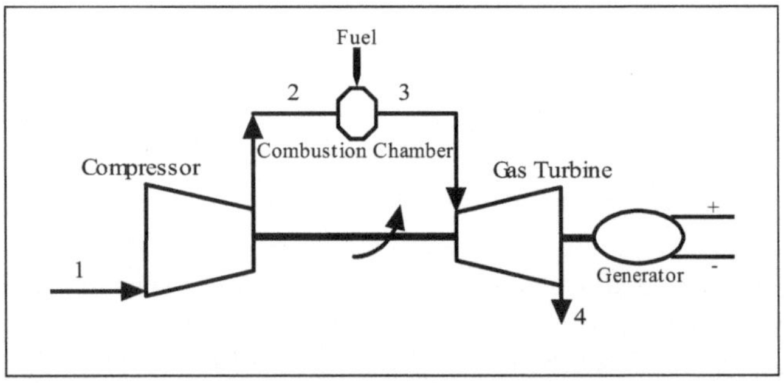

Figure 1. Schematic diagram of a simple gas turbine power plant

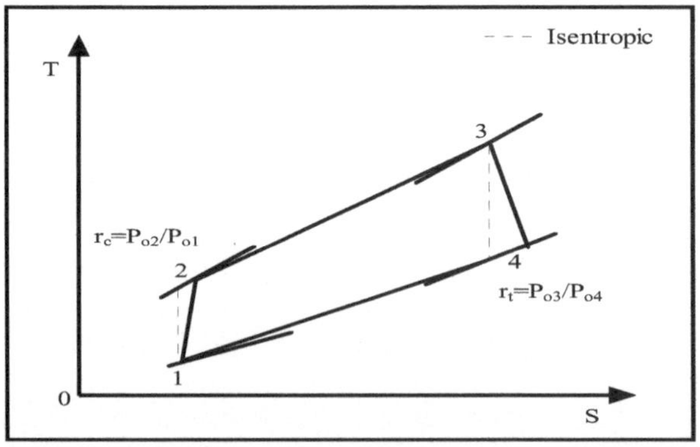

Figure 2. Temperature-Entropy diagram of the gas turbine cycle

2.1.1 Specific Heats of Air and Combustion Gases

The thermodynamic properties of combustion gases and air at various stages throughout the gas turbine cycle are calculated by considering variation of temperature but without dissociation. Tables containing the values of the specific heats against temperature variation have been published in many references such as Chappel and Cockshutt (1974). In the present work, to compute the values of specific heats at constant pressure and various temperatures for air and combustion gases, data from the tables were fitted with polynomial curves to obtain Equations 1 to 5. These equations provide details of the polynomials. Here T_a and T_g refer to the average temperatures during the compression and expansion processes in the compressor and turbine respectively.

For air at low temperature range of 200 to 800 K

$$C_{Pa} = 1.0189\times10^3 - 0.13784T_a + 1.9843\times10^{-4}T_a^{\ 2} + 4.2399\times10^{-7}\,T_a^3 - 3.7632\times10^{-10}\,T_a^4 \quad (1)$$

For air at high temperature range of 800 to 2200 K

$$C_{Pa} = 7.9865\times10^2 + 0.5339T_a - 2.2882\times10^{-4}T_a^{\ 2} + 3.7421\times10^{-8}\,T_a^3 \quad (2)$$

For specific heats of products of combustion

$$C_{Pg} = C_{Pa} + (f/(1+f))B_T \quad (3)$$

Where B_T at low temperature range of 200 to 800 K

$$B_T = -3.59494\times10^2 + 4.5164T_g + 2.8116\times10^{-3}T_g^{\ 2} - 2.1709\times10^{-5}T_g^3 + 2.8689\times10^{-8}T_g^4 - 1.2263\times10^{-11}T_g^5 \quad (4)$$

and B_T at high temperature range of 800 to 2200 K

$$B_T = 1.0888\times10^3 - 0.1416T_g + 1.916\times10^{-3}T_g^{\ 2} - 1.2401\times10^{-6}\,T_g^3 + 3.0669\times10^{-10}\,T_g^4 - 206117\times10^{-14}\,T_g^5 \quad (5)$$

2.1.2 Thermodynamic Governing Equations

i. Air intake Process

$$P_{o1} = (1-\xi_{loss})P_{atm} \quad (6)$$

ii. Air Compression Process

Compression work is given by:

$$W_c = \dot{m}_a C_{Pa}\frac{T_{o1}}{\eta_c}\left[\left(\frac{P_{o2}}{P_{o1}}\right)^{\gamma_a-1/\gamma_a} - 1\right] \quad (7)$$

Where the final stagnation temperature in the compression process (T_{o2}) equal

$$T_{o2} = T_{o1} + \frac{T_{o1}}{\eta_c}\left[\left(\frac{P_{o2}}{P_{o1}}\right)^{\gamma_a-1/\gamma_a} - 1\right] \quad (8)$$

Similarly the final stagnation temperature T_{o2} at the end of the compression process cannot be computed directly from Equation 8. This is because the air specific heat ratio γ_a is a function of the mean stagnation temperature across the compression process. Therefore, in order to compute T_{o2} , an iterative method was used.

This method, named as AIRPROP, has been developed in the form of a computer subroutine. The flowchart of AIRPROP is shown in Figure 3.

iii. Combustion Process

The fuel to air ratio (f) in the combustion chamber is given by

$$f = \frac{1}{\dfrac{\eta_{cc}(LCV)}{C_{Pg}(T_{o3}-T_{o2})} - 1} \tag{9}$$

The pressure loss in the combustion chamber (ξ) is a percentage constant value from the inlet pressure where

$$P_{o3} = (1-\xi_{cc})P_{o2} \tag{10}$$

Due to a considerable rise of the gas temperature through the combustion chamber, an assumption of constant specific heat ratio for the whole range of combustion temperatures could lead to appreciable errors in computing the combustion chamber pressure and temperature. For this reason an average specific heat ratio computed by using an average combustion chamber temperature was used in the current work.

For the same reason the fuel/air ratio f cannot be computed directly by using Equation 9 because the specific heat at constant pressure C_{pg} of the combustion gases is a function of the mean stagnation temperature across the combustion chamber. An iterative method was used to compute fuel/air ratio f . The method was developed for the purpose of this study in the form of a computer subroutine named FARATIO. The flowchart of the subroutine FARATIO is shown in Figure 4.

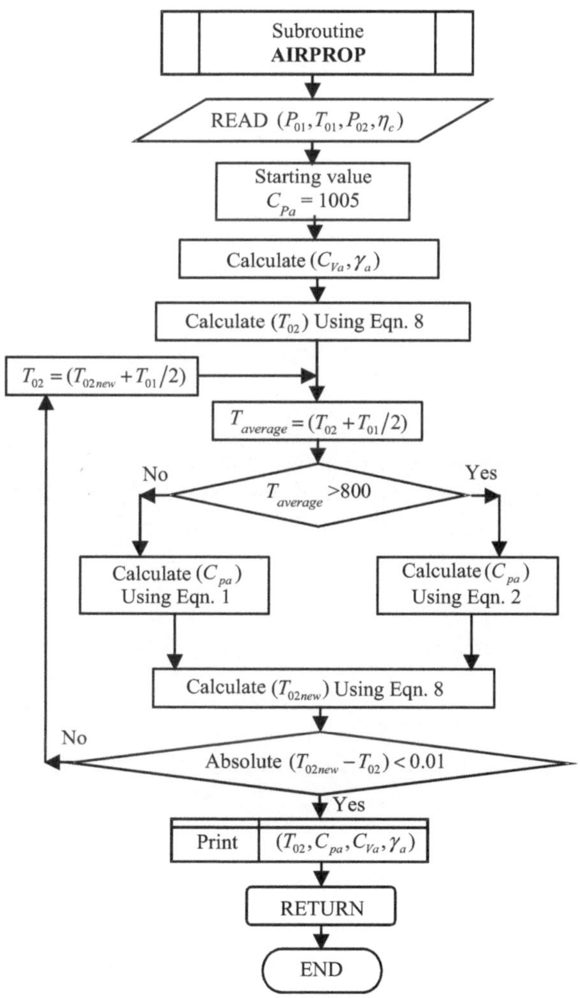

Figure 3. The flowchart for computing $(T_{o2}, C_{Pa}, C_{Va}, \gamma_a)$

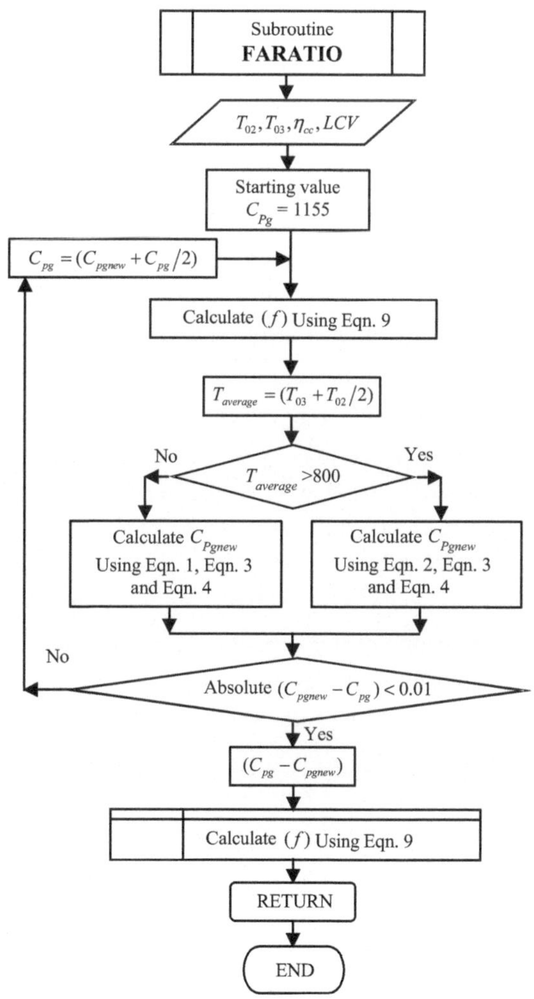

Figure 4. The flowchart for computing fuel to air ratio (f)

iv. Gas Expansion Process

The turbine power (W_t) can be described as

$$W_t = (1+f)\dot{m}_a C_{Pg} \eta_t T_{o3} \left[1 - \left(\frac{P_{o4}}{P_{o3}} \right)^{\gamma_g - 1 / \gamma_g} \right] \tag{11}$$

Where the exhaust stagnation temperature in the expansion process (T_{o4}) equal

$$T_{o4} = T_{o3} - \eta_t T_{o3} \left[1 - \left(\frac{P_{o4}}{P_{o3}} \right)^{\gamma_g - 1 / \gamma_g} \right] \tag{12}$$

For the reasons explained earlier in this section, the values of the variable specific heat ratio were computed by using the mean temperature across the turbine. A computer subroutine named GASPROP was developed to calculate mean specific heat. The flowchart of subroutine GASPROP is similar to subroutine AIRPROP.

Thermal efficiency of the gas turbine cycle η_{gt} and work output W_{gt} can be calculated from Equation 13a and Equation 13b

$$W_{gt} = W_t - W_c \tag{13a}$$

$$\dot{W}_{gt} = W_{gt} / \dot{m}_a \tag{13b}$$

$$\eta_{gt} = W_t - W_c / f\, \dot{m}_a (LCV) \tag{14}$$

For the reasons explained earlier in this section, the values of the variable specific heat ratio were computed by using the mean temperature across the turbine. A computer subroutine named GASPROP was developed to calculate mean specific heat. The flowchart of subroutine GASPROP is similar to subroutine AIRPROP.

2.2 Steam Turbine Power Plant Cycle Analysis

The schematic diagram of a simple steam power plant operating on Rankine cycle is depicted in Figure 5, and its corresponding cycle on the temperature entropy diagram is presented in Figure 6.

2.2.1 Thermodynamic Governing Equations

a. Heat recovery steam generator (HRSG)

The energy balance in the steam generator can be expressed as follows:

$$\dot{m}_a (1+f) C_{Pg} (T_{o4(gt)} - T_{o6(gt)}) \eta_B = \dot{m}_{st} (h_{2(st)} - h_{1(st)}) \tag{15}$$

The gas stack temperature ($T_{o6(gt)}$) should be kept as low as possible, but at the same time condensation should be avoided. The lowest stack temperature is determined by the fuel type used, for instance sulfuric fuels should have a higher stack temperature. The temperature-heat diagram of the heat recovery steam generator is shown in Figure 7.

The heat added to the water or the steam was supplied in three steps:

i. The economizing step where the temperature of water rises from $T_{1(st)}$ to the saturation liquid temperature at that boiler pressure.

ii. The evaporation step where the water absorbs heat at constant temperature.

iii. The superheating step where the temperature of steam increases from the saturation temperature to the desired maximum superheated temperature ($T_{2(st)}$).

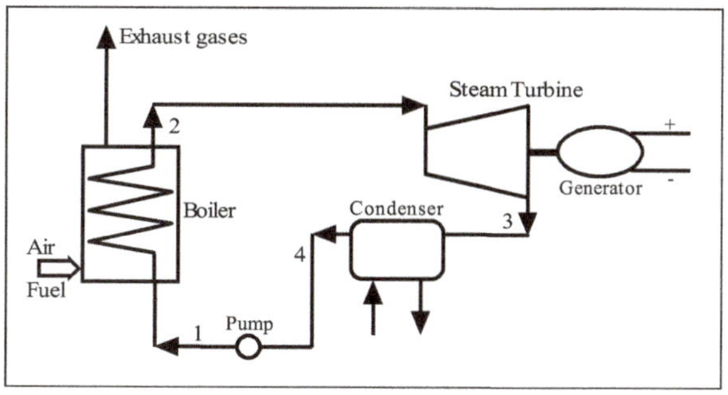

Figure 5. Schematic diagram of a simple steam power plant

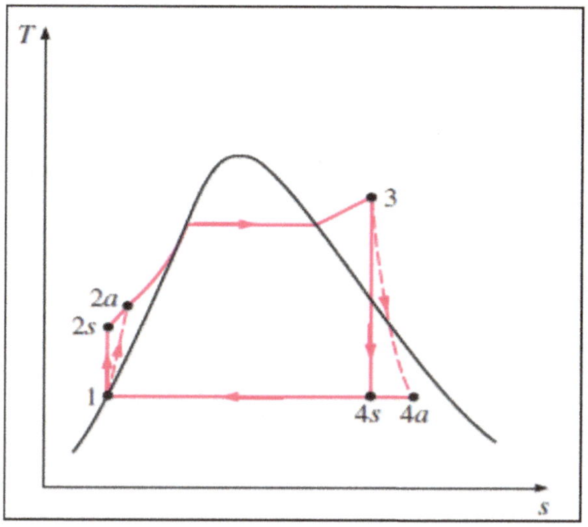

Figure 6. Temperature-Entropy diagram of the simple steam turbine cycle

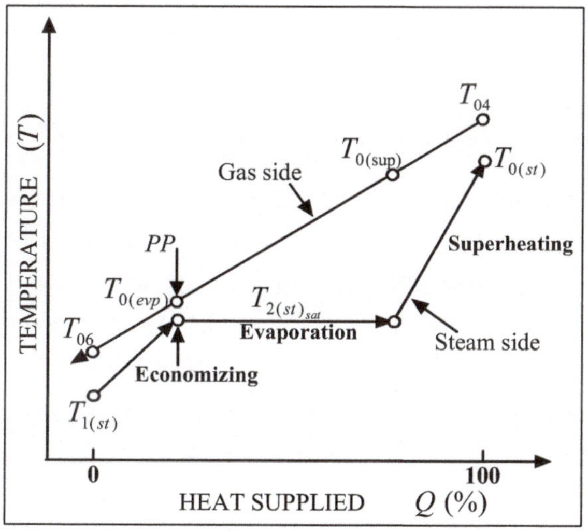

Figure 7. Temperature variation in the heat recovery boiler

$$T_{2(st)} = \varepsilon_{\sup}(T_{o4(gt)} - T_{2(st)sat}) + T_{2(st)sat} \tag{16}$$

Where $T_{2(st)sat}$ is the saturated temperature at the $P_{2(st)}$

The enthalpy of the steam at the exit of the boiler $(h_{2(st)})$ is

$$h_{2(st)} = f(T_{2(st)}, P_{2(st)}) \tag{17}$$

and

$$T_{o(evp)} = PP + T_{2(st)sat} \tag{18}$$

where $T_{o(evp)}$ is the gas temperature at the exit of the evaporator.

$$\dot{m}_{st} = \frac{\dot{m}_a(1+f)\eta_B C_{Pg}(T_{o4} - T_{o(evp)})}{h_{2(st)} - h_{2(lg)sat}}$$

$$\dot{m}_{st} = \frac{\dot{m}_a(1+f)\eta_B C_{Pg}(T_{o4} - T_{o(evp)})}{h_{2(st)} - h_{2(lg)sat}} \tag{19}$$

Where $h_{2(lg)sat}$ is the saturated liquid enthalpy at $P_{2(st)}$.

The enthalpy of the water at the exit of the pump $(h_{1(st)})$ is

$$h_{1(st)} = f(P_{1(st)}, S_4) \tag{20}$$

And

$$T_{o6} = T_{o(evp)} - \frac{\dot{m}_{st}(h_{2(lq)sat} - h_{1(st)})}{\dot{m}_a(1+f)C_{Pg}} \tag{21}$$

The enthalpy of the steam at the exit of the steam turbine ($h_{3(st)}$) is

$$h_{3(st)} = f(P_{4(st)}, S_3) \tag{22}$$

The enthalpy of the water at the exit of the condenser ($h_{4(st)}$) is

$$h_{4(st)} = f(P_{4(st)}) \tag{23}$$

i. The heat exchange process in a counter flow heat recovery steam generator must satisfy the following conditions:

ii. The gas stack temperature $(T_{o6(gt)})$ must be greater than the inlet water temperature $(T_{1(st)})$ at least by $10^\circ C$.

iii. The gas temperature at the outlet of the evaporator $(T_{o(evp)})$ must be greater than the liquid saturation temperature of the steam $(T_{2(st)sat})$ by a minimum value (pinch point temperature difference (PP)).

iv. The superheated steam temperature $(T_{2(st)})$ must be less than the gas turbine exhaust temperature $(T_{o4(gt)})$.

The gas loss in the heat recovery steam generator is a percentage value from the atmospheric pressure where the heat recovery steam generator inlt could be expressed as a function of the atmospheric pressure as:

$$P_{04} = (1 - \xi_{HRSG})P_{atm} \tag{24}$$

b. Feed Water Pump Power

The feed water pump power (W_P) can be described as

$$W_P = \frac{\dot{m}_{st}}{\eta_P}\left(h_{1(st)} - h_{4(st)}\right) \tag{25}$$

Where, the entropy at state 1 equals the entropy at state 4.

c. Steam Expansion

The steam turbine power ($W_{t(st)}$) can be described as

$$W_{t(st)} = \dot{m}_{st}\eta_{t(st)}\left(h_{2(st)} - h_{3(st)}\right) \tag{26}$$

Where the entropy at state 2 equals the entropy at state 3.

The efficiency of the steam turbine cycle (η_{st}) can be described as:

$$\eta_{st} = \frac{W_{st}}{Q_B} = \frac{W_{t(st)} - W_P}{\dot{m}_a(1+f)C_{pg}(T_{o4} - T_{o6})} \tag{27}$$

i.e. $$\eta_{st} = f(\eta_{stt}, \eta_p, P_B, P_{com}, T_{2st}) \tag{28}$$

2.3 Combined Cycle Power Plant Analysis

The efficiencies given in Eqs 14 and 28 shows that modifications and improvements to the gas turbine and steam turbine power plants would increase their efficiency. However, the cost of such modifications may be high because they invariably necessitate installation of new components. The alternative is to use the heat rejected by the gas turbine, hereafter named as the higher cycle (H), may be used to raise high-pressure steam, which is expanded in the steam turbine, hereafter known as the lower cycle (L). A block diagram of a simple combined power and power (CPP) is shown in Figure 8.

The gas turbine plant (H) had a thermal efficiency of η_{gt}, absorbs heat of Q_H to produce work W_{gt} and rejects the exhaust heat of Q_{HL}. The steam turbine (L) plant had a thermal efficiency of η_{st}, absorbs some of the heat Q_L rejected from the upper plant and produces work of W_{st} and rejects heat Q_{Latm} to the atmosphere. A supplementary heat (Q_{add}) can be added between the two power plants while a heat (Q_{loss}) can be lost at that point.

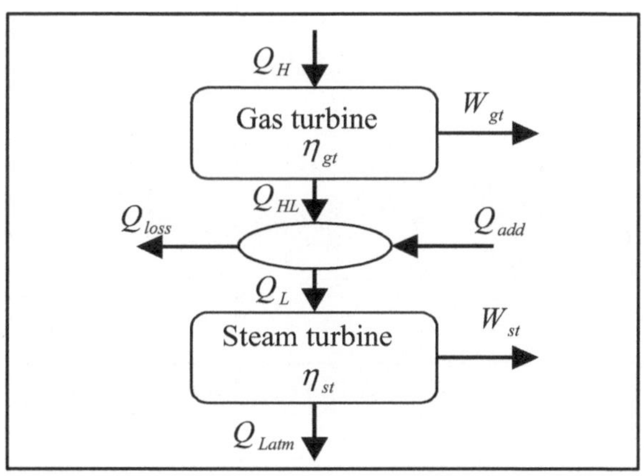

Figure 8. Two power plants in series with heat losses

Thermal efficiency of gas turbine power plant is

$$\eta_{gt} = W_{gt}/Q_H \tag{29}$$

Thermal efficiency of steam turbine power plant is:

$$\eta_{st} = W_{st}/Q_L \tag{30}$$

A schematic diagram of the CPP plant consisting of simple gas turbine combined with simple steam plant is shown in Figure 9 where the gas turbine exhaust will be used as the heat source of the steam power plant; an afterburner can be used to raise the gas turbine exhaust temperature.

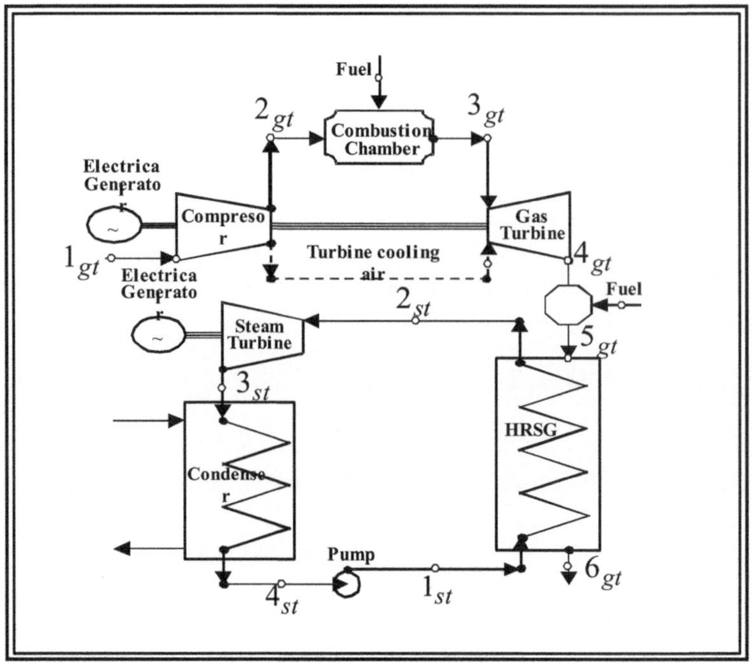

Figure 9. Schematic diagram of a combined gas-steam power plant with a waste heat recovery steam generator

The temperature-entropy diagram of the combined plant is shown in Figure 10

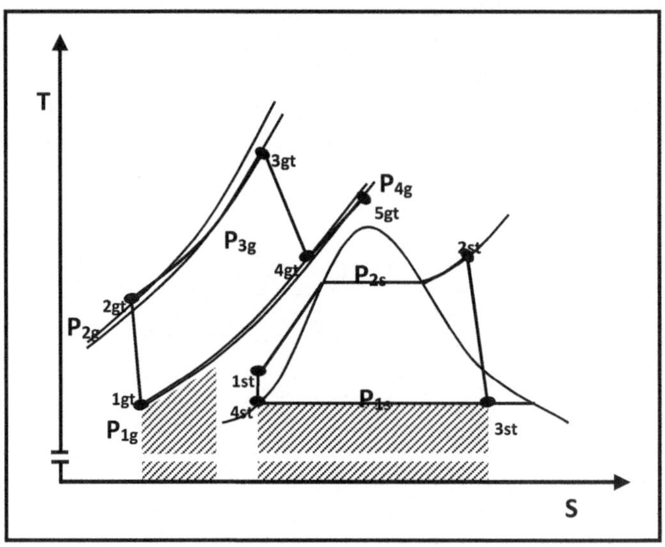

Figure 10. Temperature-Entropy diagram of a combined gas-steam power cycle

3.1 Thermodynamic Governing Equations of CPP Plant

The CPP cycle thermodynamic analysis can be simplified by making the following assumptions:

i. The air used by the gas turbine as well as the products of the combustion are perfect gases.

ii. The specific heat capacities can be constant through the process and represented at the average temperature of that process.

iii. The loss of stagnation pressure in the compressor inlet is a constant percentage of the compressor inlet pressure.

iv. The loss of stagnation pressure in the combustion chamber is a constant percentage of the combustion chamber inlet pressure.

The specific work of the gas turbine cycle $\dot{W}_{gt}$ can be described as

$$\dot{W}_{gt} = \frac{W_{gt}}{\dot{m}_a} = \frac{W_t - W_c}{\dot{m}_a} \quad (31)$$

And the heat supplied to the steam power plant is given by:

$$Q_L = Q_{HL} + Q_{add} - Q_{loss} \quad (32)$$

Thermal efficiency of the combined power plant is given by:

$$\eta_{CPP} = \frac{W_{gt} + W_{st}}{Q_H + Q_{add}} \quad (33)$$

or

$$\eta_{CPP} = \frac{W_{gt} + W_{st}}{\dot{m}_f (LCV)} = \frac{W_{gt} + W_{st}}{\dot{m}_a f (LCV)} \quad (34)$$

Using Esq. 29, 30, 31 and 32, η_{cc} can be developed to

$$\eta_{CPP} = \frac{\eta_{gt} + \eta_{st}\left(\frac{Q_{add}}{Q_H} + 1 - \eta_{gt} - \frac{Q_{loss}}{Q_H}\right)}{1 + \frac{Q_{add}}{Q_H}} \quad (35)$$

If there isn't any supplementary heating $(Q_{add} = 0)$ and no heat loss $(Q_{loss} = 0)$ then

$\eta_{CPP} = \eta_{gt} + \eta_{st} - \eta_{gt}\eta_{st}$ (36)

Also, the specific work of the combined cycle $\dot{W}_{CPP}$ can be described as

$$\dot{W}_{CPP} = \frac{W_{CPP}}{\dot{m}_a} = \frac{W_{gt} + W_{st}}{\dot{m}_a} \quad (37)$$

Based on the previous thermodynamic analysis, it can be concluded that the thermal efficiency and specific work of the CPP plant are functions of many parameters as described by Equation 38 and Equation 39 where for each set of values of the parameters in these equations, there is only one solution for the η_{CPP} and only one solution for $\dot{W}_{CPP}$.

$$\dot{W}_{CPP} = f(\theta, r, C_p, \eta_c, \eta_{cc}, \eta_{(gt)}, \eta_{mech}, P_{con}, T_{2st}, P_{2st}, \eta_{t(st)}, \eta_p, ...) \quad (38)$$

$$\eta_{CPP} = f(\theta, r, C_p, \eta_c, \eta_{cc}, \eta_{(gt)}, \eta_{mech}, P_{con}, T_{2st} P_{2st}, \eta_{t(st)}, \eta_p, ...) \quad (39)$$

There are many gas turbine /steam turbine combined cycle configurations, therefore studying the effect of each parameter on each configuration performance will be very difficult and tedious to achieve. Simultaneous variations of the main parameters in both cycles would show the effect of these parameters on the combined cycle (CPP) performance. Calculations were made by varying some parameters and holding others constant. However, precautions for the current parametric analysis were taken into consideration as stated hereafter

2.4 Precautions for Using the Parametric Analysis

The parameters can be varied within the following thermodynamic, technological and physical constraints:

i. The temperature ratio θ can have any value starting from the ratio of (T_{o2} / T_{o1}) to a maximum value limited by the metallurgical consideration.

ii. The pressure ratio (r) can have any value starting from one to a maximum value determind by mechnical and aerodynamic factors such as stress and Mach number.

iii. The steam temperature can be assumed all the values from the saturation temperature at that pressure to a maximum value where the maximum value is tied to technological factors.

iv. The steam pressure in the boiler and the condenser pressure are related to the wetness of the steam at the exit of the steam turbine, which should lie between 0.9 and 1.0. this is because wet steam can have detrimental effect on turbine blades.

v. The exhaust gas temperature in the boiler should be higher than the temperature of the steam by a minimum value where this value is dependent on the economic and the design parameters.

vi. The stack exhaust temperature should be higher than the condensation temperature of water vapour in exhaust gas in order to prevent corrosive

Table 1. Assumed parameters' values used in the parametric study

Parameter	Assumed Values	Parameter	Assumed Values
T_{o1} (K)	293	$\eta_{t(gt)}$	0.88
P_{o1} (kPa)	101.3	$\eta_{t(gt)}$	0.88
T_{o3} (K)	(1100 - 1700)	$\eta_{t(st)}$	0.87
r	(4 – 32)	η_B	0.85
$P_{2(st)}$ (bar)	60 - 150	η_P	0.85
$P_{1(st)}$ (bar)	0.05 - 0.5	η_{mec}	0.98
$T_{3(st)\max}$ (K)	800	ξ	7%
PP (K)	15	$\varepsilon_{\sup erheater}$	0.9
$D_{\min}$	0.88	LCV (kJ/kg)	42400
η_c	0.86	η_{cc}	0.98

Using the thermodynamic analysis and parameters limitations, a computer program was written in Visual Basic language to solve Equation 6 to Equation 37 incorporating AIRPROP, FARATIO and GASPROP subroutines. The flowchart of the computer program is shown in Figs. (11a-11c). Figure 11a shows the solution of the gas turbine cycle, Figure (11b) shows the solution of the steam turbine cycle, and Figure (11c) shows the solution of the combined gas steam cycle.

3. Results and Discussion

The primary purpose of the parametric study was to show the influence of either the principal design variables and cycle configurations or the operating conditions on the performance of the gas turbine engine working in series with a steam turbine engine as a part of the combined power and power or CPP plant. It was hypothesised that a combination of a particular set of the design parameters, cycle configuration or the operating conditions that might produce the optimum performance when the gas turbine engine was used of its own for electrical power generation, might not be optimum if the same engine worked in series with a steam turbine in a CPP plant. The reason which gave rise to this hypothesis was that in the CPP plant environment, the factor that must be taken into consideration would be the grade of the thermal energy of gas turbine exhaust indicated by its temperature. The outputs of the computer programs developed for the parametric study were used to generate Figs. (12, 13, 15, 16, 17, 18) for the performance of the gas turbine cycles and Figs. (19 – 34) for the performance of the combined power and power cycles.

3.1 Results of Parametric Study of the Gas Turbine Cycles

The relationship between the gas turbine efficiency η_{gt} and specific work w_{gt} at constant pressure ratio r and turbine inlet temperature T_{o3} are shown in Figure 12. It can be seen that the maximum efficiency points and the maximum specific work output points at different values of constant temperature or pressure ratio are not coincident. The design choice can be either to opt for maximum efficiency (industrial applications with low operating cost) or maximum specific work output (military applications with high power/weight ratio) or any other point that may represent the optimum choice for a particular application.

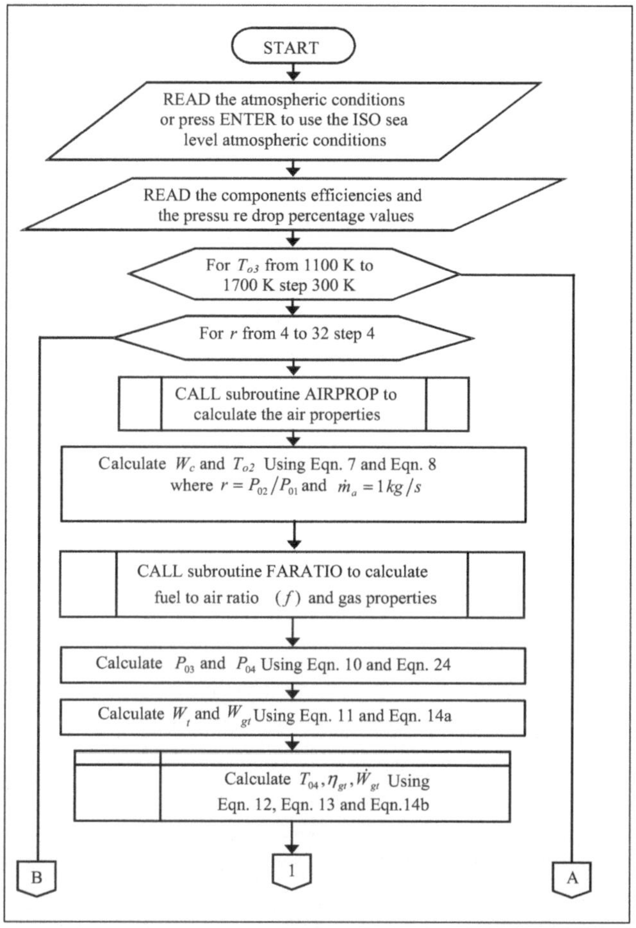

Figure (11a). Flowchart of gas turbine cycle calculations

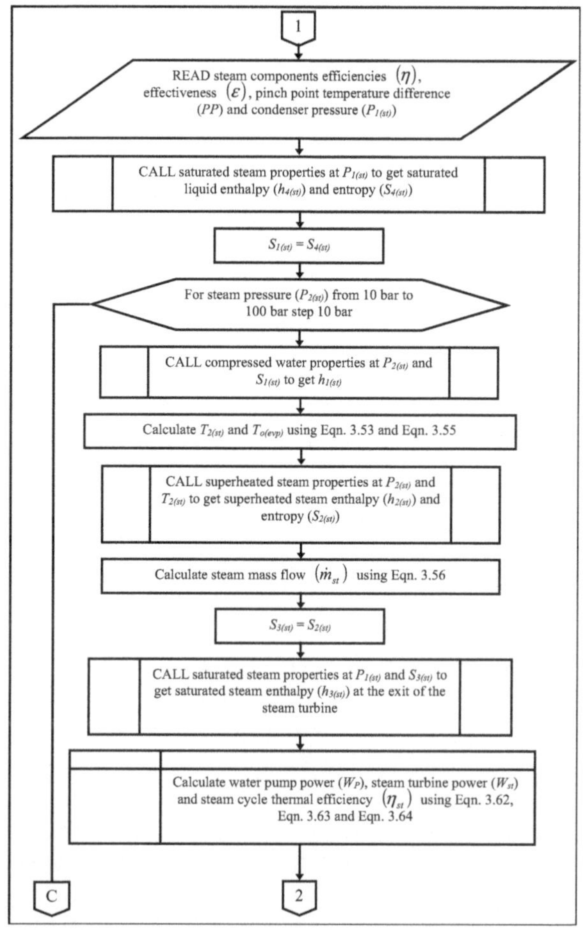

Figure (11b). Flowchart of steam turbine cycle calculations

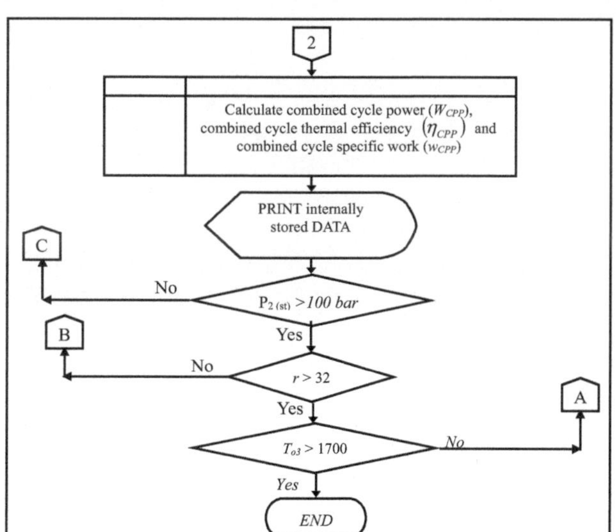

Figure (11c). Flowchart of combined cycle calculations

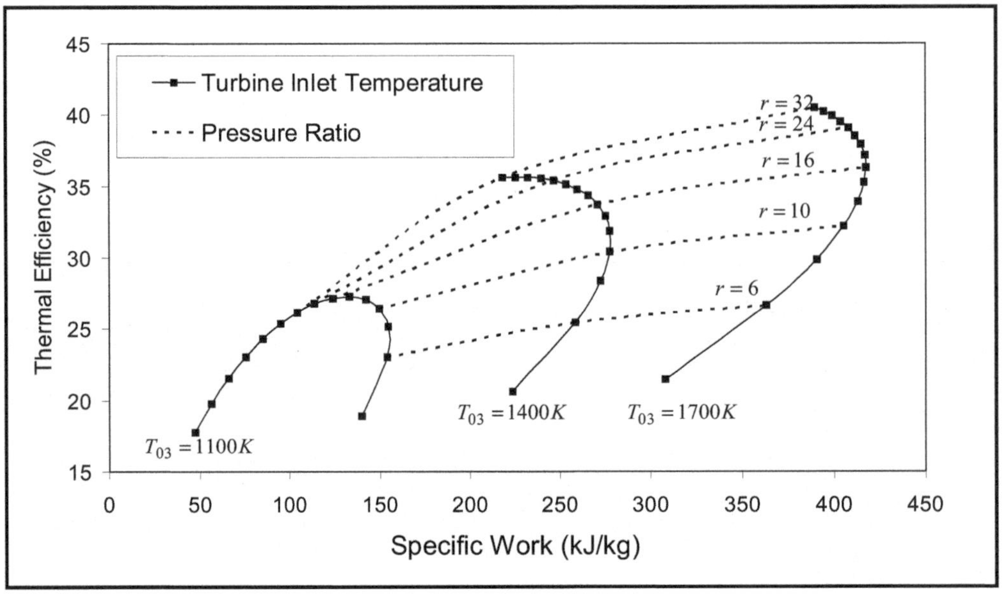

Figure 12. Gas turbine efficiency versus specific work at constant turbine inlet temperatures and pressure ratios

Figure 13 depicts the relationship between the pressure ratio r and the gas turbine exhaust temperature T_{o4} at constant turbine inlet temperature T_{o3}. It can be seen that raising the turbine inlet temperature T_{o3} as well as lowering the compressor pressure ratio r can increase the gas turbine exhaust temperature T_{o4}. It is worth noting that the maximum specific work output points have a higher turbine exhaust temperatures than the maximum efficiency points. Furthermore, $T_{o3} = 1100K$ the higher the pressure ratio the greater would be the difference between the points of maximum thermal efficiency and maximum specific work output. This is because of the increasing divergence between the maximum thermal efficiency and maximum specific work output lines with increasing cycle pressure ratio.

Reheating in the gas turbine cycle increases the specific work output with some reduction of thermal efficiency. This is because reheating of exhaust gas after the first expansion does not contribute in any way to the cycle pressure ratio, as shown in the T-S diagram, Figure 14, consequently the thermal efficiency reduces.

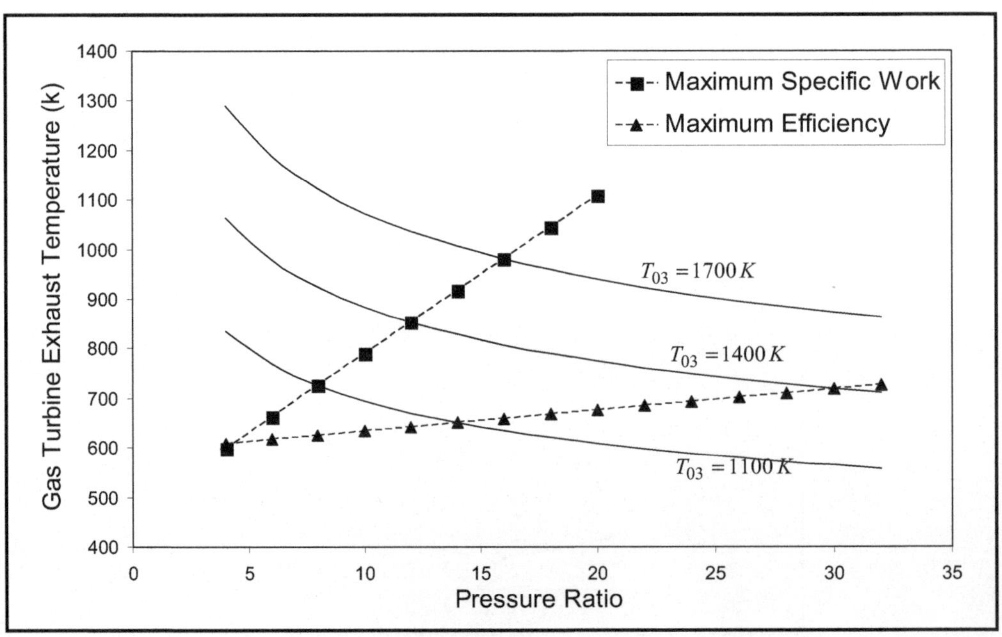

Figure 13. Gas turbine exhaust temperature versus pressure ratio at constant turbine inlet temperatures

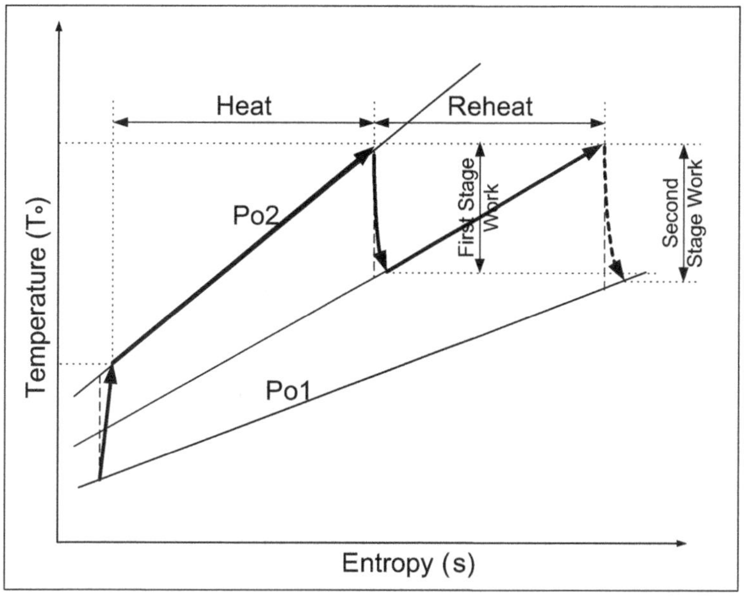

Figure14. Temperature-Entropy diagram of reheat gas turbinecycle

Reheating also tends to reduce the difference between the maximum efficiency points and the maximum specific work points. Hence, for the reheated gas turbine cycle at constant turbine inlet temperature there is only one point for both maximum thermal efficiency and maximum specific work output at each cycle pressure ratio as illustrated in Figure 15.

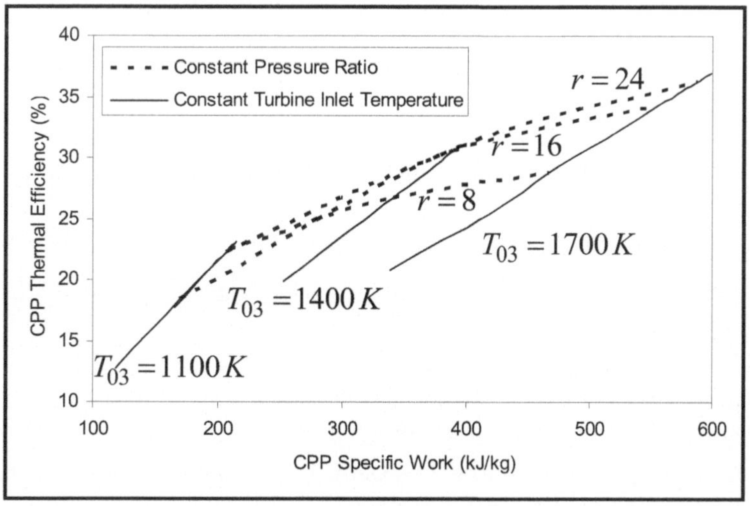

Figure 15. Gas turbine reheat cycle efficiency versus specific work at constant turbine inlet temperatures and pressure ratios

The exhaust gas temperature of the reheat gas turbine cycle can also be plotted against pressure ratio, Figure16. An increase in the exhaust temperature is expected due to reheating but the change in exhaust temperature with the pressure ratio is totally different than that for the simple gas turbine cycle especially at lower values T_{o3}.

The effect of cooling the air before it enters the compressor on the gas turbine exhaust gas temperature is shown in Figure 17. As would be expected, at a constant value of the turbine entry temperature the exhaust gas temperature decreases as the cycle pressure ratio is increased. From the point of view of the combined power and power cycle this is an undesirable feature. However, both the specific work output and the thermal efficiency increase with increasing cycle pressure ratio. This can be seen in Figure 18, which gives a comparison between the simple gas turbine cycle, reheat cycle and the pre-cooling cycle.

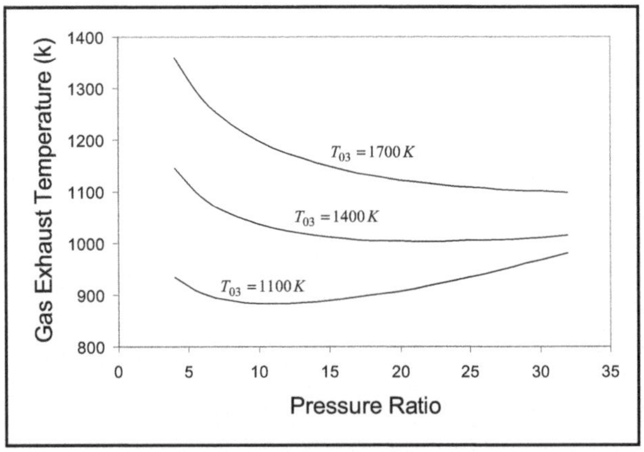

Figure 16. Exhaust temperature of gas turbine reheat cycle versus pressure ratio at constant turbine inlet temperatures

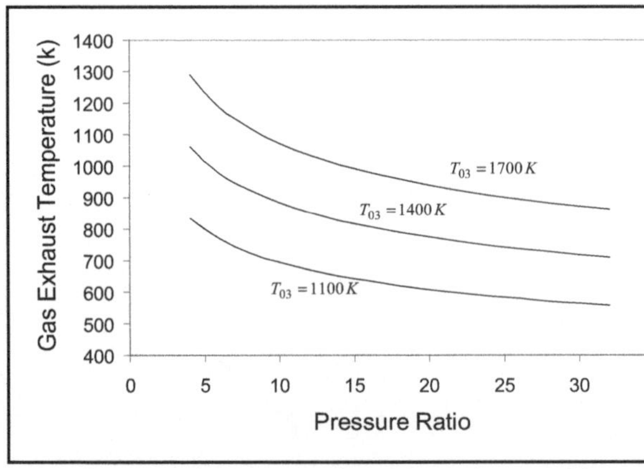

Figure 17. Exhaust temperature of gas turbine pre-cooling cycle versus pressure ratio at constant turbine inlet temperature

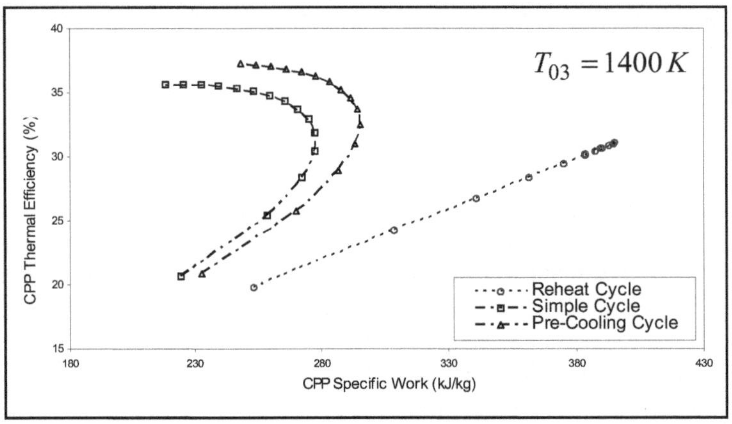

Figure 18. Efficiency vs. specific work output for different gas turbine cycles

3.2 Results of Parametric Study of the CPP Plant Cycles

Several configurations of the gas turbine plant and the steam turbine plant that comprise the CPP plant were studied as follows:

i. Simple gas turbine cycle combined with simple steam cycle

The first set of results, which covers the simple gas turbine cycle combined with simple steam turbine cycle, is given in Figs. 19 to 23. The conditions for calculating the relevant data are shown on each figure.

Figure 19 shows the relationship between the CPP thermal efficiency η_{CPP} and the pressure ratio r at constant turbine inlet temperature T_{o3}. It can be seen that increasing the gas turbine inlet temperature will increase the combined cycle efficiency and for each turbine inlet temperature T_{o3}, there is an optimum pressure ratio value.

Figure 20 depicts the relationship between the CPP specific work output $\dot{W}_{CPP}$ and the pressure ratio r at constant turbine inlet temperature T_{o3}. It can be seen that increasing the gas turbine inlet temperature will increase the specific work output. The maximum specific work output at each turbine inlet temperature will be at low pressure ratio values. This is due to the fact at these values of pressure ratio, the gas turbine cycle will have a higher exhaust temperatures; hence the steam cycle will produce more power output. Consequently the CPP specific work output increases.

Figure 21 is a combination of Figure 19 and Figure 20, which shows the relationship between the CPP of thermal efficiency η_{CPP} and specific work output $\dot{W}_{CPP}$ at various turbine inlet temperature T_{o3}. It is noticeable that the maximum efficiency points and the maximum specific work output points do not coincide.

Figure 22 shows again the relationship between the CPP thermal efficiency η_{CPP} and the pressure ratio r at constant turbine inlet temperature T_{o3}. Furthermore, it shows the maximum CPP efficiency lines at different boiler pressure values of 50 bar and 90 bar. It can be seen that increasing the turbine inlet temperature should be accompanied with an increasing in gas turbine cycle pressure ratio in order to achieve the maximum combined efficiency.

Figure 23 shows the relationship between the CPP thermal efficiency η_{CPP} and the heat recovery boiler pressure P_{2st} at various cycle pressure ratios r and a constant turbine inlet temperature T_{o3} of 1400 K. It can be noted that the heat recovery boiler pressure can be selected from a wide range. Furthermore, it is worth noting that decreasing the heat recovery boiler pressure will have an economic advantage of lowering the combined plant total cost. In particular, the capital cost of building the heat recovery steam generator.

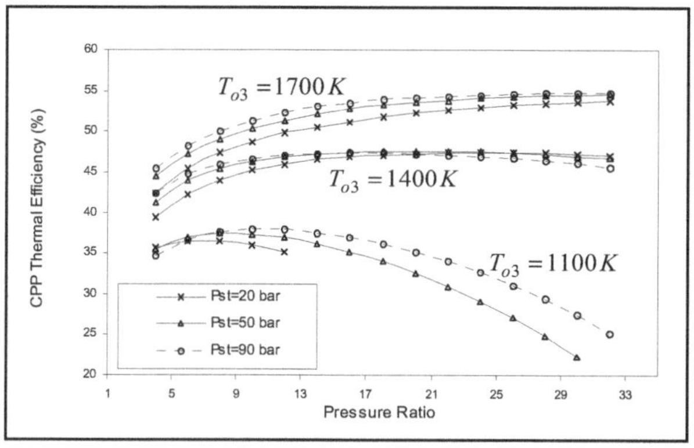

Figure 19. Combined efficiency versus pressure ratio at constant turbine inlet temperatures and boiler pressures (simple combined cycle)

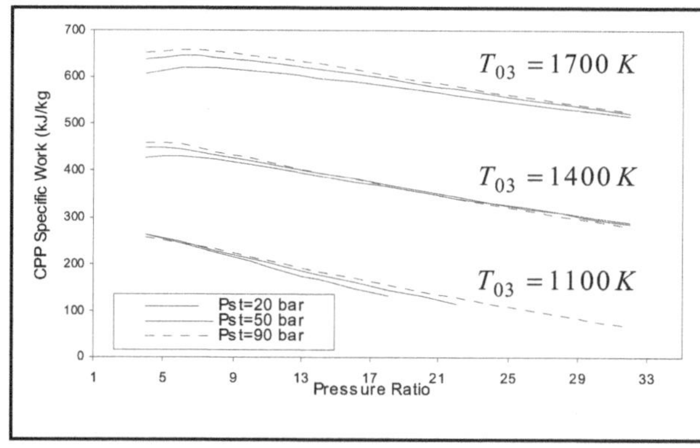

Figure 20. Combined specific work versus pressure ratio at constant turbine inlet temperatures and boiler pressures (simple combined cycle)

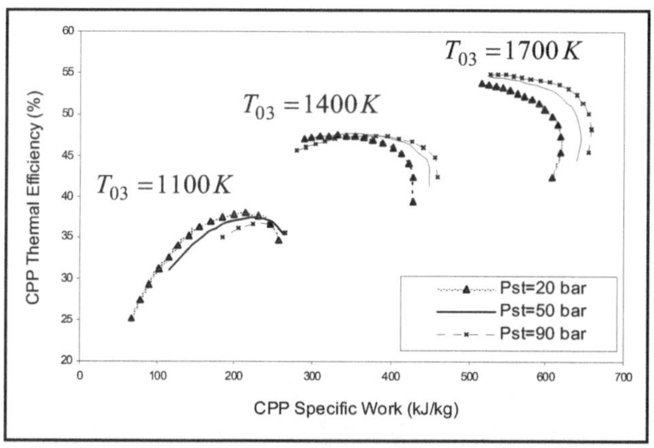

Figure 21. Combined specific work versus combined efficiency at constant turbine inlet temperatures and boiler pressures (simple combined cycle)

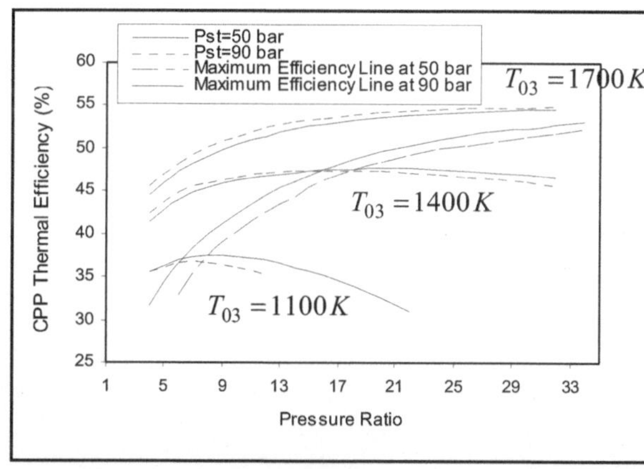

Figure 22. Combined efficiency versus pressure ratio at constant turbine inlet temperatures with maximum efficiency lines (simple combined cycle)

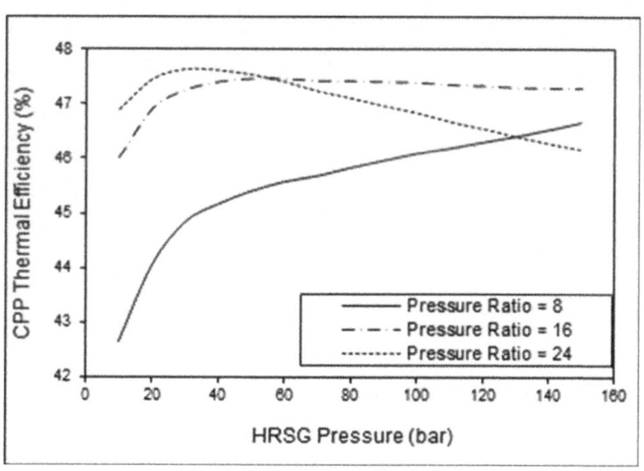

Figure 23. Combined efficiency versus steam boiler pressure at different gas turbine pressure ratios

ii. Reheat gas turbine cycle combined with simple steam cycle analysis.

The second set of results, which covers the reheat gas turbine cycle combined with simple steam turbine cycle, is given in Figures 24 to 26. The conditions for calculating the relevant data are shown on each figure. Reheating the gas turbine cycle reduces the CPP thermal efficiency, see Figure 24 and increases the CPP specific work output, see Figure 25. The reason for that, reheating of gas turbine exhaust after the first gas expansion lower the gas turbine cycle efficiency, as explained in the previous section.

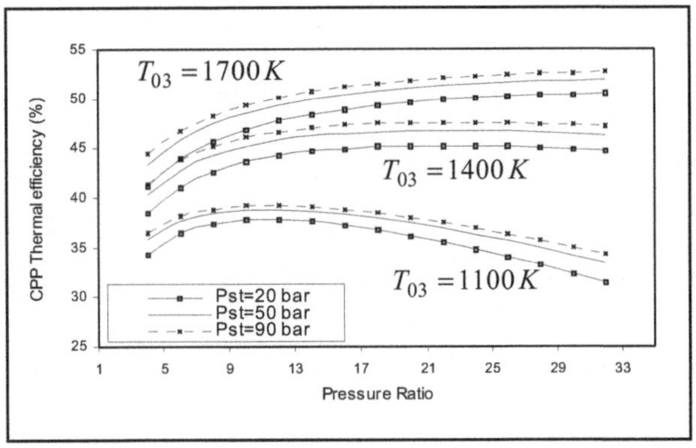

Figure 24. Combined efficiency versus pressure ratio at constant turbine inlet temperatures and steam boiler pressures (gas reheat combined cycle)

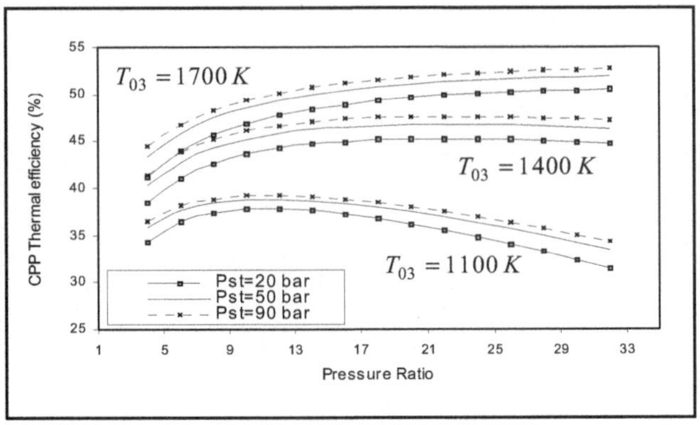

Figure 25. Combined specific work versus pressure ratio at constant turbine inlet temperatures and steam boiler pressures (gas reheat combined cycle)

On the other hand, reheating will increase the gas turbine cycle specific work output and the exhaust gas temperature. Consequently, higher gas turbine exhaust temperature will increase both the steam turbine thermal efficiency and specific work output. Hence, the CPP specific work output increases. However, the increase in the steam turbine efficiency will not compensate for the reduction of gas turbine efficiency, consequently the CPP thermal efficiency reduces, refer to Figure 26.

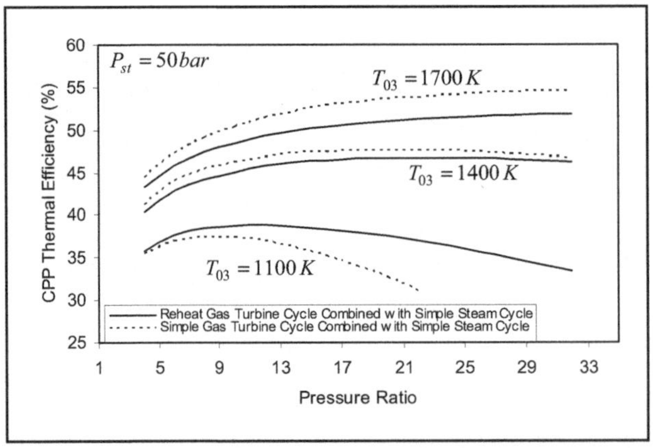

Figure 26. Combined efficiency versus pressure ratio at constant turbine inlet temperatures for simple combined cycle and gas reheat combined cycle

iii. Simple gas turbine cycle combined with dual pressure steam cycle analysis.

The third set of results, which covers the simple gas turbine cycle combined with dual pressure steam turbine cycle, is given in Figs. 27 to 29. The conditions for calculating the relevant data are shown on each figure. Figs. 27 and 28 show that the dual pressure steam cycle will increase both the CPP thermal efficiency and CPP specific work output. It can be also noted that raising the turbine inlet temperature will also increase both the CPP thermal efficiency and CPP specific work output. Furthermore, the effect of higher values of pressure ratio on CPP thermal efficiency is fairly small at constant T_{o3} higher than 1400 K, while the CPP specific work output showed similar trends at all temperatures of T_{o3}.

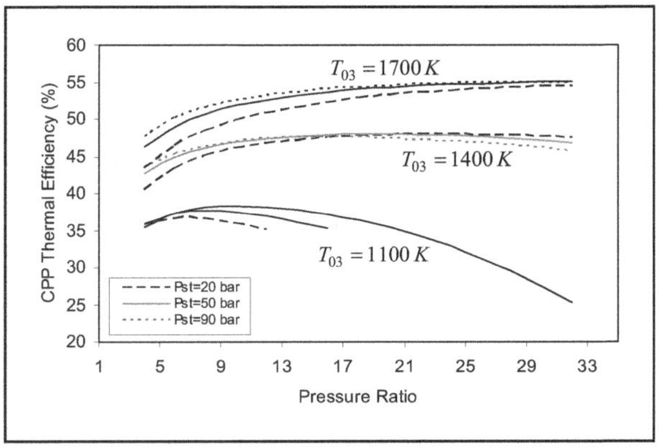

Figure 27. Combined efficiency versus pressure ratio at constant turbine inlet temperatures (simple gas and dual pressure steam combined cycle)

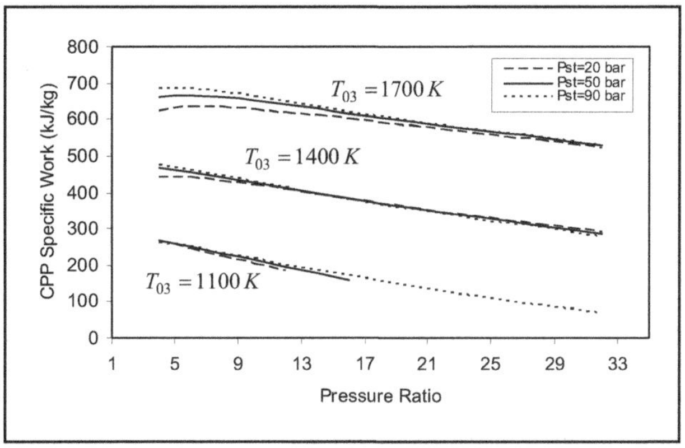

Figure 28. Combined specific work versus pressure ratio at constant turbine inlet temperatures (simple gas and dual pressure steam combined cycle)

Figure 29 shows the comparison of various cycles; (a) Simple gas turbine cycle combined with simple steam cycle; (b) Simple gas turbine cycle combined with dual pressure steam cycle; and (c) Reheat gas turbine cycle combined with simple steam cycle. The trends of CPP thermal efficiency lines for theses cycles are similar. It also shows that Simple gas turbine cycle combined with simple steam cycle has the highest thermal efficiency.

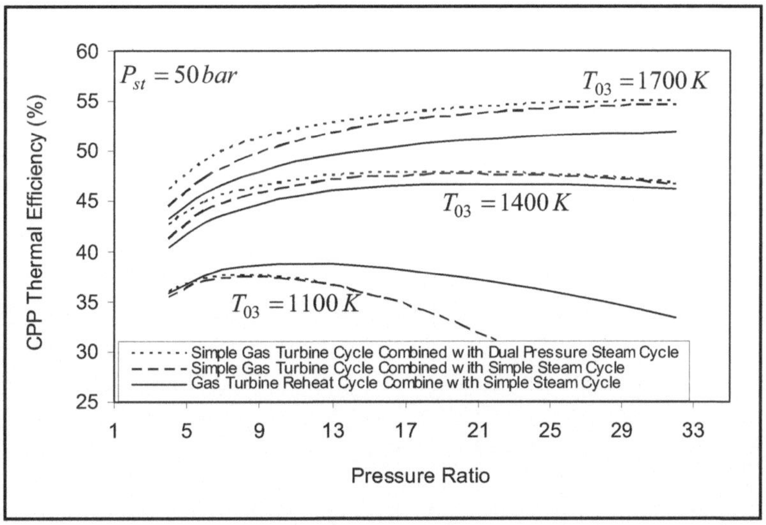

Figure 29. Combined efficiency versus pressure ratio for simple combined cycle, simple gas combined with dual pressure steam cycle and gas reheat combined cycle

vi. Gas turbine Intercooling cycle combined with dual pressure steam cycle analysis.

The fourth set of results, which covers the pre-cooling gas turbine cycle combined with dual pressure steam turbine cycle, is given in Figs. 30 and 31. The conditions for calculating the relevant data are shown on each figure. Figures 30 to 31 show that the pre-cooling gas turbine cycle combined with dual pressure steam turbine cycle will have a slight increase on both the CPP thermal efficiency and CPP specific work output. The trends of CPP for efficiency and specific work output are similar to the other combined cycles, i.e. increasing T_{o3} will increase both the CPP for thermal efficiency and specific work output.

For comparison purposes of various studied cycles, Figure 33 above was drawn at T_{o3} of 1400 K and steam boiler pressure P_{2st} of 50 bar. It shows that the maximum efficiency can be attained with the gas turbine pre-cooling cycle combined with dual pressure steam cycle. However, it is worth mentioning that pre-cooling cycles require the addition of new component, this means adding complexity to the system as well as increasing the capital cost of the plant. Therefore, the selection of such a cycle might incur a heavy economic penalty.

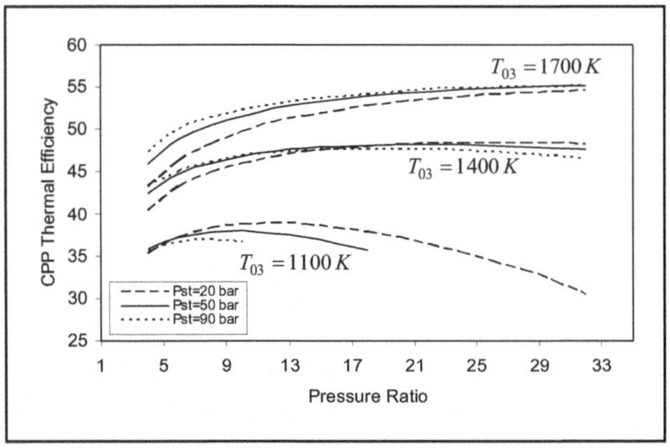

Figure 30. Combined efficiency versus pressure ratio at constant turbine inlet temperatures (gas turbine intercooling cycle combined with dual pressure steam cycle)

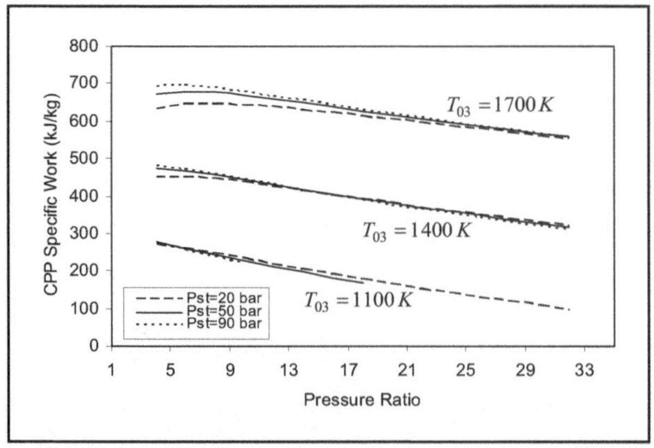

Figure 31. Combined specific work versus pressure ratio at constant turbine inlet temperatures (gas turbine intercooling cycle combined with dual pressure steam cycle)

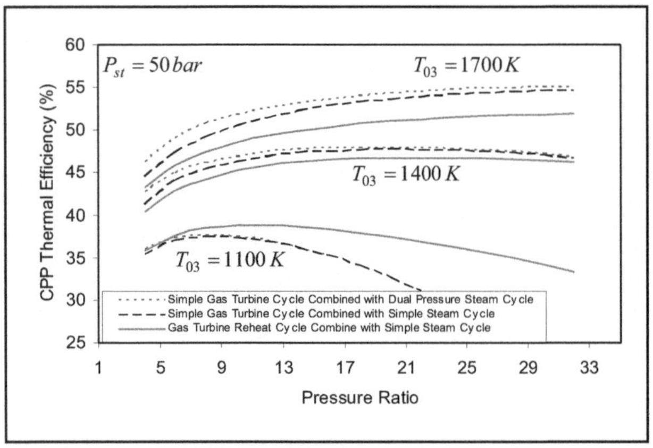

Figure 32. Combined efficiency versus pressure ratio for simple combined cycle, gas intercooling combined with dual pressure steam cycle and gas reheat cycle

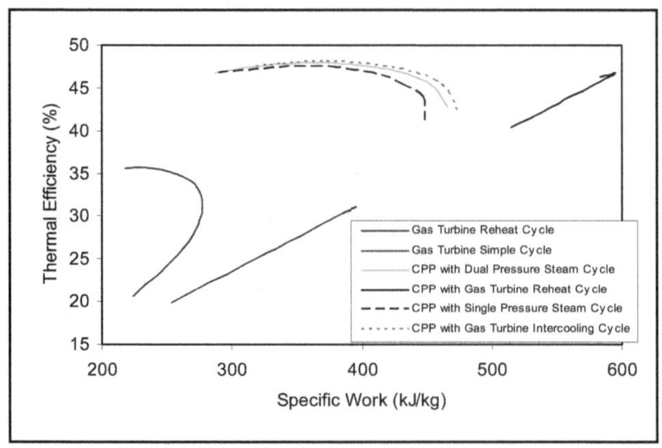

Figure 33. Efficiency versus specific work for different gas turbine cycles at 1400 K turbine inlet temperature and 50 bar steam boiler pressure

v. The Effect of Supplementary Heating on CPP Plant

From Eq. 35 and with a constant higher cycle efficiency η_H of 0.3, the values of the combined efficiency η_{CPP} were calculated for a range of values of the lower cycle efficiency η_L at different values of supplementary heat ratios Q_{add}/Q_{AH} . The results have plotted in Figure 34. Without supplementary heating $(Q_{add}/Q_H = 0)$ and by varying the gas turbine thermal efficiency η_{gt} between 0.5 and 0.6, Equation 35 has been used to calculate the steam turbine thermal efficiency η_{st} at two constant combined cycle efficiencies η_{CPP} of 0.5 and 0.6. The calculated results versus the gas turbine thermal efficiency is plotted in Figure 35 at two constant heat loss percentages of 0% and 10%.

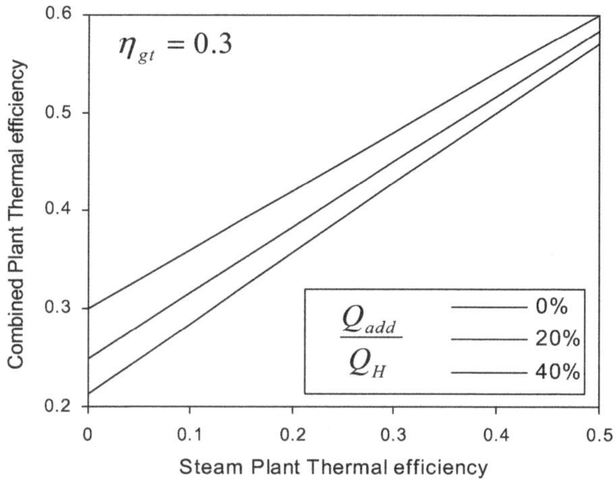

Figure 34. Combined power plant thermal efficiency versus steam plant thermal efficiency at constant supplementary heating ratios and 0.30 gas turbine plant thermal efficiency

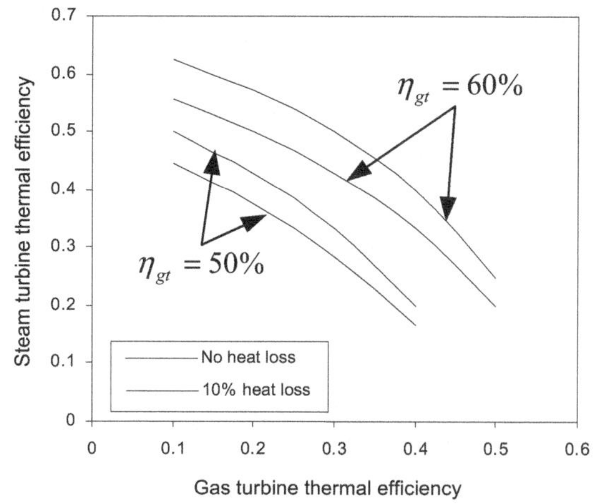

Figure 35. steam plant thermal efficiency versus gas turbine plant thermal efficiency at constant heat loss ratios for 0.50 and 0.60 Combined power plant thermal efficiency

From Figures 34 and 35, the following observations can be stated:

i. Supplementary heating would decrease the overall combined power plant thermal efficiency except when the supplementary heating results in a significant increase the steam turbine thermal efficiency.

ii. To reach specified combined power plant thermal efficiency, correct combination between the two

iii. plant's efficiencies is necesary.

iv. The heat loss between the two plants Q_{loss} increases the importance of gas turbine thermal efficiency in the combined power plant performance.

3. Concluding Remarks

The gas turbine plant represents a complex system where its performance depends on many parameters. A parametric study of the gas turbine cycle was carried out to assess the influence of each parameter on engine performance in order to identify the design point. The design point in this case is defined as that point, which would give the optimum performance. Achieving the design point parameters depends on factors such as economic, technological, operational and environmental, etc. The design of gas turbine plants often requires a trade off between these factors.

The gas turbine engine operating as part of the combined power and power plant does not only produce power but also the necessary thermal energy to operate the steam turbine plant. Therefore, what might be considered as the optimum performance for the gas turbine plant may not necessarily be the optimum performance of the combined power and power or the CPP plant. The results of the parametric study confirm this hypothesis.

The results of the current work, led to the following conclusions:

i. For a simple gas turbine cycle, the maximum efficiency and the maximum specific work depend on different performance parameters, such as cycle pressure ratio or the dimensionless mass flow and speed parameters, while for a reheat gas turbine cycle, the maximum efficiency and the maximum specific work depend on identical performance parameters.

ii. Supplementary heating will always decrease the combined cycle efficiency except in the case that the supplementary heating significantly increases the steam turbine cycle efficiency.

iii. Increasing the gas turbine inlet temperature at constant cycle pressure ratio will increase the combined cycle efficiency and the combined specific work.

iv. The higher the gas turbine inlet temperature the greater the influence of the pressure ratio difference on the combined cycle efficiency of any combined cycle configurations.

v. Gas turbine reheating can be justified only if the turbine inlet temperature is low (low gas exhaust temperatures) and/or higher combined specific work output.

vi. Although gas turbine pre-cooling improves the gas turbine performance, it has a slight effect on the combined cycle efficiency and the combined specific work output.

vii. In the combined power and power cycle, the combined cycle maximum efficiency depends on neither the gas turbine maximum efficiency parameters nor the gas turbine maximum specific work parameters, but on new parameters that are closer to the gas turbine maximum specific work parameters.

viii. Increasing the gas turbine inlet temperature will always increase the combined cycle efficiency and the combined specific work.

Nomenclature

AIRPROP	Air properties	P	Pressure (kPa)
FARATIO	Fuel to air ratio	T	Temperature (K)
GASPROP	Gas properties	S	Entropy $(kJ/kg.K)$
HRSG	Heat recovery steam generator	r	Pressure ratio
C_P	Specific heat at constant pressure $(kJ/kg.K)$	η	Efficiency
C_V	Specific heat at constant volume $(kJ/kg.K)$	θ	Ratio of maximum to minimum temperature
γ	Ratio of specific heats	h	Enthalpy (kJ/kg)
$\dot{m}$	Mass flow rate (kg/s)	LCV	Lower calorific value (kJ/kg)
Q	Heat supplied or rejected (kJ)	PP	Pinch point temperature difference
q	Specific heat supplied or rejected (kJ/kg)	f	Fuel to air ratio, function
W	Work output (kJ)	ε	Effectiveness
w	Specific work output (kJ/kg)	D	Dryness factor
		ξ	Pressure loss in combustion chamber

Subscripts

1, 2, 3	State points in the cycles	*H*	Higher plant (Gas turbine)
gt	Gas turbine	*HL*	Higher to lower (Gas turbine)
st	Steam turbine	*L*	Lower plant(Steam plant)
s	Isentropic	*Latm*	Lower to atmosphere
o	Stagnation	*evp*	Evaporator
g	Gas	*sup*	Super heater
a	Air	*sat*	Saturation
c	Compressor	*CPP*	Combined cycle power plant
t	Turbine	*max*	Maximum
cc	Combustion chamber	*min*	Minimum
P	Pump	*atm*	Minimum
B	Boiler (Heat recovery steam generator)		

Superscripts

.	Rate

References

Andreades C., Dempsey L., & Peterson P. F. (2014), Reheat Air–Brayton combined cycle power conversion off-nominal and transient performance. *J. Eng. Gas Turbines Power, 136*(7), 11 pages. http://dx.doi.org/10.1115/1.4026612

Bannister, R. L., Briesch, M. S., Diakunchak, I. S., & Huber, D. J. (1995). A combined cycle designed to achieve greater than 60 % efficiency. Trans. *ASME Journal of Engineering for gas turbines and power, 117*(4), 734-741. http://dx.doi.org/10.1115/1.2815459.

Bannister, R. L., Cheruvu, N. S., Little, D. A., & McQuiggan, G. (1995). Development requirements for an advanced gas turbine system. Trans. *ASME Journal of Engineering for gas turbines and power, 117*(4), 724-733. http://dx.doi.org/10.1115/1.2815458

Bassily A. M. (2015). The application of novel techniques for gas turbine inlet-cooling that improve both the power and efficiency of the modern commercial steam-air-cooled gas turbine combined cycle power plants in hot and humid climates. *Proceedings of the Institution of Mechanical Engineers, Part A: Journal of Power and Energy, 229*(4), 406-430. http://dx.doi.org/10.1177/0957650915570348

Bhinder, F. S., & Mango, O. I. K. (1995). A parametric study of the combined power and power (CPP) plant for generating electricity. ASME Turbo Cogeneration. Vienna, Austria.

Bianci, M., Melino, F., & Peretto, A. (20060. Parametric studies of combined cycles equipped with inlet fogging. *J. Eng. Gas Turbines Power, 128*(2), 326-335. http://dx.doi.org/ 10.1115/1.1765122

Cerri, G. (1987). Parametric analysis of combined gas-steam cycles. Trans. *ASME Journal of Engineering for gas turbines and power, 109*(1), 46-55. http:// dx.doi:10.1115/1.3240005

Chappel, M. S., & Cockshutt, E. P. (1974). Gas turbine cycle calculations: Thermodynamic data tables for air and combustion products for three systems of units. NRC No. 14300, Ottawa.

Chiesa P., & Macchi E. (2004). A thermodynamic analysis of different options to break 60% electric efficiencies in combined cycle power plant. *J. Eng. Gas Turbines Power, 126*(4), 770-785. http://dx.doi.org/10.1115/1.1771684

Chin, W. W., & El-Masri, M. A. (1987). Exergy analysis of combined cycles: Part 2 –Analysis and optimisation of two-pressure steam bottoming cycles. Trans. *ASME Journal of Engineering for gas turbines and power, 109*(2), 237-243. http:// dx.doi:10.1115/1.3240030.

Chodkiewicz R., Krysinski J., & Porochnicki J., (2002). A recuperated gas turbine incorporating external heat sources in the combined gas-turbine cycle. *J. Eng. Gas Turbines Power, 124*(2), 263-269. http://dx.doi.org/10.1115/1.1448325

Da Cunha Alves, M. A., de Franca Mendes Carneiro, H. F., Barbosa, J. R., Travieso, L. E., Pilidis, P., & Ramsden, K. W. (2001). An insight on intercooling and reheat gas turbine cycles. *Proceedings of the IMechE, Part A: Journal of Power and Energy, 215*(2), 163-171. http://dx.doi.org/10.1243/0957650011538433

El-Masri, M. A. (1985). On thermodynamics of gas turbine cycles: Part 1 –Second law analysis of combined cycles. Trans. *ASME Journal of Engineering for gas turbines and power, 107*(4), 880-236. http://dx.doi.org/10.1115/1.3239832

El-Masri, M. A. (1987). Exergy Analysis of Combined Cycles: Part 1 –Air cooled Brayton cycle gas turbines. Trans. *ASME Journal of Engineering for gas turbines and power, 109*(2), 228-236. http://dx.doi.org/10.1115/1.3240029

Gülen S. C. (2011). Importance of auxiliary power consumption for combined cycle performance. *J. Eng. Gas Turbines Power, 133*(4). http://dx.doi.org/10.1115/1.4002254

Gülen S. C., & J Joseph, J. (2011). Combined cycle off-design performance estimation: A second law perspective. *J. Eng. Gas Turbines Power, 134*(1). http://dx.doi.org/ 10.1115/1.4004179

Gülen S. C., & Smith, R. W. (2010). Second law efficiency of the Rankine bottoming cycle of a combined power plant. *J. Eng. Gas Turbines Power, 132*(1). http://dx.doi.org/ 10.1115/1.3124787

Hawthorne, W. R., & Davis, G. de V. (1956). Calculating gas turbine performance. *Engineering, 181*, 461.

Horlock, J. H. (1995). Combined power plants: past, present and future. Trans. *ASME Journal of Engineering for gas turbines and power, 117*(4), 608-616. http:// dx.doi.org/10.1115/1.2815448

Horlock, J. H. (1997). Aero-engine derivative gas turbines for power generation: Thermodynamic and economic perspectives. Trans. *ASME Journal of Engineering for gas turbines and power, 119*(1), 119-123, http://dx.doi.org/10.1115/1.2815534

Kehlhofer, R. (1991). *Combined cycle gas turbine and steam turbine power plants.* Lilburn GA: Fairmont Press.

Korakianitis T., Grantstrom J., Wassingbo P., & Massardo, A. F. (2005). Parametric performance of combined –cogeneration power plants with various power and efficiency enhancements. *J. Eng. Gas Turbines Power, 127127*(1), 65-72. http://dx.doi.org/10.1115/1

Palestra, N., Barigozzi G., & Perdichizzi A., (2008). Inlet air cooling applied to combined cycle power plants: Influence of site climate and thermal storage systems. *J. Eng. Gas Turbines Power, 130*(2). http://dx.doi.org/ 10.1115/1.2771570

Rahim, M. A. (2012). Performance and sensitivity analysis of a combined cycle gas turbine power plant by various inlet air-cooling systems. *Proceedings of the Institution of Mechanical Engineers, Part A: Journal of Power and Energy, 226*(7), 922-931. http://dx.doi.org/10.1177/0957650912456657

Ruffled, P. (1987). A systematic analysis of the combined gas/steam cycle. ASME COGEN-TURBO: International Symposium on Turbomachinery, Combined-Cycle Technologies and Cogeneration held in Montreux, Switzerland (IGTI).

Sarabchi, K. & Polley, G. T. (1994). Thermodynamical optimisation of a combined cycle plant performance. *ASME paper 94-GT-304*.

Shi, X., Agnew, B., & She, D. (2011). Analysis of a combined power plant integrated with a liquid natural gas gasification and power generation system. *Proceedings of the IMechE, part A: Journal of Power and Energy, 225*(1), 1-11. http://dx.doi.org/10.1243/09576509JPE805

15

Solutions of the Problems of a Viscoelastic Dynamic Contact between Smooth Curvilinear Surfaces of two Solid Bodies by the Application of the "Method of the Specific Forces"

Nikolay Goloshchapov[1]

[1] 71 Grosvenor Road, London, N3 1EY, UK

Correspondence: Nikolay Goloshchapov, 71 Grosvenor Road, London, N3 1EY, UK. E-mail: glshchpv@yahoo.co.uk

Abstract

Solutions of the problems of a viscoelastic dynamic contact between smooth curvilinear surfaces of two solid bodies by the application of the "Method of the specific forces" have been given in the article, and the new conception for the definition of the elastic and the viscous forces in the common case of dynamics of a viscoelastic contact is proposed here by the further development of this method. Essence of this method is that, the forces of viscosity and the forces of elasticity can be found by integration of the specific forces acting inside an elementary volume of the contact zone. It is shown here, that this method allows finding the viscoelastic forces for any theoretical or experimental dependencies between the distance of mutual approach of two solid bodies and the diameter of the contact area. Also, the derivation of the integral equations of the viscoelastic forces, the equations for pressure in the contact is presented. Viscoelastic dynamic contacts between two spherical bodies, and between a spherical solid body and a semi-space at impact have been examined. Work and Energy in the phases of compression and restitution, and at the rolling shear have been derived. Approximate solutions for the differential equations of movement (displacement) by using the method of equivalent work have been derived. Equations for the normal contact stresses have been obtained. Also, equations for kinematic and dynamic parameters of the viscoelastic collision have been derived in this article. Examples of the comparison of theoretical results and conclusions have been given in the paper.

Keywords: viscoelastic forces, method specific forces, elementary distributed axial loads, geometry contact area, dynamic modules, dissipative energy, viscoelastic parameters, method equivalent work

1. Introduction

The objective of this paper is the further development of the application of the "Method of the specific forces (MSF)" for solutions to the problems of a viscoelastic dynamic contact between smooth curvilinear surfaces of two solid bodies. The new conception is proposed here, how to find the elastic and the viscous forces by an integration of the specific forces in the boundaries of the contact area, which can be found by considering of the geometry of the contact. This method has been already used by author for case of the collision between a spherical solid body and a semi-space (Goloshchapov, 2003, 2015).

It is assumed here that the surfaces of contact are smooth and in this case we are not considering the influence of roughness on the contact forces, and the initial velocities of contact V_x and $V\tau$ is less than the effective sound speed in the volume of deformation (Figure 1). Also the effect of the adhesive forces and a plastic deformations have not been considered in this paper.

As we know, the mechanics of an elastic contact problem between two smooth surfaces have been studied yet in the 19-th century by Herts (1882, 1896) and Boussinesq (1885), and then later, for example, it was examined by many others researchers, such as: Bowden and Tabor (1939); Landau and Lifshits (1944); Timoshenko and Goodier (1951); Archard (1957); Galin (1961); Sneddon(1965); Greenwood and Williamson (1966); Johnson, Kendall and Roberts (1971); Derjaguin, Muller and Toporov (1975); Bush, Gibson and Thomas (1975); Tabor (1977); Johnson (1985); Webster and Sayles, 1986; Stronge (2000); Persson, Bucher and Chiaia (2002); Wriggers (2006); Hyun and Robbins (2007). Also a viscoelastic contact between smooth and rough curvilinear surfaces of two solids already have been researched very widely and their results was published in many different manuscripts (Mindlin,1949; Radok. 1957; Hunter,1960; Goldsmith, 1960; Galin, 1961; Lee, 1962; Graham, 1965; Ting, 1966; Greenwood & Williamson, 1966; Simon, 1967; Jonas, 1982; Padovan, & Paramadilok, 1984; Johnson, 1985; Brilliantov, 1996; Brilliantov, Spahn, Hertzsch, & Poeschel, 1996; Ramírez, Poeschel, Brilliantov, & Schwager, 1999; Stronge, 2000; Barber & Ciavarella, 2000; Goloshchapov, 2001, 2003; Laursen, 2002; Dintwa, 2006; Carbone, Lorenz, Persson, & Wohlers, 2009; Harrass, Friedrich, & Almajid, 2010; Persson, 2010; Cummins, Thornton, & Cleary, 2012; Carbone & Putignano, 2013; Popov, 2015). In all these researches for a finding of the viscoelastic forces and stresses, the traditional theories and methods usually have been applied. But, in this paper, the novel theoretical and practical principals have been used for finding these forces and stresses.

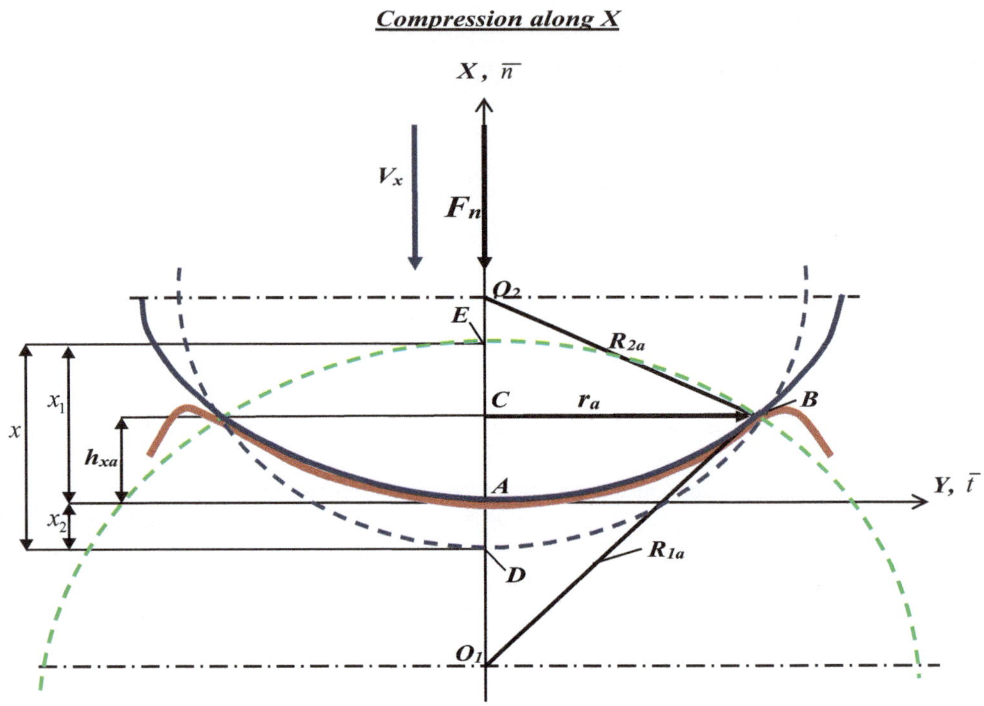

a.

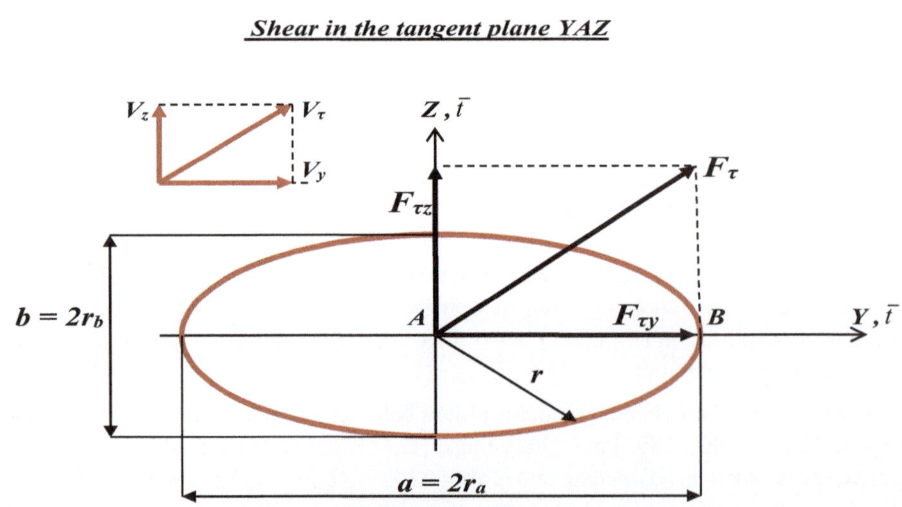

b.

Figure 1. Illustration of the mutual approach between two curvilinear surfaces of two solid bodies along the axis X relative to the initial point of contact A

The illustration of the mutual approach between two curvilinear surfaces of two solid bodies along the axis X relative to the initial point of contact A is depicted in the Figure 1. Here is not shown the displacement of the bodies along the axes Z and Y.

It is obviously that during the time of a dynamic contact, in the initial point A, the curvilinear surfaces are moving relative to each other with the normal relative velocity V_x and the tangential relative velocity V_τ between them under action of the normal force F_n and the tangential shear force F_τ. Also let: $F_{\tau y}$ and $F_{\tau z}$ are the tangential forces along axes Y and Z; x is the size of the mutual approach of the bodies along the axis X; r is a current radius of the contact area; h_{xa} is the depth of the contact surface in the plane XAY , or other words it is the depth of indentation of the more hard body into the surface of more soft body; h_{xb} is the depth of the contact surface in the plane XAZ *(This size is not shown in the Figure 1)*; x_1 and x_2 are normal compression deformations of the contacting surfaces relative to the initial point of the contact A; R_{1a} and R_{2a} are the radiuses of curvature of the contacting surfaces on the border of contact area in the plane XAY; $V_{\tau y}$ and $V_{\tau z}$ are the projections of the tangential relative velocity V_τ by axes Y and Z, or they are the tangential relative velocities of the displacement of surfaces of bodies along axes Y and Z; a is the big axis of an elliptic (oval) contact area; b is the small axis of an elliptic (oval) contact area; O_1 and O_2 are the centres of curvature of the contacting surfaces in the initial point of contact A; R_{1b} and R_{2b} are the radiuses of curvature of the contacting surfaces on the border of contact area in the plane XAZ.

It is obvious, that in a general case the area of contact is taking the shape of an oval or an ellipse, and in the same time, the contact surface takes the identical shape with the harder surface, and the surface, which has less hardness slips by a surface of the harder body. It is a micro-slip and we do not take in account the energy of this dissipative process in this paper. You can say that in process of sliding deformation under action of the tangential force, the area of contact does not take the elliptical shape, but it is the oval or other asymmetric figure.

But indeed, if the tangential force equal zero the contact area takes the right elliptical shape, but in the case of sliding deformation under the action of the tangential force, the elliptic shape of the contact area is transformed in the oval. Therefore, we can suppose, since the normal force stays same independently of the action of the tangential force, the normal contact pressure (or the normal contact stress) is not changed and stays same, and therefore the size of square of the contact is not changed too. According to this statement, we approximately can consider the area of contact like having the elliptical shape during all the contact time. Also in the case of an impact between the contact surfaces, when the rolling shear between them has place, we can state that the area of contact approximately keeps the elliptical shape too. Moreover, as we will see further, it does not matter, the area of contact is an ellipse or it is an oval, in the both cases, the derivation of the sizes of the contact area can be done in the same manner.

The viscoelastic forces can be found as the sums of the elastic forces and the viscous forces:

$$\begin{cases} F_n = F_{bn} + F_{cn} \\ F_{\tau y} = F_{b\tau y} + F_{c\tau y} \\ F_{\tau z} = F_{b\tau z} + F_{c\tau z} \end{cases} \tag{1}$$

Where: F_{cn} is the normal elastic force; F_{bn} is the normal viscous force; $F_{b\tau y}$ is the tangential viscous force by axis ***Y***; $F_{c\tau y}$ is the tangential elastic force by axis ***Y***; $F_{b\tau z}$ is the tangential viscous force by axis ***Z***; $F_{c\tau c}$ is the tangential elastic force by axis ***Z***.

The general tangential force can be found as sum

$$F_\tau = \sqrt{F_{\tau y}^2 + F_{\tau z}^2} \tag{2}$$

As it is known (Brilliantov et al., 1996; Goloshchapov, 2003, 2015; Jonson, 1985; Mindlin, 1949; Ramirez et al., 1999; Schafer et al., 1996; Schwager & Poschel, 2007; Stronge, 2000; Thornton, 2009 we can write all these viscoelastic forces, simply as:

$$\begin{cases} F_{bn} = b_x\dot{x} & F_{cn} = c_x x \\ F_{b\tau y} = b_y\dot{y} & F_{c\tau y} = c_y y \\ F_{b\tau z} = b_z\dot{z} & F_{c\tau z} = c_z z \end{cases} \tag{3}$$

Where c_y, c_z, b_x, b_y, b_z are variable parameters depending of the displacement ***x, y*** and ***z***, and where: b_x is the effective parameter of viscosity (damping parameter), c_x is the effective parameter of elasticity (stiffness); b_y, b_z are the effective parameters viscosity at a shift; c_y, c_z are the effective parameters of elasticity (stiffness) at a shift. x, y, z are the relative displacements (deformations) between contacting bodies or between contacting surfaces in the initial point ***A***; $\dot{y}, \dot{x}, \dot{z}$ are relative velocities of displacements (deformations) between contacting bodies or between contacting surfaces in the initial point ***A***.

In the past many old papers and others published recently (Mindlin, 1949; Simon, 1967; Johnson, 1985 ; Goloshchapov, 2003, 2015; Cundall & Strack,1979; Hertzsch, Spahn , Schafer, Dippel, & Wolf, 1996; Ramírez, Poeschel, Brilliantov, & Schwager, 1999; Stronge, 2000; Roylance, 2001; Brilliantov & Poeschel, 2004; Makse, Gland, Jnohnson, & Schwartz, 2004; Schwager & Poschel, Van Zeebroeck, 2005; Dintwa, 2006; Schwager & Poschel, 2007, 2008; Cheng, Subic, & Monir Takla, 2008; Becker, Schwagerand, & Pöschel, 2008; Schwager & Poschel, 2008; Thornton, 2009; Cummins, Thornton, & Cleary, 2012) have been used traditional theoretical rheological models, such as the "Linear Spring Dashpot Model"- (LS+D), the "Hertz Mindlin Spring Dashpot Model"- (HM+D), and the "Discrete Elements Method" - (DEM) and others . In all of these methods and models, for the definition of the effective parameter of elasticity c_x (Some authors name it like a stiffness, or spring parameter), the Hertz's theory of elastic contact between two surfaces (Landau and Lifshitz, 1944, 1965) has been used. Also, for the purpose of finding the tangential forces, the coefficient of friction was taken as a constant value. The more comprehensive analysis and review of these already known models and methods can be found, for example, in the monographs of the authors, such as: Stronge (2000); Van Zeebroeck (2005); Dintwa (2006); Li (2006). But, the most basic problem in the finding of the mechanical dynamic parameters of viscoelasticity c_y, c_z, b_x, b_y in the equations (3) is that, they are not the constant values. They all are variable magnitudes, because all dynamic mechanical and physical properties of the materials depend on dynamic conditions of loading (displacements, a velocity and a frequency) and temperature. But, the Hertz theory allows only the finding the normal elastic force. The existing methods still cannot give the complete answer, how these nonlinear parameters of viscoelasticity can be found for the practical application by using the dynamic modules of elasticity and viscosity, which usually can be found by using the known methods (Ferry, 1948, 1963; Moore, 1975; Van Krevelen 1972; Nilsen, 1978, 1994) For example as we know according the Hertz theory (Landau & Lifshtz, 1944, 1965) for the contact of two spherical surfaces $c_x = \frac{4}{3} E R^{1/2} x^{1/2}$. Where E is effective elasticity modulus, R is the effective radius of contact curvature. As we can see, the stiffness c_x is the nonlinear function of displacement ***x***, but we still have a problem in definition of others parameters c_y, c_z, b_x, b_y, b_z, which are the variable nonlinear functions too.

And also in already existing researches, the coefficient of friction usually is taken like a constant value for the purpose of finding the tangential forces, but according to Equations (1), (2) and (3) it can be defined as follows:

$$f = \frac{F_\tau}{F_n} = \frac{\sqrt{(F_{b\tau y} + F_{c\tau y})^2 + (F_{b\tau z} + F_{c\tau z})^2}}{F_{cn} + F_{bn}} = \frac{\sqrt{(b_y \dot{y} + c_y y)^2 + (b_z \dot{z} + c_z z)^2}}{c_x x + b_x \dot{x}} \tag{4}$$

As we can see from this equation, the coefficient of friction is not a constant value, but it changes during the time of contact, because the dynamic contact between two bodies is a non-equilibrium process, and all dynamic mechanical and physical properties of the materials depend on dynamic conditions of loading and temperature. For example, the dynamic elasticity modules are very yieldable to a changing of a velocity and temperature of the matter in the area of deformations (Ferry, 1948, 1963; Lee, 1962; Van Krevelen, 1972; Moor 1978; Nilsen, 1978, 1994; Lakes, 1998; Meyers, 1994; Menard, 1999; Roylance, 2001; Goloshchapov, 2001, 2003, 2015; Hosford, 2005; Popov, 2010, Popov & Hess, 2015).

Also, it is necessary to mention that, the researches in the field of the collision of viscoelastic particles (granules) with identical mechanical properties have been made by Brilliantov, Spahn, Hertzsch and Pöschel (1996). They have obtained the equation for the normal viscous force with variable viscosity parameter

$$F_{bn} = F_{dis}^{N} = 4 \frac{Y}{(1-\nu^2)} \sqrt{R^{eff}} A \sqrt{\xi} \dot{\xi}^{1/2}, \tag{4*}$$

where $\xi = x$, $R^{eff} = R$, Y is the Young modulus or the elasticity modul, ν is the Poisson ratio,

$A = \frac{1}{3}\frac{(3\eta_2 - \eta_1)^2}{(3\eta_2 + 2\eta_1)}\left[\frac{(1-\nu^2)(1-2\nu)}{Y\nu^2}\right]$ is the damping viscous parameter, and where η_1 and η_2 are the viscous constants. But this theoretical result can only be used for the contact of the bodies with the same physical-mechanical properties, and in this case we have the problem of finding the viscous constants "η_1" and "η_2". If the contacting surfaces have different physical-mechanical properties this conception does not give the answer, because this is a yet more difficult problem.

Also the interesting method - "Method of Dimensionality Reduction (MDR)" has been presented by Popov (2015), but it can be used only in the case of contact between three-dimensional, axial-symmetric bodies and a foundation (a semi-space). In all these papers, to find the equations for tangential forces, the coefficient of friction again was taken as a constant value.

Thus, as we can see, the many problems still exist now in these research areas. Therefore, especially for the solving of these problems, such as the definition of the normal viscous force and the all tangential viscoelastic forces, and for the finding of the kinematic and the dynamic mechanical parameters between two contacting surfaces, such as the elasticity modulus and the viscosity modulus, the theoretical and experimental ways have been developed and represented in this article below.

2. Derivation of the Equations for the Viscoelastic Forces by the "Method of the Specific Forces (MSF)"

Let us assume that in the infinitesimal period of the time *dt*, when the mutual approach between a body and a semi-space is the infinitesimal magnitude *dx* (Figure 2), inside the elementary infinitesimal volume *dV*, which is arising around the initial point of the contact *A* (Figure1 and Figure2), the infinitesimal viscoelastic forces dF_n, $dF_{\tau y}$ and $dF_{\tau z}$ are beginning to act.

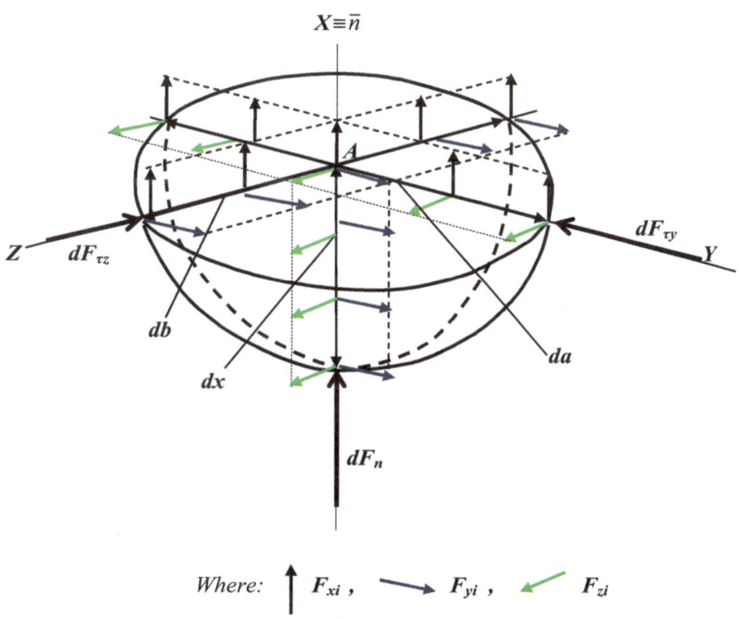

Figure 2. Illustration of the action of the specific elementary viscoelastic forces inside the infinitesimal volume ***dV*** in the vicinity of point ***A***

These forces can be found by the differentiation of the normal F_{xi}, and the tangential F_{yi} and F_{zi} specific forces by sizes *da, db* and *dx*:

$$dF_n = F_{xi}da + F_{xi}db\ ,\ dF_{\tau y} = F_{yi}db + F_{yi}dx\ ,\ dF_{\tau z} = F_{zi}da + F_{zi}dx \qquad (5)$$

Where: *da* and *db* the big and the small axes of the infinitesimal contact area; F_{xi} is the normal effective specific viscoelastic force; F_{zi} is the tangential effective specific viscoelastic force by axis ***Z***; F_{yi} is the tangential effective specific viscoelastic force by axis ***Y***.

Also it is very important to understand indeed that, these specific forces *Fxi*, *Fyi* and *Fzi* are the elementary axial loads distributed on the infinitesimal sizes, *da*, *db* and *dx* parallel to axes *X*, *Y* and *Z* (Figure2). The name "specific force" already was applied by author of article (Goloshchapov, 2015), and this name is the short name, which usually is used to denote the elementary distributed axial loads.

2.1 The Effective Dynamic Modules and the Effective Dynamic Viscosities

According to the "Newton's Third Law" the effective specific forces and the specific forces between bodies have to be equal: $F_{xi} = F_{x1} = F_{x2}$, $F_{yi} = F_{y1} = F_{y2}$, $F_{zi} = F_{z1} = F_{z2}$. Where: F_{x1} and F_{x2} are the normal specific viscoelastic forces of the bodies by axis *X*; F_{y2} and F_{y1} is the tangential viscoelastic specific forces of the bodies by axis *Y*; F_{z2} and F_{z1} is the tangential viscoelastic specific forces by axis *Z*. (Here and further in this paper the subscript *1* is used for more soft body, and *2 is* used for more solid body). On the other hand, the specific viscoelastic forces can be found as the sum of the specific elastic forces and the specific viscous forces:

$$\begin{cases} F_{xi} = F_{xb} + F_{xc},\ F_{x1} = F_{x1b} + F_{x1c},\ \ F_{x2} = F_{x2b} + F_{x2c} \\ F_{yi} = F_{yb} + F_{yc},\ F_{y1} = F_{y1b} + F_{y1c},\ \ F_{y2} = F_{y2b} + F_{y2c} \\ F_{zi} = F_{zb} + F_{zc},\ F_{z1} = F_{z1b} + F_{z1c},\ \ F_{z2} = F_{z2b} + F_{z2c} \end{cases} \tag{6}$$

Where: F_{xb} is the normal effective specific viscous force; F_{xc} is the normal effective specific elastic force; F_{x1b} and F_{x2b} are the normal specific viscous forces; F_{x1c} and F_{x2c} are the normal specific elastic forces; F_{yb} and F_{zb} are the tangential effective specific viscous forces; F_{yc} and F_{zc} is the tangential effective specific elastic forces; F_{y1b}, F_{y2b} and F_{z1b}, F_{z2b} are the tangential viscous specific forces; F_{y1c}, F_{y2c} and F_{z1c}, F_{z2c} are the tangential elastic specific forces.

Also let us suppose that the volume of deformation is the system of an infinitely large number of elementary discrete elements (Figure 3.) connected among themselves definitely. And also, in this case let us assume, that for the infinitesimal period of the contact time *dt* all deformations inside of each elementary discrete element are changing linearly and therefore all specific forces are changing linearly too. Based on this, the equations for all specific forces can be written as the linear functions:

$$\begin{cases} F_{xb} = \eta'_E \dot{x},\ F_{xc} = E'x,\ F_{yb} = \eta'_G \dot{y},\ F_{yc} = G'y,\ F_{zb} = \eta'_G \dot{z},\ F_{zc} = G'z \\ F_{x1b} = \eta'_{1E}\dot{x}_1,\ F_{x1c} = E'_1 x_1,\ F_{y1b} = \eta'_{1G}\dot{y}_1,\ F_{y1c} = G'_1 y_1,\ F_{z1b} = \eta'_{1G}\dot{z}_1,\ F_{z1c} = G'_1 z_1, \\ F_{x2b} = \eta'_{2E}\dot{x}_2,\ F_{x2c} = E'_2 x_2,\ F_{y2b} = \eta'_{2G}\dot{y}_2,\ F_{y2c} = G'_2 y_2,\ F_{z2b} = \eta'_{2G}\dot{z}_2,\ F_{z2c} = G'_2 z_2 \end{cases} \tag{7}$$

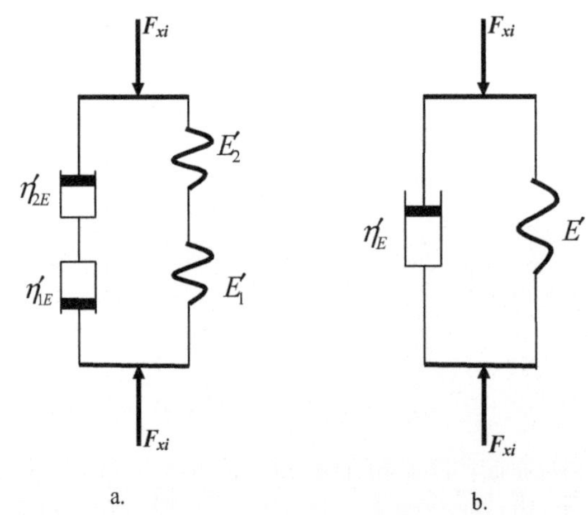

Figure 3. Illustration of the "Elementary discrete elements model (EDEM)": a. the elementary discrete element of the normal contact between two bodies; b. the effective elementary discrete element of the normal contact

Where: E' is the effective dynamic elasticity modulus; η'_E is the effective dynamic viscosity; G' is effective dynamic elasticity modulus at the shear; η'_G is the effective dynamic viscosity at the shear; E'_1 and E'_2 are the dynamic elasticity modules. η'_{1E} and η'_{2E} are the dynamic viscosities; G'_1 and G'_2 is the dynamic elasticity modules at the shear. η'_{1G} and η'_{2G} is the dynamic viscosities at the shear.

In the proposed model, each elementary deformation between two bodies develops analogically like the deformation of the elementary discrete element, which is depicted in Figure 3.a. It is a simple case of the linear model of deformations of elementary discrete elements, and instead this model with four elements we can use its analogy - the model with two effective elements depicted in Figure 3.b. Also the “Elementary discrete elements model” for the normal forces can be used for the tangential forces in the same manner.

It is obvious that the according to the initial conditions, when $t = 0$, $x = 0$, the specific elastic forces acting by X are equal at the initial instant of the contact (in this point they are equal zero); and the according to the boundary conditions, when $x = x_m$, they are equal at the instant of the maximum compression between two surfaces (in this point they reach the maximum value). But at the same time, the specific viscous forces are equal at the initial instant of the contact, when $t = 0$, $x = 0$ (in this point they are equal to the maximum value, because the velocity $\dot{x}$ has the maximum value), and they are equal at the instant of the maximum compression between two surfaces, when $x = x_m$ (in this point they are equal zero, because the velocity $\dot{x}=0$). All of these forces in the Equation 7 are linear continuous functions and if they are equal for these two values of the argument x, they have to be equal for any other values as well, or by other words, they are equal in any instant of the time of the contact. Analogically this conclusion is valid for other specific forces acting by axes Y and Z. Thus consequently we can write that

$$F_{xc}=F_{x1c}=F_{x2c},\ F_{xb}=F_{x1b}=F_{x2b},\ F_{yc}=F_{y1c}=F_{y2c},\ F_{yb}=F_{y1b}=F_{y2b} \text{ and } F_{zb}=F_{z1b}=F_{z2b}, \quad (7^*)$$

and hence we get respectively

$$\begin{cases} E'x = E'_1x_1=E'_2x_2, & \eta'_E\dot{x}=\eta'_{1E}\dot{x}_1=\eta'_{2E}\dot{x}_2 \\ G'y=G'_1y_1=G'_2y_2; & \eta'_G\dot{y}=\eta'_{1G}\dot{y}_1=\eta'_{2G}\dot{y}_2 \\ G'z=G'_1z_1=G'_2y_2; & \eta'_G\dot{z}=\eta'_{1G}\dot{z}_1=\eta'_{2G}\dot{z}_2 \end{cases} \quad (8)$$

Since as $x = x_1 + x_2$, $y = y_1 + y_2$, $z = z_1 + z_2$ according to (8) we can write the equations for the effective dynamic modules and the effective dynamic viscosities as

$$E' = \frac{E'_1E'_2}{E'_1+E'_2},\ \eta'_E = \frac{\eta'_{1E}\eta'_{2E}}{\eta'_{1E}+\eta'_{2E}},\ G' = \frac{G'_1G'_2}{G'_1+G'_2},\ \eta'_G = \frac{\eta'_{1G}\eta'_{2G}}{\eta'_{1G}+\eta'_{2G}} \quad (9)$$

And also we can write that

$$x_1 = D_1x \text{ and } x_2 = D_2x \quad (10)$$

Where: $D_1 = \dfrac{E'_2}{E'_1+E'_2}$ and $D_2 = \dfrac{E'_1}{E'_1+E'_2}$ are the coefficients of deformation.

2.2 Integral Equations for the Viscoelastic Forces

Now, since according to (5) and (6) we can write that

$$\begin{cases} dF_n = (F_{xb} + F_{xc})da + (F_{xb} + F_{xc})db\ , \\ dF_{\tau y} = (F_{yb} + F_{yc})db + (F_{yb} + F_{yc})dx\ , \\ dF_{\tau z} = (F_{zb} + F_{zc})da + (F_{zb} + F_{zc})dx\ , \end{cases} \quad (10^*)$$

and according to (7), since $F_{xb} = \eta'_E \dot{x}$, $F_{xc} = E'x$, $F_{yb} = \eta'_G \dot{y}$, $F_{yc} = G'y$, $F_{zb} = \eta'_G \dot{z}$, $F_{zc} = G'z$, the twelve expressions for the infinitesimal viscoelastic forces can be written separately as:

$$\begin{cases} dF_{abn} = \eta'_E \dot{x}da\ ,\ dF_{acn} = E'xda\ ,\ dF_{bbn} = \eta'_E \dot{x}db\ ,\ dF_{bcn} = E'xdb\ , \\ dF_{bby} = \eta'_G \dot{y}db,\ dF_{bcy} = G'ydb,\ dF_{hby} = \eta'_G \dot{y}dx,\ dF_{hcy} = G'ydx, \\ dF_{abz} = \eta'_G \dot{z}da,\ dF_{acz} = G'zda,\ dF_{hbz} = \eta'_G \dot{z}dx,\ dF_{hcz} = G'zdx \end{cases} \quad (11)$$

And since the limits of integration h_{xa} and h_{xb} are known, the integral equations for all viscoelastic forces can be written respectively as

$$\begin{cases} F_{abn} = \eta'_E \dot{x} \int da\ ,\ F_{acn} = E' \int xda\ ,\ F_{bbn} = \eta'_E \dot{x} \int db\ ,\ F_{bcn} = E' \int xdb\ , \\ F_{bby} = \eta'_G \dot{y} \int db\ ,\ F_{bcy} = G'y \int db\ ,\ F_{hby} = \eta'_G \dot{y} \int_0^{h_{xa}} dx\ ,\ F_{hcy} = G'y \int_0^{h_{xa}} dx\ , \\ F_{abz} = \eta'_G \dot{z} \int da\ ,\ F_{acz} = G'z \int da\ ,\ F_{hbz} = \eta'_G \dot{z} \int_0^{h_{xb}} dx\ ,\ F_{hcz} = G'z \int_0^{h_{xb}} dx \end{cases} \quad (12)$$

2.3 The Geometry of the Area of the Contact and the Pressure Distribution

And now, the important moment, it can be seen that for a finding of the solutions for all Equations (12) we have to know only the equations or the formulas for $a = f(x)$, $b = f(x)$, and for $h_{xa} = f(x)$ and $h_{xb} = f(x)$. For example, we can use that $r = (Rx)^{1/2}$ according to the Hertz theory, but according to this theory, the area of contact is a flat surface and the depth of indentation (the depth of the contact surface) $h_x = 0$. But in reality the area of contact usually is not a flat, it is a curvilinear surface. In Hertz's theoretical models has been taken that the contacting surfaces deform together without of the micro-sliding, but in reality each surface deforms independently. Therefore, to find the radiuses of the contact area r_a and r_b in reality, first of all, let us consider the geometry of contact between two curvilinear surfaces in the normal section in the plane *XAY*, like it is depicted in the illustration in Figure1. It is obviously that, in the time of indentation of more hard surface into a soft surface, the contact surface takes a curvilinear shape, where the point *B* (see Figure1) is a special point where the deformations always equal zero, and the border of the area of contact always pass through this point *B*. According to this statement, the distance O_2B between this point and the centre of curvature O_2 of the surface of more hard body will not be changed in the period of time of contact. This distance always equals to the radius of curvature R_2. Also the distance O_1B between this point and the centre of curvature O_1 of the surface of less hard body will not be changed in the period of time of contact too. This distance always equals to the radius of curvature R_1. Hence, obviously that $O_2B = O_2D = R_{2a}$ and $O_1B = O_1E = R_{1a}$, and also we can write that $O_1C + O_2C = (R_{1a} + R_{2a}) - x$, and since as $O_1C = (R_{1a}^2 - r_a^2)^{1/2}$ and $O_2C = (R_{2a}^2 - r_a^2)^{1/2}$, after a simple calculation, if to neglect by members of smallest order, we get the next equation for the radius of contact area $r_a = f(x)$ equals $a/2$, and then analogically in same way we can find the radius of contact area $r_b = f(x)$ equals $b/2$ as follows:

$$r_a^2 = 2R_a x - x^2\ ,\ r_b^2 = 2R_b x - x^2 \quad (13)$$

Where $R_a = \dfrac{R_{1a}R_{2a}}{R_{1a} + R_{2a}}$ is the effective initial radius of contact curvature in the plane *XAY* and $R_b = \dfrac{R_{1b}R_{2ab}}{R_{1b} + R_{2b}}$ is the effective initial radius of contact curvature in the plane *XAZ*. The Equation (13) is not convenient for using and therefore, let us rewrite them as

$$r_a^2 = k_{pa}^2 R_a x\,,\ r_b^2 = k_{pb}^2 R_b x \tag{14}$$

where

$$k_{pa} = \sqrt{2 - \frac{x}{R_a}}\,,\ k_{pb} = \sqrt{2 - \frac{x}{R_b}} \tag{15}$$

are the correlation coefficients. Practically for the solution of the contact problems, the correlation coefficient can be found by the method of iterations and a consecutive approximation.

Obvious from (13) that the surface of the contact takes the elliptical shapes in the planes *XAY* and *XAZ*, but for a simplification, its shape can be approximated by the parabolic functions (14) respectively as

$$x = \frac{1}{k_{pa}^2 R_a} r_a^2\,,\ x = \frac{1}{k_{pb}^2 R_b} r_b^2 \tag{16}$$

Since the surface of the contact has approximately the parabolic shape, let us to take that the radial distribution of the pressure inside of this area changes analogically according to the parabolic function as

$$P = P_c\left(1 - \frac{r_y^2}{r_a^2}\right),\ P = P_c\left(1 - \frac{r_z^2}{r_b^2}\right) \tag{17}$$

Where: r_y is a current radius of the contact area by axis *Y*; r_z is a current radius of the contact area by axis *Z*; P_c is the maximum magnitude of the pressure in the centre of the contact area.

Further since the square under these functions in the Equation17 and the square under the linear function of the mean pressure P_m in the contact area are equal, we can write that

$$P_c\int_0^{r_a}\left(1 - \frac{r_y^2}{r_a^2}\right)dr_y = P_m r\,,\ P_c\int_0^{r_b}\left(1 - \frac{r_z^2}{r_b^2}\right)dr_z = P_m r \tag{18}$$

Then after the integration follows

$$P_c\left(r - \frac{1}{3}r\right) = P_m r\,, \tag{19}$$

and finally the ratio between maximum and the mean pressure in the contact zone can be found as

$$P_c = \frac{3}{2}P_m \tag{20}$$

Now let us to define the depths of the contact surface (see Figure 1) in the planes *XAY* and *XAZ*. The expressions for the radiuses of contact area can be found also as follows

$$r_a^2 = R_{2a}^2 - \left(R_{2a}^2 - (x_2 + h_{xa})\right)^2,\ r_b^2 = R_{2b}^2 - \left(R_{2b}^2 - (x_2 + h_{xb})\right)^2 \tag{21}$$

After a simple geometric calculation, if to neglect by members the smallest order, we obtain the next equations for the radiuses of contact area:

$$r_a^2 = 2R_{2a}(x_2 + h_{xa}) - (x_2 + h_{xa})^2\,,\ r_b^2 = 2R_{2b}(x_2 + h_{xb}) - (x_2 + h_{xb})^2 \tag{22}$$

Then after the comparison equations (22) and (13) we can write that

$$2R_{2a}(x_2 + h_{xa}) \approx 2R_a x\ ,\ 2R_{2b}(x_2 + h_{xb}) \approx 2R_b x \tag{23}$$

Finally since $x_2 = D_2 x$, the formulas for h_{xa} and h_{xb} can be written as follows

$$h_{xa} = \left(\frac{R_a - D_2 R_{2a}}{R_{2a}}\right)x = k_{ha}x\,,\ h_{xb} = \left(\frac{R_b - D_2 R_{2b}}{R_{2b}}\right)x = k_{hb}x \tag{24}$$

Where $k_{ha} = \left(\frac{R_a - D_2 R_{2a}}{R_{2a}}\right)$ and $k_{hb} = \left(\frac{R_b - D_2 R_{2b}}{R_{2b}}\right)$ are the coefficients of depth of the contact surface.

Also let us remark that, in the case a very small deformations, when $R_a >> x$ and $R_b >> x$ according to Equations (15) follows $k_{pa} \approx k_b \approx \sqrt{2}$, and we can write that

$$r_a = (2R_a x)^{1/2},\ r_b = (2R_b x)^{1/2} \tag{25}$$

But, according to the Hertz theory (Landau and Lifshitz 1944, 1965) $r = (Rx)^{1/2}$, but it is possible only in the one case when $k_{pa} = k_{pb} = 1$.

2.4 The Equations for all Viscoelastic Forces

Since as $a = 2r_a$ and $b = 2r_b$, according to the Equation (14) we can write that

$$a = 2k_{pa}R_a^{1/2}x^{1/2}\ ,\ b = 2k_{pb}R_b^{1/2}x^{1/2} \tag{26}$$

The taking of the derivatives *da/dx* and *db/dx* gives us

$$da = \frac{k_{pa}R_a^{1/2}}{x^{1/2}}dx\ ,\ db = \frac{k_{pb}R_b^{1/2}}{x^{1/2}}dx\,, \tag{27}$$

and then after an integration of the integral equations for the normal forces from (12), we get

$$F_{acn} = k_{pa}E'R_a^{1/2}\int x^{1/2}dx = \frac{2}{3}k_{pa}E'R_a^{1/2}x^{3/2},\ F_{bcn} = k_{pb}E'R_b^{1/2}\int x^{1/2}dx = \frac{2}{3}k_{pb}E'R_b^{1/2}x^{3/2} \tag{28}$$

Since as $F_{cn} = F_{acn} + F_{bcn}$ we get

$$F_{cn} = \frac{2}{3}E'x^{3/2}(k_{pa}R_a^{1/2} + k_{pb}R_b^{1/2}) = \frac{2}{3}E'\psi x^{3/2}, \tag{29}$$

where $\psi = (k_{pa}R_{pa}^{1/2} + k_{pb}R_{pb}^{1/2})$, [$m^{1/2}$] is the parameter of curvature.

If $k_{pa} = k_{pb} = 1$ and if the contact area is a circle $R_a = R_b = R$ we have the same solution that have been obtained for the contact between spherical surfaces by using the Hertz theory (Landau & Lifshitz, 1944, 1965):

$$F_{cn} = \frac{4}{3}E'R^{1/2}x^{3/2} \tag{30}$$

Thus, it is obvious that the proposed method of the finding of the normal elastic forces definitely is valid and correct. It can be seen that, if we know a functional dependency between ***r*** and ***x***, we can always find the elastic force. But, if this method is correct for the definition this force, hence it should be valid for the definition of all viscoelastic forces in the equations (12). The equation for the normal viscous forces can be found in the same way by integration:

$$\begin{cases} F_{abn} = k_{pa}\eta'_E\dot{x}(t)\int \frac{R_a^{1/2}}{x^{1/2}}dx = 2k_{pa}\eta'_E R_a^{1/2}\dot{x}(t)x^{1/2} \\ F_{bbn} = k_{pb}\eta'_E\dot{x}(t)\int \frac{R_b^{1/2}}{x^{1/2}}dx = 2k_{pb}\eta'_E R_b^{1/2}\dot{x}(t)x^{1/2} \end{cases} \tag{31}$$

Here $\dot{x}(t)$ are the function linear independent from x, and it cannot be integrated by x, and it stays outside of integrals. And since as $F_{bn} = F_{abn} + F_{bbn}$ follows

$$F_{bn} = 2\eta'_E \dot{x} x^{1/2} (k_{pa} R_a^{1/2} + k_{pb} R_b^{1/2}) = 2\eta'_E \psi \dot{x} x^{1/2} \tag{32}$$

In the case when the contact area is a circle $k_{pa} = k_{pb} = k_p$ and $R_a = R_b = R$ we get

$$F_{bn} = 4k_p \eta'_E R^{1/2} \dot{x} x^{1/2} \tag{33}$$

Since as x and y are linearly independent and $da = \frac{k_{pa} R_a^{1/2}}{x^{1/2}} dx$, $db = \frac{k_{pb} R_b^{1/2}}{x^{1/2}} dx$, and $h_{ax} = k_{ha} x$, $h_{xb} = k_{hb} x$, after an integration of the equations for the tangential forces from Equation (12) their solutions can be written as follows:

$$\begin{cases} F_{bby} = \eta'_G \dot{y} \int db = 2k_{pb} \eta'_G R_b^{1/2} x^{1/2} \dot{y}, \quad F_{bcy} = G' y \int db = 2k_{pb} G' R_b^{1/2} x^{1/2} y \\ F_{hby} = \eta'_G \dot{y} \int_0^{h_{xa}} dx = k_{ha} \eta'_G x \dot{y}, \quad F_{hcy} = G' y \int_0^{h_{xa}} dx = k_{ha} x y, \\ F_{abz} = \eta'_G \dot{z} \int da = 2k_{pa} \eta'_G R_a^{1/2} x^{1/2} \dot{z} \quad F_{acz} = G' z \int da = 2k_{pa} G' R_a^{1/2} x^{1/2} z \\ F_{hbz} = \eta'_G \dot{z} \int_0^{h_{xb}} dx = k_{hb} x \dot{z}, \quad F_{hcz} = G' z \int_0^{h_{xb}} dx = k_{hb} x z \end{cases} \tag{34}$$

Here $\dot{y}, y, \dot{z}, z$ are the functions linear independent from x, and they cannot be integrated by x, and they stay outside of integrals.

The equations for the tangential elastic and viscous forces can be written now, as the sum of the elastic and the viscous tangential forces from Equation (34):

$$\begin{cases} F_{b\tau y} = F_{hby} + F_{bby} = \eta'_G P_{bx} \dot{y}, \quad F_{c\tau y} = F_{hcy} + F_{bcy} = G' P_{bx} y \\ F_{b\tau z} = F_{hbz} + F_{bbz} = \eta'_G P_{ax} \dot{z}, \quad F_{c\tau z} = F_{hcz} + F_{acz} = G' P_{ax} z \end{cases} \tag{35}$$

Where:

$$P_{bx} = k_{ha} x + 2k_{pb} R_b^{1/2} x^{1/2}, \quad P_{ax} = k_{hb} x + 2k_{pa} R_a^{1/2} x^{1/2} \tag{36}$$

Thus finaly, according to Equations (1), (31), (34) and (35) the next system of equations for general viscoelastic forces can be written respectively:

$$\begin{cases} F_n = F_{bn} + F_{cn} = 2\eta'_E \psi \dot{x} x^{1/2} + \dfrac{2}{3} E' \psi x^{3/2} \\ F_{\tau y} = F_{b\tau y} + F_{c\tau y} = \eta'_G P_{bx} \dot{y} + G' P_{bx} y \\ F_{\tau z} = F_{b\tau z} + F_{c\tau z} = \eta'_G P_{ax} \dot{z} + G' P_{ax} z \end{cases} \tag{37}$$

And thus, we can write the expressions for the variable viscoelasticity parameters as follows:

$$b_x = 2\eta'_E \psi x^{1/2},\ c_x = \frac{2}{3} E' \psi x^{1/2},\ b_y = \eta'_G P_{bx},\ c_y = G' P_{bx},\ b_z = \eta'_G P_{ax},\ c_z = G' P_{ax} \tag{38}$$

It is not possible the finding of the viscoelastic forces separately for each contacting body if to use the Hertz theory and others already existing theories, but it is possible by using the "Method of the specific forces (MSF)". The integral Equation (12) separately for each contacting body can be written as follows:

$$\begin{cases} F_{abni} = \eta'_{iE} \dot{x}_i \int da,\ F_{acni} = E'_i \int x_i da,\ F_{bbni} = \eta'_{iE} \dot{x}_i \int db,\ F_{bcni} = E'_i \int x_i db, \\ F_{hbyi} = \eta'_{iG} \dot{y}_i \int_0^{h_{xa}} dx,\ F_{hcyi} = G'_i y_i \int_0^{h_{xa}} dx,\ F_{hbzi} = \eta'_{iG} \dot{z}_i \int_0^{h_{xb}} dx,\ F_{hczi} = G'_i z_i \int_0^{h_{xb}} dx, \\ F_{bbyi} = \eta'_{iG} \dot{y}_i \int db,\ F_{bcyi} = G'_i y_i \int db,\ F_{abzi} = \eta'_{iG} \dot{z}_i \int da,\ F_{aczi} = G'_i z_i \int da \end{cases} \tag{39}$$

Where usually the index ***i*** = *1* use for a soft body and ***i*** = *2* use for a hard body.

Since as $x = \dfrac{x_1}{D_1} = \dfrac{x_2}{D_2}$, we can write that $a = 2k_{pa} R_a^{1/2} \dfrac{x_1^{1/2}}{D_1^{1/2}}$ and as well $a = 2k_{pa} R_a^{1/2} \dfrac{x_2^{1/2}}{D_2^{1/2}}$, and $b = 2k_{pb} R_b^{1/2} \dfrac{x_1^{1/2}}{D_1^{1/2}}$ and as well $b = 2k_{pb} R_b^{1/2} \dfrac{x_2^{1/2}}{D_2^{1/2}}$, then after differentiation we get:

$$da = \frac{k_{pa} R_a^{1/2}}{D_1^{1/2} x_1^{1/2}} dx_1,\ da = \frac{k_{pa} R_a^{1/2}}{D_2^{1/2} x_2^{1/2}} dx_2,\ db = \frac{k_{pb} R_b^{1/2}}{D_1^{1/2} x_1^{1/2}} dx_1,\ db = \frac{k_{pb} R_b^{1/2}}{D_2^{1/2} x_2^{1/2}} dx_2 \tag{40}$$

After a substituting the expressions (40) into (39) and then after their integration we can get the equations for the viscoelastic forces separately for each body. If we can find the viscoelastic forces separately for each body hence we can find separately the viscoelastic stresses for each body too. It is not possible to do by using the Hertz theory or by using others already existing theories and methods.

The normal viscoelastic stresses, which equal to the mean pressure P_m in the contact area, can be found as

$$P_m = \frac{F_n}{S_x}, \tag{41}$$

where $S_x = \dfrac{\pi}{4} ab$ is the contact area, and according to Equation (26) follows $S_x = \pi k_{pa} k_{pb} R_a^{1/2} R_b^{1/2} x$. In correspondence with Equations (20), (37) and (41) the expression for maximum value of the normal contact pressure (stresses) can be written as follows:

$$P_c = \frac{\psi(3\eta'_E \dot{x} + E'x)}{\pi k_{pa} k_{pb} R_a^{1/2} R_b^{1/2} x^{1/2}} \tag{42}$$

3. Dynamic Contact between two Spherical Bodies at Impact

In the case of the contact between two spheres, when the contact area is a circle $R_a = R_b = R$, $r_a = r_b = r$, $a = b = 2r$, $F_{\tau z} = 0$, $F_\tau = F_{\tau y}$, $F_{b\tau} = F_{b\tau y}$, $F_{c\tau} = F_{c\tau y}$ (see Figure1) follows that $k_p = k_{pa} = k_{pb}$, $h_x = h_{xa} = h_{xb}$ and $k_h = k_{ha} = k_{hb}$, $\psi = 2k_p R^{1/2}$, and then according to Equations (1) and (37) we get the next system of equations for the main viscoelastic forces:

$$\begin{cases} F_n = F_{bn} + F_{cn} = b_x \dot{x} + c_x x = 4k_p \eta'_E R^{1/2} \dot{x} x^{1/2} + \frac{4}{3} k_p E' R^{1/2} x^{3/2} \\ F_\tau = F_{b\tau} + F_{c\tau} = b_y \dot{y} + c_y y = \eta'_G P_x \dot{y} + G' P_x y \end{cases} \tag{43}$$

Where $P_x = k_h x + 2k_p R^{1/2} x^{1/2}$, and where the expressions for the variable viscoelasticity parameters in the Equation (3) can be written as follows

$$b_x = 4\eta'_E R^{1/2} x^{1/2}; \quad c_x = \frac{2}{3} E' R^{1/2} x^{1/2}, \quad b_y = \eta'_G P_{bx}; \quad c_y = G' P_{bx} \tag{44}$$

Also, for this case, the expression for the depth of the contact surface can be written as follows

$$h_x = \left(\frac{R - D_2 R_2}{R_2} \right) x = k_h x \tag{44*}$$

4. Contact between a Spherical Solid Body and a Semi-Space at Impact

This example is given here, because the results, which can be obtained in this case, can be used also for others cases of a dynamic contacts, such as for the dynamic contact at impact between two spherical bodies or between two cylindrical bodies.

Let a spherical body, having the average statistical mass m_2, the radius R and the initial velocity V_0, comes into viscoelastic contact under an arbitrary angle of attack α to the surface of semi-space at the initial instant of the time $t = 0$, at the initial point of contact 0 (Figure 4). And let the vectors of velocities are applied to the centre of mass of the body (the point C_0). Also in Figure4 are designated: $V_{0x} = V_0 sin\alpha$, $V_{0y} = V_0 cos\alpha$ are the initial normal and tangential velocities of a body; $M = F_\tau l_x$ is the reactive moment; l_x is the shoulder of tangential force; ω is the angular velocity and ε is the angular acceleration around of the centre of mass of a body; V_d is the volume of deformations, which is forming in the course of contact. Also the geometry of surface of the contact zone is characterised by the geometrical parameters, such as (Figure 4): x_1 is the normal deformation of the surface of semi-space in the middle of the contact area, which as well, is the approach of a semi-space relative to a body; x_2 is the normal deformation of the surface of a body in the middle of the contact area, which as well, is the approach of a semi-space relative to a semi-space; r - the radius of the contact area; h_x is the depth of the contact surface, or other words, h_x is the depth of indentation of a body into the surface of semi-space.

Also, it is seen here (Figure4) that at the initial instant of the time, the body with the centre of mass in the point C_0 comes into contact with the surface of semi-space at the initial point of the contact 0 with coordinates $x = 0$ and $y = 0$, but at the instant of the time t, the centre of mass of a body (the point C_t) takes the position with coordinates x and y.

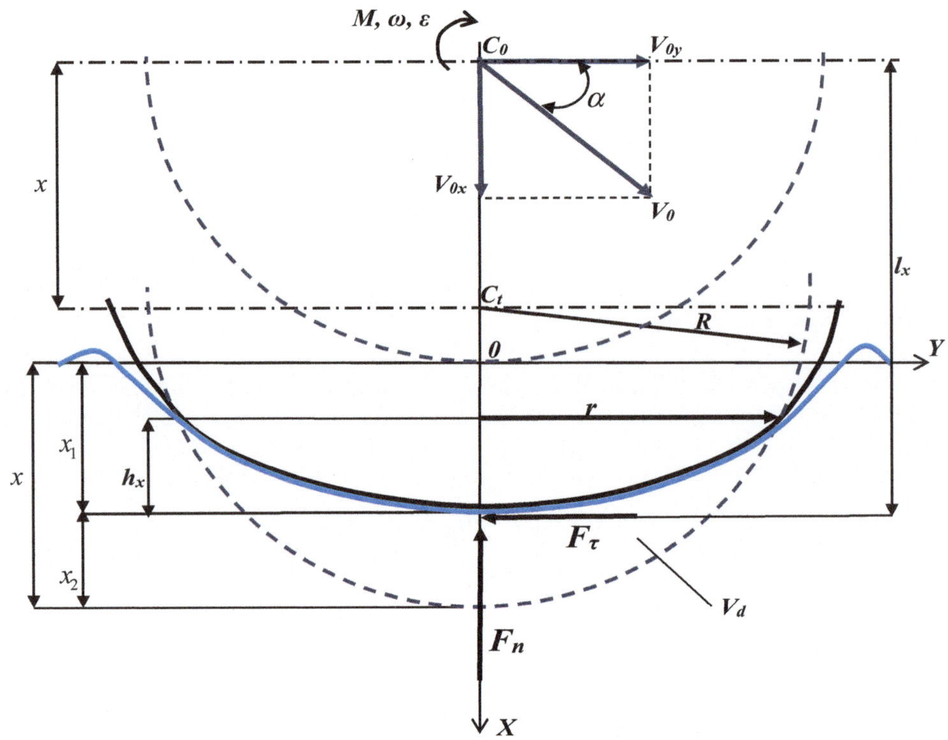

Figure 4. Schematic illustration of the indentation of a spherical body into a semi-space

It is obviously that the viscoelastic forces F_n and F_τ are acting in the contact area between the surfaces of the contact and according to Newton's Second Law we can write:

$$\begin{cases} F_n = -m\ddot{x} \\ F_\tau = -m\ddot{y} \\ M = -J_z\ddot{\varphi} \end{cases} \quad (45)$$

Where: m is the effective or reduced mass of the contacting bodies; $\ddot{y}, \ddot{x}$ - the accelerations of the centre of mass of the body; J_z is the moment of inertia of a body; φ the angle of rotation of the body around the centre of mass; $\ddot{\varphi}$ is the angular acceleration of a body around the centre of mass.

Remark: The term "effective mass" already have been used by Stronge (2000), Dintwa, (2006), Bordbar, Hyppänen (2007), Antypov, Elliott, Hancock (2011), and by many others authors. Also the mass m was called like the reduced mass by Landau (1944, 1965), Brilliantov (1996).

We can see here that $x = x_1 + x_2$ is the distance of the mutual approach (the total deformation) between a body and a semi-space, and as well, in the same instant of the time, it is the displacement of the centre of mass of a body relative to the initial point of contact *0* by axis *X*. At impact of two bodies, the effective mass m enters usually like the mass of the third body, and the movement (the displacement) x of the centre of mass of this third body takes equal to the distance of the mutual approach (or a compression, an overlapping) of the colliding bodies. At impact of two bodies, according to the second law of Newton, we can write that

$$F_n = -m\frac{dV_{0x}}{dt} = -m_1\frac{dV_{1x}}{dt} = -m_2\frac{dV_{2x}}{dt},$$

where $V_{0x} = V_{1x} + V_{2x}$. These equations are valid only for the movement of the centres of mass of the bodies. All authors, who use these equations, for example, Stronge (2000), Dintwa, (2006), Bordbar, Hyppänen (2007), Antypov, Elliott, Hancock (2011), Landau (1944, 1965), Brilliantov (1996) take that x is the displacement of centre mass of this third body and x is the mutual approach (a compression, an overlapping) between the bodies too. From these two expressions follows that $\frac{1}{m} = \frac{1}{m_1} + \frac{1}{m_2}$ and since as in the case of the collision of a body and a semi-space $m_1 >> m_2$, and hence we can take that $m = m_2$. As well, if the semi-space is immovable, when the velocity of a semi-space $V_{1x} = 0$, follows that $m = m_2$. Hence it is proved that in the case of collision between a body and a semi-space, the mass of a body m_2 is equal to the effective mass *m*. Further in this paper, the mass of body is designated by the symbol *m*. Consequently, the distance of the mutual approach between a body and a semi-space is equal to the displacement *x* of the centre of mass of a body. Analogically as for *x* we can write that $y = y_1 + y_2$ is the displacement of the centre of mass of a body relative to the initial point of contact *0* by axis *Y*, where y_1 - is the tangential deformation of the surface of a semi-space y_2 is the tangential deformation of the surface of a body.

It is obvious that, in the case of the contact between a spherical body and a semi-space, the effective radius *R* and the radius of a body are equal, and if take in account (25) and that $M = F_\tau l_x$, then if the dynamic viscosities replace by the dynamic viscosity modulus according to the known expressions (Ferry, 1948, 1963; Lee, 1962; Van Krevelen, 1972; Moor 1978; Nilsen, 1978, 1994; Lakes 1998; Meyers 1994; Menard 1999; Goloshchapov, 2001, 2003, 2015; Hosford 2005; Popov 2010, Popov and Hess 2015)

$$\frac{E''}{\omega_x} = \eta'_E \text{ and } \frac{G''}{\omega_y} = \eta'_G, \tag{46}$$

the equations for the normal and the tangential viscoelastic forces and for the reactive moment can be written as follows:

$$\begin{cases} F_n = F_{bn} + F_{cn} = \dfrac{4k_p E''}{\omega_x} R^{1/2} \dot{x} x^{1/2} + \dfrac{4}{3} k_p E' R^{1/2} x^{3/2} \\ F_\tau = F_{b\tau} + F_{c\tau} = \dfrac{G''}{\omega_y} P_x \dot{y} + G' P_x y \\ M = \left(\dfrac{G''}{\omega_y} \dot{y} + G' y \right) P_x l \end{cases} \tag{47}$$

Where E'' is the effective viscosity modulus under, G'' is the effective viscosity modulus at shear, ω_x is the normal angular frequency of damped oscillations of the volume of deformation V_d by the axis ***X***, ω_y is the tangential angular frequency of damped oscillations of the volume of deformation V_d by the axis ***Y***.

We have to mark that in the case of contact between a spherical body and a semi-space, when $R_2 = R$ (see Equation 44*) follows that $k_h = (1 - D_2) = D_1$, and hence

$$h_x = x_1 = D_1 x \tag{47*}$$

Viscosity modules can be found by using the known (Ferry,1948; Moore,1975; Van Krevelen, 1972; Nilsen, 1978; Landel, 1994) formula

$$\frac{E''}{E'} = \frac{G''}{G'} = tg\beta, \tag{48}$$

where β is the angle of mechanical losses.

Let us notice that often the dynamic modulus of elasticity is named yet as the accumulation (storage) modulus, and the dynamic modulus of viscosity is named yet as the loss modulus.

And now, according to (45) and (47), the differential equations of the movement (displacement) of the centre of mass of a body can be written as follows:

$$\begin{cases} m\ddot{x} + 4k_p \dfrac{E''R^{1/2}}{\omega_x}\dot{x}x^{1/2} + \dfrac{4}{3}k_p E'R^{1/2}x^{3/2} = 0 \\ m\ddot{y} + \dfrac{G''}{\omega_y}P_x\dot{y} + G'P_x y = 0 \\ J_z\ddot{\varphi} + \left(\dfrac{G''}{\omega_y}P_x\dot{y} + G'P_x y\right)l_x = 0 \end{cases} \qquad (49)$$

Or it can be also written in the canonical form as

$$\begin{cases} m\ddot{x} + b_x\dot{x} + c_x x = 0 \\ m\ddot{y} + b_y\dot{y} + c_y y = 0 \\ J_0\ddot{\varphi} + l_x \cdot \left(b_y\dot{y} + c_y y\right) = 0 \end{cases} \qquad (50)$$

Where, the formulas for the variable viscoelasticity parameters in the system of equation (50) can be written as

$$b_x = \frac{4E''R^{1/2}}{\omega_x}x^{1/2}, \quad c_x = \frac{4}{3}E'R^{1/2}x^{1/2}, \quad b_y = \frac{G''}{\omega_y}P_x, \quad c_y = G'P_x \qquad (51)$$

4.1 Work and Energy

As we know the period of time at impact includes two principally different phases such as, the phase of the compression and the phase of the restitution. Also in the duration of a collision, the full initial kinetic energy of a body $W_0 = \frac{mV_0^2}{2}$ is divided into the two independent parts such as, the normal initial kinetic energy of a body $W_{0x} = \frac{mV_{0x}^2}{2}$ and the tangential initial kinetic energy of a body $W_{0y} = \frac{mV_{0y}^2}{2}$. On the other hand, the full kinetic energy of a body at the instant of rebound $W_t = \frac{mV_t^2}{2}$ (where V_t is the velocity of the centre of mass of a body in the instant of rebound) is included two independent parts such as, the normal kinetic energy of a body at the instant of rebound $W_{tx} = \frac{mV_{tx}^2}{2}$ (where V_{tx} is the normal velocity of the centre of mass of a body in the instant of the rebound) and the tangential kinetic energy of a body at the instant of rebound $W_{ty} = \frac{mV_{ty}^2}{2}$ (where V_{ty} is the tangential velocity of the centre of mass of a body in the instant of rebound). Therefore, the description of the processes of the compression and the restitution along the axis *X*, and the shear along the axis *Y* are given independently in this part of the paper. The basic problems here are the finding the equations for the work of viscoelastic forces, the dissipative energy in the phases of the compression, the restitution and at the shear. Also it is necessary to determine the coefficients of restitution, the maximum size of the compression between a body and a semi-space, the dynamic modules of elasticity and viscosity.

4.1.1 Work and Energy in the Phases of Compression and Restitution

The graphical illustration of the functional dependences between the normal viscoelastic forces and the displacement of the centre of mass of a body is depicted in Figure 5: (a). Also the "Rheological model of Kelvin-Vogt", which usually is used for the viscoelastic contact, is represented in Figure 5: (b). As we can see, the viscosity force has the extremum in some point of the time equal τ_b.

It is obvious that the normal initial kinetic energy of the body W_{0x} is spent for the work A_{xm} of the normal viscoelastic force F_n in the compression phase. But on other hand, A_{xm} can be found as the sum of the works A_{xcm} and A_{xbm}, where A_{xcm} is the work of the normal elastic force F_{cn} and A_{xbm} is the work of the normal viscous force F_{bn} in the compression phase. Also we can say that the part of the kinetic energy W_{0x} is transformed into the potential energy of the nonlinear elastic element (spring) (Figure5: (b)) and the other part of this kinetic energy is dissipated during the time of deformation at the compression of the nonlinear viscous element (dashpot). However, on the other hand, the work A_{xt} of the normal viscoelastic force F_n in the restitution phase is equal to the normal energy of a body W_{tx} at the instant of rebound, and also A_{xt} can be found as the difference between A_{xct} and A_{xbt}, where A_{xct} is the work of the normal elastic force F_{cn} and A_{xbt} is the work of the normal viscous force F_{bn} in the restitution phase. Consequently, we can write that

$$\begin{cases} A_{xm} = A_{xcm} + A_{xbm} = W_{0x} = \dfrac{mV_{0x}^2}{2} \\ \\ A_{xt} = A_{xct} - A_{xbt} = W_{tx} = \dfrac{mV_{tx}^2}{2} \end{cases} \tag{52}$$

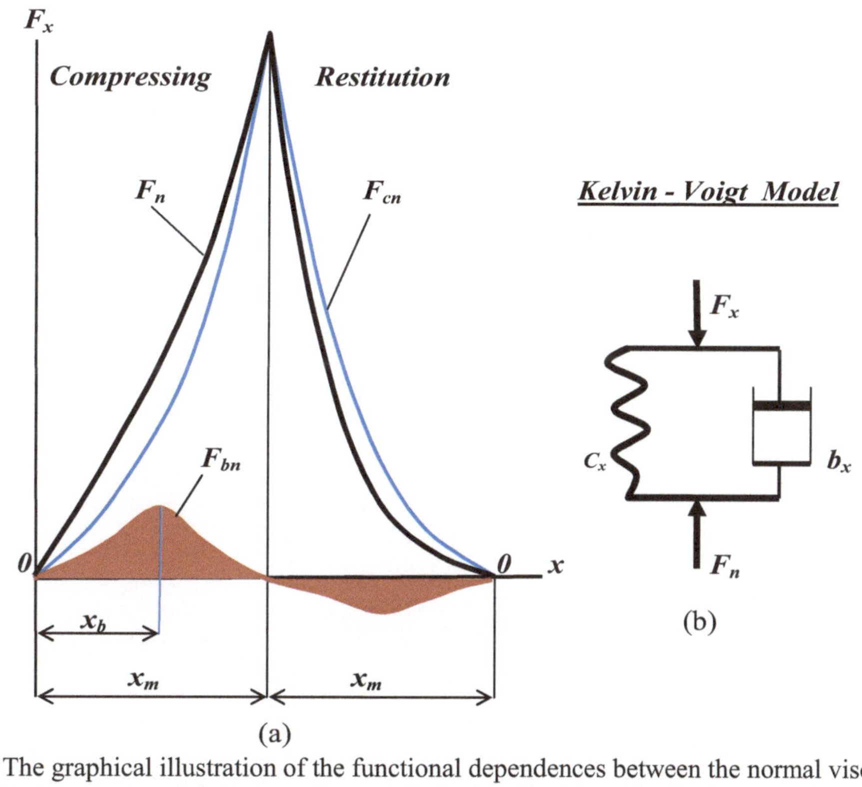

Figure 5: (a) - The graphical illustration of the functional dependences between the normal viscoelastic forces and the displacement x of the centre of mass of a body; (b) - The "Nonlinear Rheological Model of Kelvin-Vogt", where c_x and b_x are not the constant magnitudes

It is obvious that $A_{xcm} = A_{xct}$ and hence the potential energy which has been accumulated inside of the elastic element (spring) fully returns back to the body in the instant of rebound. The works A_{xcm} and A_{xbm} at the compression can be found by integration:

$$A_{xcm} = \int_0^{x_m} F_{cn} dx = \frac{4}{3} k_p E' R^{1/2} \int_0^{x_m} x^{3/2} dx = \frac{8}{15} k_p E' R^{1/2} x_m^{5/2} \tag{53}$$

and

$$A_{xbm} = \int_0^{x_m} F_{bn} dx = \int_0^{x_m} \frac{4k_p E'' R^{1/2}}{\omega_x} \dot{x} x^{1/2} dx = \left(\frac{4k_p E'' R^{1/2}}{\omega_x} \right) \frac{\int_0^{x_m} \int dx x^{1/2} dx}{\int_0^{\tau_1} dt} = \frac{8k_p E'' R^{1/2} x_m^{5/2}}{5\omega_x \tau_1} \tag{54}$$

Analogically the works A_{xct} and A_{xbt} in the restitution phase can be found as follows:

$$A_{xct} = -\int_{x_m}^{0} F_{cn} dx = -\int_{x_m}^{0} \frac{4}{3} k_p E' R^{1/2} x^{3/2} dx = \frac{8}{15} k_p E' R^{1/2} x_m^{5/2} \tag{55}$$

and

$$A_{xbt} = -\int_{x_m}^{0} F_{bn} dx = -\int_{x_m}^{0} \frac{4k_p E'' R^{1/2}}{\omega_x} \dot{x} x^{1/2} dx = -\left(\frac{4k_p E'' R^{1/2}}{\omega_x} \right) \frac{\int_{x_m}^{0} x^{1/2} dx \int dx}{\int_{\tau_1}^{\tau_x} dt} = \frac{8k_p E'' R^{1/2} x_m^{5/2}}{5\omega_x \tau_2} \tag{56}$$

Where: $\tau_x = \tau_1 + \tau_2$ is the period time of the contact; τ_1 is the period time of the compression; τ_2 is the period time of the restitution; x_m is the maximum magnitude of the compression between a body and a semi-space (also it is the maximum displacement of the centre of mass of a body, which is equal to the maximum of mutual approach between a body and a semi-space).

According (52), (53), (54), (55) and (56) the equations for the work of the compression and the restitution can be written as follows:

$$\begin{cases} A_{xm} = A_{xcm} + A_{xbm} = \dfrac{8}{15} k_p R^{1/2} x_m^{5/2} (E' + \dfrac{3E''}{\omega_x \tau_1}) \\ \\ A_{xt} = A_{xct} - A_{xbt} = \dfrac{8}{15} k_p R^{1/2} x_m^{5/2} (E' - \dfrac{3E''}{\omega_x \tau_2}) \end{cases} \tag{57}$$

and according (48) and (57) we can write

$$\begin{cases} A_{xm} = \dfrac{8}{15} k_p E' R^{1/2} x_m^{5/2} \left(1 + \dfrac{3tg\beta}{\omega_x \tau_1} \right) \\ \\ A_{xt} = \dfrac{8}{15} k_p E' R^{1/2} x_m^{5/2} \left(1 - \dfrac{3tg\beta}{\omega_x \tau_2} \right) \end{cases} \tag{58}$$

Since as $A_{xm} = W_{0x} = \dfrac{mV_{0x}^2}{2}$, $A_{xt} = W_{tx} = \dfrac{mV_{tx}^2}{2}$, and by using the first of the equations (58), we get the formula for x_m respectively

$$x_m = \left[\frac{15 m \omega_x \tau_1 V_{0x}^2}{16 (3tg\beta + \omega_x \tau_1) k_p E' R^{1/2}} \right]^{2/5} \tag{59}$$

Also, we can define the energetic coefficient of restitution e_x, which equals to the square of the kinematic coefficient of restitution k_x (further it will be named simply the coefficient of restitution), like the ratio between W_{tx} and W_{0x}:

$$e_x = k_x^2 = \frac{V_{tx}^2}{V_{0x}^2} = \left(\frac{\omega_x \tau_2 - 3tg\beta}{\omega_x \tau_1 + 3tg\beta}\right)\frac{\tau_1}{\tau_2} \tag{60}$$

Since as we can take that

$$x_m \approx \frac{/V_{0x}/}{2}\tau_1 = \frac{/V_{tx}/}{2}\tau_2, \tag{61}$$

we get that

$$k_x = \frac{\tau_1}{\tau_2}, \tag{62}$$

and using (60) and (62) we get that

$$tg\beta = \frac{\omega_x \tau_1}{3} \times \frac{1-k_x}{k_x} \tag{63}$$

Thus, we have got the equation, which binds the coefficient of restitution and the tangent of the angle of mechanical losses. So, if $k_x = 1$, $tg\beta \to 0$ we get the totally elastic impact, but if $k_x = 0$, $tg\beta \to \infty$ then we get the totally viscous impact. Using (63) we can write the formula for the restitution coefficient as

$$k_x = \left[\frac{\omega_x \tau_1}{(3tg\beta + \omega_x \tau_1)}\right] \tag{64}$$

If to compare the Equations (59) and (64) we can finally get the expression for the maximum magnitude of the compression between a body and a semi-space respectively as

$$x_m = \left[\frac{15\, mV_{0x}^2}{16\, k_p E' R^{1/2}} k_x\right]^{2/5} \tag{65}$$

In the case of a totally elastic impact, when $k_x = 1$ and $k_p = 1$ we get the same result, as it has been obtained by L. Landau (1944, 1965) according to the Hertz theory for the absolutely elastic contact.

4.1.2 Work and Energy at the Rolling Shear

It is obvious that, in the during time of the displacement and the rolling shear along axis Y, the tangential initial kinetic energy of a body $\boldsymbol{W_{0y}}$ is spent for the work A_y of the tangential viscoelastic force F_τ. The work A_y can be found as the sum of the works A_{yb} and A_{yc}, where A_{yb} is the work of the tangential viscous force $F_{b\tau}$ and A_{yc} is the work of the tangential elastic force. But on other hand, it is obvious as well, that the work A_{yb} is transformed into the dissipative energy Q_ω and the work A_{yc} is transformed into the work A_ω of the rotation of the body around the centre of mass of a body. Thus, according to the "Law of preservation of energy for a non-conservative (dissipative) mechanical systems", we can write the equations for the displacement of the centre of mass of a body and for the rotation of a body relative to the centre of mass of a body, as follows below:

$$\begin{cases} \dfrac{m}{2}\left(\dfrac{dy}{dt}\right)^2 + A_y = \dfrac{mV_{oy}^2}{2} \\ \dfrac{J_z}{2}\left(\dfrac{d\varphi}{dt}\right)^2 + A_\omega + Q_\omega = \dfrac{J_z \omega_0^2}{2} \end{cases} \tag{66}$$

Where: $A_y = \int F_\tau dy$; $A_\omega = -\int M d\varphi$; $Q_\omega = \int F_{b\tau} dy$, and where $M = F_\tau l_x$.

Since $F_\tau = F_{c\tau} + F_{b\tau}$ and if $R >> x$, we can take that $M = F_\tau R = (F_{c\tau} + F_{b\tau})R$, and since as $d\varphi = dy/R$, hence

$$A_\omega = -\int M d\varphi = -\int (F_{c\tau} + F_{b\tau})dy \tag{67}$$

Also since, if the initial angular velocity ω_0 equals zero we can write the Equation (66) for the boundary conditions in the instant of the time $t=\tau_1$ of the maximum compression $x=x_m$ and $y=y_1$ as follows:

$$\begin{cases} \dfrac{mV_{my}^2}{2} + \displaystyle\int_0^{y_1}(F_{c\tau} + F_{b\tau})dy = \dfrac{mV_{oy}^2}{2} \\ \dfrac{J_z\omega_m^2}{2} - \displaystyle\int_0^{y_1}(F_{c\tau} + F_{b\tau})dy + \int_0^{y_1} F_{b\tau}dy = 0 \end{cases} \tag{68}$$

Where: V_{my} is the velocity at the instant of the time $t=\tau_1$; ω_m is the angular velocity relative to the centre of mass of a body at the instant of the time $t=\tau_1$; y_1 is displacement of the centre of mass of a body along axis Y at the instant of the time $t=\tau_1$. The Equation (68) can be rewritten as

$$\begin{cases} \displaystyle\int_0^{y_1}(F_{c\tau} + F_{b\tau})dy = \dfrac{mV_{oy}^2}{2} - \dfrac{mV_{my}^2}{2} \\ \dfrac{J_z\omega_m^2}{2} = \displaystyle\int_0^{y_1} F_{c\tau}dy \end{cases} \tag{69}$$

Also at the point of the rebound, when $t=\tau_x$ we get

$$\begin{cases} \dfrac{mV_{ty}^2}{2} + \displaystyle\int_{y_1}^{y_t}(F_{c\tau} + F_{b\tau})dy = \dfrac{mV_{my}^2}{2} \\ \dfrac{J_z\omega_t^2}{2} - \displaystyle\int_{y_1}^{y_t}(F_{c\tau} + F_{b\tau})dy + \int_{y_1}^{y_t} F_{b\tau}dy = \dfrac{J_z\omega_m^2}{2} \end{cases} \tag{70}$$

Then we can write that

$$\begin{cases} \dfrac{mV_{ty}^2}{2} + \displaystyle\int_{y_1}^{y_t}(F_{c\tau} + F_{b\tau})dy = \dfrac{mV_{my}^2}{2} \\ \dfrac{J_z\omega_t^2}{2} - \dfrac{J_z\omega_m^2}{2} = \displaystyle\int_{y_1}^{y_t} F_{c\tau}dy \end{cases} \tag{71}$$

The summation of the systems (69) and (71) together yields the following result

$$\begin{cases} \dfrac{mV_{ty}^2}{2} = \dfrac{mV_{0y}^2}{2} - \displaystyle\int_0^{y_1}(F_{c\tau} + F_{b\tau})dy - \int_{y_1}^{y_t}(F_{c\tau} + F_{b\tau})dy \\ \dfrac{J_z\omega_t^2}{2} = \displaystyle\int_0^{y_1} F_{c\tau}dy + \int_{y_1}^{y_t} F_{c\tau}dy \end{cases} \tag{72}$$

We can rewrite Equation (72) in the next order

$$\begin{cases} A_y = \dfrac{mV_{0y}^2}{2} - \dfrac{mV_{ty}^2}{2} = \int_0^{y_1}(F_{c\tau} + F_{b\tau})dy + \int_{y_1}^{y_t}(F_{c\tau} + F_{b\tau})dy \\ \dfrac{J_z\omega_t^2}{2} = \int_0^{y_1} F_{c\tau}dy + \int_{y_1}^{y_t} F_{c\tau}dy \end{cases} \tag{73}$$

Finally we get

$$\begin{cases} A_y = \dfrac{mV_{0y}^2}{2} - \dfrac{mV_{ty}^2}{2} = A_{ybm} + A_{ycm} + A_{ybt} + A_{yct} \\ A_{yc} = \dfrac{J_z\omega_t^2}{2} = A_{ycm} + A_{yct} \end{cases} \tag{74}$$

Where: $A_{ybm} = \int_0^{y_1} F_{b\tau}dy$ is the work of the tangential viscous force $\overline{F_{b\tau}}$ in the compression period τ_1 ;

$A_{ycm} = \int_0^{y_1} F_{c\tau}dy$ is the work of the tangential elastic force $\overline{F_{c\tau}}$ in the compression period τ_1 ;

$A_{ybt} = -\int_{y_1}^{y_t} F_{b\tau}dy$ is the work of the tangential viscous force $\overline{F_{b\tau}}$ in the restitution period τ_2 ;

$A_{yct} = -\int_{y_1}^{y_t} F_{c\tau}dy$ is the work of the tangential elastic force $\overline{F_{c\tau}}$ in the restitution period τ_2 .

Since as from Equation 47 follows that $F_{b\tau} = \dfrac{G''}{\omega_y}P_x\dot{y}$ and $F_{c\tau} = G'P_x y$, all these works in (74) can be found by integration, as follows:

$$\begin{cases} A_{ybm} = \dfrac{G''}{\omega_y}\int_0^{P_m} dP_x \int_0^{y_1} \dot{y}dy = \dfrac{G''}{\omega_y}P_m \dfrac{\int_0^{y_1}\int dydy}{\int_0^{\tau_1} dt} = \dfrac{G''}{2\omega_y}P_m \dfrac{y_1^2}{\tau_1} \\ A_{ycm} = G'\int_0^{P_m} dP_x \int_0^{y_1} ydy = \dfrac{G'}{2}P_m y_1^2 \\ A_{ybt} = -\dfrac{G''}{\omega_y}\int_{P_m}^{0} dP_x \int_{y_1}^{y_t} \dot{y}dy = -\dfrac{G''}{\omega_y}(0 - P_m)\dfrac{\int_{y_1}^{y_t}\int dydy}{\int_{\tau_1}^{\tau_x} dt} = \dfrac{G''}{2\omega_y}P_m \dfrac{y_t^2 - y_1^2}{\tau_x - \tau_1} \\ A_{yct} = -G'\int_{P_m}^{0} dP_x \int_{y_1}^{y_t} ydy = \dfrac{G'}{2}P_m(y_t^2 - y_1^2) \end{cases} \tag{75}$$

Where

$$P_m = D_1 x_m + 2k_p R^{1/2} x_m^{1/2} \tag{76}$$

The full changing of the energy of the dissipative system at the rolling shear can be found as the difference between A_y and A_{yc} from (74):

$$\Delta W_y = A_y - A_{cy} = \frac{mV_{0y}^2}{2} - \frac{mV_{ty}^2}{2} - \frac{J_z\omega_t^2}{2} = A_{ybm} + A_{ybt} = A_{yb} \quad (77)$$

According to the Equation (74) the conclusion can be drawn that the work A_{yc} is transformed into the kinetic energy of the rotation of the body relative its centre of mass, but on the other hand the work A_{yb} is transformed into dissipative energy Q_ω in the process of the internal friction. Accordingly, using (74) and (75) we have

$$A_{yc} = \frac{J_z\omega_t^2}{2} = \frac{G'}{2}P_m y_t^2 \quad (78)$$

Hence the equation for the angular velocity at the instant time of rebound can be written as follows

$$\omega_t = \left(\frac{G'P_m}{J_z}\right)^{\frac{1}{2}} y_t \quad (79)$$

Since the work A_{yb} of the viscous tangential force $F_{b\tau}$ is equal to the dissipative energy Q_ω, using Equation (75) we get

$$A_{by} = Q_\omega = A_{ybm} + A_{ybt} = \frac{G''}{2\omega_y}P_m\left(\frac{y_1^2}{\tau_1} - \frac{y_1^2}{\tau_2} + \frac{y_t^2}{\tau_2}\right) \quad (80)$$

Since as $\tau_1 = k_x\tau_2$, finally we get

$$A_{by} = Q_\omega = A_{ybm} + A_{ybt} = \frac{G''}{2\omega_y\tau_2 k_x}P_m\left[y_1^2(1-k_x) + k_x y_t^2\right] \quad (81)$$

5. Approximate solution to the Differential Equations of the Displacement by using the Method of the Equivalent Works

For practical application of the differential Equation (50) with the variable viscoelasticity parameters, we can find their approximate solutions in the same manner as for the equations with the equivalent constant viscoelasticity parameters, if we choose the equivalent constant parameters B_x, C_x and B_y, C_y so that the work A_{xcm} and A_{xbm}, A_{ycm} and A_{ybm} with the variable viscoelasticity parameters c_x, b_x and c_y, b_y will be equal to the works with the constant viscoelasticity parameters. Thus, according to this statement, and since as the work A_{xcm} and A_{xbm} , A_{ycm} and A_{ybm} are known from Equations (53), (54) and (75), we can write next equations in the phase of the compression

$$\begin{cases} A_{xcm} = C_x\int_0^{x_m} x\,dx = \frac{1}{2}C_x x_m^2 = \frac{8}{15}k_p E'R^{1/2}x_m^{5/2} \\ \\ A_{xbm} = B_x\int_0^{x_m}\dot{x}\,dx = B_x\frac{\int_0^{x_m} x\,dx}{\int_0^{\tau_1} dt} = B_x\frac{x_m^2}{2\tau_1} = \frac{8k_p E''R^{1/2}}{5\Omega_x\tau_1}x_m^{5/2} \end{cases} \quad (82)$$

and also in the phase of the rolling shear for the period of the compression time

$$\begin{cases} A_{ycm} = C_y \int_0^{y_1} y dy = \frac{1}{2} C_y y_1^2 = \frac{1}{2} G' P_m y_1^2 \\ \\ A_{ybm} = B_y \int_0^{y_1} \dot{y} dy = B_y \frac{\int_0^{y_1} \int dy dy}{\int_0^{\tau_1} dt} = B_y \frac{y_1^2}{2\tau_1} = \frac{G''}{2\omega_y \tau_1} P_m y_1^2 \end{cases} \tag{83}$$

Hence, according to the results obtained in (82) and in (83), we can write the expressions for the equivalent constant viscoelasticity parameters, respectively as:

$$B_x = \frac{16 E'' k_p R^{1/2}}{5 \omega_x} x_m^{1/2}, \quad C_x = \frac{16}{15} k_p E' R^{1/2} x_m^{1/2}, \quad B_y = \frac{G''}{\omega_y} P_m, \quad C_y = G' P_m \tag{84}$$

Thus, the Equation (50) with variable parameters can be rewritten as the equations with constant parameters as follows:

$$\begin{cases} m\ddot{x} + B_x \dot{x} + C_x x = 0 \\ m\ddot{y} + B_y \dot{y} + C_y y = 0 \\ J_z \ddot{\varphi} + l_x \cdot (B_y \dot{y} + C_y y) = 0 \end{cases} \tag{85}$$

The Equation (85) are the equations of the damped oscillations and the solutions to these equations are known:

$$\begin{cases} x = \frac{\upsilon_{0x}}{\omega_x} e^{-\delta_x t} \sin(\omega_x t) \\ \\ y = \frac{\upsilon_{0y}}{\omega_y} e^{-\delta_y t} \sin(\omega_y t) \end{cases} \tag{86}$$

Where: $\omega_x = \sqrt{\omega_{0x}^2 - \delta_x^2}$; $\delta_x = \frac{B_x}{2m}$ is the normal damping factor; $\omega_{0x} = \sqrt{\frac{C_x}{m}}$ is the angular frequency of the harmonic oscillations by axis X ; $\omega_y = \sqrt{\omega_{0y}^2 - \delta_y^2}$; $\delta_y = \frac{B_y}{2m}$ is the tangential damping factor; $\omega_{0y} = \sqrt{\frac{C_y}{m}}$ is the angular frequency of the harmonic oscillations by axis Y.

It is obviously that the period of time of the contact τ_x is equal to the semi-period of damped oscillations $T_x/2$ by axis X.

$$\tau_x = \frac{T_x}{2} = \frac{\pi}{\omega_x} \tag{87}$$

Since as $\tau_x = \tau_1 + \tau_2$ and also by using Equations (60), (62),(63) and (87) we get:

$$tg\,\beta = \frac{\pi}{3} \times \frac{(1 - k_x)}{(1 + k_x)} \tag{88}$$

The equation for the restitution coefficient we can write now as follows:

$$k_x = \frac{(\pi - 3tg\beta)}{(\pi + 3tg\beta)} \tag{89}$$

If $tg\beta = 0$ hence $k_x = 1$, it is a totally elastic impact, but if $tg\beta = \pi/2$ hence $k_x = 0$ and $x = 0$, it is absolutely plastic impact. Both of these two cases are not possible in nature.

Finally, from (65) and (89) follows that

$$x_m = \left[\frac{15 mV_{0x}^2}{16 k_p E'R^{1/2}} \times \frac{(\pi - 3tg\beta)}{(\pi + 3tg\beta)} \right]^{2/5} \tag{90}$$

Thus we have a very simple way to calculate x_m, if we know the value of $tg\beta$. According to the Equation (9), (46) and (48) $tg\beta$ can be calculated by formula

$$tg\beta = \frac{E''}{E'} = \frac{E_1''E_2''(E_1' + E_2')}{E_1'E_2'(E_1'' + E_2'')} \tag{91}$$

The equations for the velocities of the centre of mass of a body can be received by differentiation of (86):

$$\begin{cases} \dot{x} = \dfrac{V_{0x}}{\omega_x} e^{-\delta_x t} [\omega_x \cos(\omega_x t) - \delta_x \sin(\omega_x t)] \\ \\ \dot{y} = \dfrac{V_{0y}}{\omega_y} e^{-\delta_y t} [\omega_y \cos(\omega_y t) - \delta_y \sin(\omega_y t)] \end{cases} \tag{92}$$

Using (92) for the velocity, the duration of the time of the impact equals to the period of the time of the contact can be found now from the conditions $\dot{x} = V_{tx}$ and $t = \tau_x$ as

$$\tau_x = -\frac{\ln k_x}{\delta_x} , \tag{93}$$

where

$$\delta_x = \frac{B_x}{2m} = \frac{8k_p E''R^{1/2}}{5m\,\omega_x} x_m^{1/2} = \frac{8k_p E'tg\beta}{5\pi m} \tau_x R^{1/2} x_m^{1/2} , \tag{94}$$

and since $tg\beta$ is known from (88), by using (65),(93) and (94) we get

$$\tau_x^2 = -\frac{2(1 + k_x)\ln k_x}{V_{0x}^{2/5}(1 - k_x)k_x^{1/5}} \times \left(\frac{5m}{8k_p E'R^{1/2}} \right)^{4/5} \tag{95}$$

6. Determination of the Dynamics Modules by the Method of the "Temperature -Time Superposition"

The dynamic elasticity and viscosity modules for high velocities of the collision can be found, if to follow the principles of the "Time-temperature superposition" according to the equation of the "WLF" Williams - Landel - Ferry or Arrhenius (Ferry, 1963; Van Krevelen, 1972; Moore, 1975; Nilsen and Landel, 1994). First of all we have to define experimentally the effect of temperature for the period of the contact time τ_x, and for the coefficient of restitution k_x at the fixed initial velocity of impact. For example, if we define these parameters for velocity at 2 m/s, then using the principles of the "Time-temperature superposition" we can determine their values for any velocities interesting for us, for example for velocity 100 m/c and for temperature 100 0C. After this, when τ_x and k_x will be known, we can find the value of $tg\beta$ and the dynamic modules E'' and E'. If to use the equation (95), the expression for the calculation of the effective dynamic elasticity module can be written as follows

$$E' = \frac{5m}{8k_p V_{0x}^{1/2} R^{1/2} \tau_x^{5/2}} \times \left(\frac{(-2\ln k_x)(1+k_x)}{k_x^{1/5}(1-k_x)} \right)^{5/4} \tag{96}$$

And, if to use (48), (88) and (96) we get the formula for the calculation of the effective dynamic viscosity module

$$E'' = \frac{15\pi m}{24 k_p V_{0x}^{1/2} R^{1/2} \tau_x^{5/2}} \times (-2\ln k_x)^{5/4} \left(\frac{(1+k_x)}{k_x(1-k_x)} \right)^{1/4} \tag{97}$$

Obviously, if $k_x = 0$, then $E'' = 0$ too. We can find G' and G'' in the analogical way.

8. Analyse and Conclusion

It is a very important now to confirm of the correctness of the offered theories and methods, obtained in this article, if to compare them with others already available. For example, the equation for the elastic force $F_{cn} = \frac{4}{3} k_p E' R^{1/2} x^{3/2}$ have been obtained by the using of the "Method of the specific forces". In the case, when $k_p = 1$ we have the same solution like in the Hertz theory. Hence the Hertz theory gives the partial results in comparisons with proposed "MSF" method, because MSF" let to find the viscoelastic forces for any curvilinear contact between two surfaces, but Hertz case can be using only for the flat contact. The obtained result proves us that, the "Method of the specific forces" is definitely valid for finding of the normal elastic force, because if we know a functional dependency between r and x, we can always find the elastic force. It is obviously that, if it gives the correct way for the definition elastic force, and also as it was represented, it is valid for the definition of the viscous force and the tangential viscoelastic forces. We cannot find viscous force and the tangential viscoelastic forces by using the Hertz's theoretical model, but we can do this by using the "Method of the Specific Forces". It is obviously that, for the finding the normal viscous and the tangential viscoelastic forces, we can take $k_p = 1$, like according the Hertz theory, but we should be aware that, in this case, the contact area is a flat surface according to 2D tensor of deformations and h_x – the depth of indentation has to be equal to zero. But in reality, the contact surface takes the curvilinear shape, therefore, alternatively in this paper, the way of the finding of the radius of the contact area, by considering the geometry of contact between two curvilinear surfaces, have been proposed. It was received that the radius of contact area $r = f(x)$can be find by the equation $r^2 = 2Rx - x^2$. Since this equation is not convenient in using, and therefore it was proposed the finding $\boldsymbol{r}$ as $r^2 = k_p^2 Rx$, where $k_p = \sqrt{2 - \frac{x}{R}}$ is the correlation coefficient, which can be found by the method of iterations and consecutive approximations. If a deformation is small, when $R >> x$, hence we can take $k_p = \sqrt{2}$. And, if contact area is a flat, when $h_x = 0$, follows from Equation 47* that $D_1 = 0$. Practically the area of contact can be considered as a flat surface only in the case, when the surface of a semi-space in many times harder than the surface of a body. For example, it is possible in the case of impact between a rubber ball and a steel plane. Hence, it is obviously, that the "MSF" is the universal method, which can be used for any functional dependencies between the radius or the diameter of the contact area and the distance of the mutual approach (the total deformation) between two curvilinear surfaces. But nevertheless, we still have the question: What kind of the equation is better to take for finding of the radius of contact area, by the Hertz theory or directly by the way of consideration the geometry of the contact, like it is proposed in this article? Objectively to answer this question, we have to analyse simply logically the way as these equations were received. It was taken according to the Hertz theory that the contact surface is a flat, and the deformations are very small, the contact pressure is distributed analogically as an electrical potential (Remark: an electrical potential is the scalar function, but a pressure is the vector function), and then, on the basis of this main statements, the equation between the radius of the contact area and the normal elastic force, and the equation between the distance of the mutual approach and the normal elastic force as the effect have been obtained. Then only after that, in result of the comparison of these two equations by excluding the normal force (Landau, 1944, 1965), the expression $r^2 = Rx$ have been received. But in this article the analogical functional dependence have been proposed as the cause, in the result of the direct consideration the geometry of the contact. It was shown that, in the time of indentation of more hard surface into a soft surface, the contact surface takes a curvilinear elliptical shape (the function $r^2 = 2Rx - x^2$ is the elliptical function, which can be approximated by the parabolic function $r^2 = k_p^2 Rx$), where the point B (see Figure1) is a special point where the deformations always equal zero, and the border of the area of contact always pass through this point B. Therefore, the proposed geometrical method is more exact, than according to the Hertz's theoretical model.

On the other hand, the equation for the normal viscous force $F_{bn} = 4k_p \frac{E''}{\omega_x} R^{1/2} \dot{x} x^{1/2}$ gives the similar result as it has place in the contact between two bodies with identical mechanical properties in the equation (4*), which have been obtained by Brilliantov, N. V., Spahn, F., Hertzsch, J.-M., and Poeschel, T. (1996). After the comparison of two these equations, since as $x = \xi$ and $R = R^{eff}$ we obtain the following

$$A = \frac{E'' k_p (1 - \nu^2)}{Y \omega_x} \tag{98}$$

But since as $\omega_x = \frac{\pi}{\tau_x}$ and in the quasi-static conditions $E' = Y$ we get

$$A = \frac{k_p tg\beta}{\pi} \times (1 - \nu^2) \times \tau_x \tag{99}$$

If $\beta = 0$ hence $A = 0$ too, it is a totally elastic impact. Thus we can find the parameter A by a very simple way using the "Method of the Specific Forces".

Also the equation (65) to determine the maximum displacement x_m have been derived. It is obvious, that in the case of $k_x = 1$ and $k_p = 1$ we have the same result, as was obtained by Landau (1944,1965) for a totally elastic impact by using the Hertz Theory. It proves the correctness of the way of finding the Equation 65. But we have to understand that, this equation has the borders of application which can be found if to solve the next equation

$\omega_x = \sqrt{\omega_{0x}^2 - \delta_x^2}$. First of all since as $\omega_{0x}^2 = \frac{C_x}{m}$ and $\delta_x = \frac{B_x}{2m}$, we can write that $\delta_x = \frac{3\omega_{0x}^2}{2\omega_x} tg\beta$ and we get the next algebraic equation $\omega_x^4 - \omega_{0x}^2 \omega_x^2 + \omega_{0x}^4 tg^2\beta = 0$. This equation has only the one valid solution $\omega_x^2 = \frac{\omega_{0x}^2}{2}\left(1 + \sqrt{1 - 9tg^2\beta}\right)$ and it has the valid root only when $1 - 9tg^2\beta \geq 0$, therefore $tg\beta = \frac{E''}{E'} \leq \frac{1}{3}$, and according to Equation 88 we get for a viscoelastic contact that

$$k_x \geq \frac{\pi - 1}{\pi + 1} \tag{100}$$

In the case when $k_x < \frac{\pi - 1}{\pi + 1}$ the plastic deformations will be have place in the zone of the contact.

Also it is necessary to proof the correctness of the definition of the work for the normal viscous force in Equation (54) and in Equation (82), because two ways in the order of integrations are possible to apply here such as, that have been taken for finding the solution in Equation (54) and in Equation (82) and like it is shown below

$$A_{xbm} = \left(\frac{4k_p E'' R^{1/2}}{\omega_1}\right) \frac{\int_0^{x_m} \int x^{1/2} dx dx}{\int_0^{\tau_1} dt} = \frac{16 k_p E'' R^{1/2} x_m^{5/2}}{15 \omega_1 \tau_1} \tag{101}$$

It is simple to proof that the 1-st variant of the order of integration in Equation (54) and in Equation (82) is correct and the second variant in Equation (101) is not correct. It obviously that the attitude between the normal viscous force and the normal elastic force can be find as follows

$$\frac{b_x \dot{x}}{c_x x} = \frac{B_x \dot{x}}{C_x x} = \frac{3tg\beta}{\omega_1} \frac{\dot{x}}{x} \tag{102}$$

Or hence we can write respectively

$$\frac{b_x}{c_x} = \frac{B_x}{C_x} = \frac{3tg\beta}{\omega_1} \tag{103}$$

As we can see this attitude is a constant value and it should not change at dependence by the order of integration. If we take the second variant of integration according to Equation (101), the values of the constants of viscoelasticity C_x and B_x will be changed. Thus therefore, in the first variant of integration as it is taken in Equation (54) and in Equation (82) we have a valid solution.

In conclusion, first of all, let us to mark, that the method of specific viscoelastic forces allows to find the equations for all viscoelastic forces. The proposed method is a principally different with others in which are using the Hertz's theory, the classical theory of elasticity and the tensor algebra. In this method the new conception is proposed, how to find the elastic and viscous forces by an integration of the specific forces in the infinitesimal boundaries of the contact area. The radius of contact area can be taken according the Hertz theory or can be found by the considering the geometry of the contact. This method can be used in researches of the contact dynamics of any shape of contacting surfaces. Also in the article the method of the solution of the differential equations of a movement has been proposed and they have been solved. This method also can be used for determination of the dynamic mechanical properties of materials, and it can be used in the design of wear-resistant elements and coverings for components of machines and equipment, which are working in harsh conditions where they are subjected to the action of flow or jet abrasive particles. Also the theoretical and experimental statements which are presented here can be useful in the design of elements and details machines and mechanism which are being in the conditions of the dynamic contact. The results of the experimental and theoretical research and the method of the specific forces presented in this article can be used for the determination of the viscoelastic forces, contact stresses, durability and fatigue life for a wide spectrum of the tasks relevant to collisions between solid bodies under different loading conditions. Opportunities exist to use the obtained results practically in the design and development of new advanced materials, wear-resistant elastic coatings and elements for pneumatic and hydraulic systems, stop valves, fans, centrifugal pumps, injectors, valves, gate valves and in other installations. Also the using of this theory gives an opportunity for the development of analytical and experimental methods allowing optimising the basic dynamic and mechanical visco-elastic qualities already existing materials and in the development new advanced materials and elements of machines. Also this theory can be used not only for visco-elastic contact and also for any other kind of contacts, such as the elasto- plastic contact and for the elasto-visco-plastic contact too.

References

Antypov, D., Elliott, J. A., & Bruno, C., & Hancock, B. C. (2011). Effect of particle size on energy dissipation in viscoelastic granular collisions. *Phys. Rev. E, 84, 021303.* Retrieved from http://PhysRevE.84.021303

Archard, J. F. (1957). Elastic deformation and the laws of friction. *Proceedings of the Royal Society of London. Series A, Mathematical and Physical Sciences, 243*(1233), 190--205.

Barber, J. R., & Ciavarella, M. (2000). Contact mechanics. *International Journal of Solids and Structures, 37.* http://dx.doi.org/10.1016/S0020-7683(99)00075-X

Becker, V., Schwager, T., & Pöschel, T. (2008). Coefficient of tangential restitution for the linear dashpot model. *Phys. Rev. E, 77, 011304.* http://dx.doi.org/10.1103/PhysRevE.77.011304

Bowden, F. P., & Tabor, D. (1939). The area of contact between stationary and between moving surfaces. *Proceedings of the Royal Society of London. Series A, Mathematical and Physical Sciences, 169*(938), 391-413

Brilliantov, N. V., Spahn, F., Hertzsch, J.-M., & Pöschel, T. (1996). Model for collisions in granular gases. *Phys. Rev. E,* 53, 5382. Retrieved from arxiv.org

Bordbar, M. H., & Hyppänen, T. (2007). Modeling of Binary Collision between Multisize Viscoelastic Spheres. *Journal of Numerical Analysis in Sciences and Engineering (ESCMSE), 2*(3-4), 115-128.

Boussinesq J. (1885). Application des potentiels à l'étude de l'équilibre et du mouvement des solides élastiques. Paris, 1885, Gauthier-Villars.

Bush, A. W., Gibson, R. D., & Thomas, T. R. (1975). The elastic contact of a rough surface. *Wear, 35*(1), 87-111.

Carbone, G., Lorenz, B., Persson, B. N. J., & Wohlers, A. (2009). Contact mechanics and rubber friction for randomly rough surfaces with anisotropic statistical properties. *The European Physiscal Journal E – Soft Matter, 29*(3), 275– 284. http://dx.doi.org/10.1140/epje/i2009-10484-8

Carbone, G., & Putignano, C. (2013). A novel methodology to predict sliding/rolling friction in viscoelastic materials: theory and experiments. *Journal of Mechanics and Physics of Solids.* http://dx.doi.org/10.1016/j.jmps.2013.03.005

Cheng, N., Subic, A., & Takla, M. (2008). Development of a fast-solving numerical model for the structural analysis of cricket balls. *Sports Technol., 1*(2–3), 132–144. Retrieved from http:// publication/227591545

Cummins, S. J., Thornton, C., & Cleary, P. W. (2012). CONTACT FORCE MODELS IN INELASTIC COLLISIONS, Ninth International Conference on CFD in the Minerals and Process Industries CSIRO, Melbourne, Australia. Retrieved from http://www.cfd.com.au/cfd_conf12/PDFs/175CUM.pdf

Cundall, P. A., & Strack, O. D. L. (1979). A discrete numerical model for granular assemblies. *Geotechnique, 29*, 47-65.

Derjaguin, B. V., Muller, V. M., & Toporov, Y. P. (1975). Effect of contact deformations on the adhesion of particles. *Journal of Colloid and Interface Science, 53*(2), 314-326.

Dintwa, E. (2006). Development of accurate contact force models for use with Discrete Element Method (DEM) modelling of bulk fruit handling processes. *DISSERTATIONES DE AGRICULTURA, Doctoraatsproefschrift nr. 726 aan de faculteit Bio-ingenieurswetenschappen van de K.U.Leuven, Katholieke Universiteit Leuven, Faculteit Bio-ingenieurswetenschappen.*

Ferry, J. D. (1948). Viscoelastic Properties of Polymer Solutions. *Journal of Research of the National Bureau of Standards, Research Paper, RP1903Volume, 41, US.*

Ferry, J. D. (1963). Viscoelasticpropertiesof polymers. New York: John Wiley & Sons, Inc.

Galin, L. A. (1961). Contact problems in the theory of elasticity. *Dept. of Mathematics, School of Physical Sciences and Applied Mathematics, North Carolina State College.*

Greenwood, J. A., & Williamson, J. B. P. (1966). Contact of nominally flat surfaces. *Proceedings of the Royal Society of London. Series A, Mathematical and Physical Sciences, 295*, 300-319.

Goldsmith, W. (1960). *Impact. The Theory and Physical Behaviour of Colliding Solids*. Adward Arnold Publisher Ltd, 371 pages.

Goloshchapov, N. N. (2001). Determination methods of cycle-thermo-kinetics viscoelastic and relaxation characteristic of elastomers under impact. *Cycles, Materials of the 3-d International conference, North-Caucasian State Technical University, Stavropol, Russia.*

Goloshchapov, N. N. (2003). Tribo-cyclicity. Cycles. Materials of 5th International conference. North-Caucasian State Technical University, Russia.

Goloshchapov, N. N. (2003). Processes of abrasive wear elastomers and their connection with viscoelasticity qualities. *2-d International scientific conference "Tribology in Environmental Design", Collected articles*. London: Professional Engineering Publishing l.

Goloshchapov, N. (2015). Application of the method of the specific forces in the mechanics of a viscoelastic collision between a solid body and a semi-space. *International Journal of Damage Mechanics, 24*(6), 915-943. Retrieved from http://ijd.sagepub.com/content/24/6/915

Graham, G. A. C. (1965). The contact problem in the linear theory of viscoelasticity. *International Journal of Engineering Science, 3*, 27. http://www.sciencedirect.com/science/article/pii/0020722565900182

Harrass, M., Friedrich, K., & Almajid, A. A. (2010). Tribological behavior of selected engineering polymers under rolling contact, *Tribology International, 43*, 635-646.

Hertz H. (1882). Über die Berührung fester elastischer Körper (On the contact of elastic solids*). J. Reine und angewandte Mathematik,* 92, 156-171.

Hertz, H. (1896). Über die berührung fester elastischer Körper (On the contact of rigid elastic solids). *In: Miscellaneous Papers.* J. reine und angewandte Mathematik 92, Macmillan, London, 156 pages. English translation.

Hosford, W. F. (2005). Mechanical Behaviour of Materials. New York: Cambridge University Press. Retrieved from http://assets.cambridge.org/97805218/46707/frontmatter/9780521846707_frontmatter.pdf

Hunter, S. C. (1960). The Hertz problem for a rigid spherical indenter and a viscoelastic half space. *Journal of the Mechanics and Physics of Solids, 8*, 219-234.

Hyun, S., & Robbins, M. O. (2007). Elastic contact between rough surfaces. Effect of roughness at large and small wavelengths. *Trobology International, 40*, 1413-1422. K. L.

Johnson, K. L, Kendall, K., & Roberts, A. D. (1971). Surface energy and the contact of elastic solids. *Proc. R. Soc. London A, 324*, 301-313. Retrieved from http://rspa.royalsocietypublishing.org/content/324/1558/30

Johnson, K. L. (1985). Contact mechanics. *Cambridge University Press.*

Jonas, A. Z. (1982). Impact dynamics. *Wiley, - Technology & Engineering - 452 pages.*

Lakes, R. S. (1998). Viscoelastic Solids. *II Series: Mechanical Engineering*. CRC Press.

Landau, L. D., & Lifshits, E. M. (1944). *Theory of Elastisity. Theoretical Physics* (vol.4). Moscow.

Landau, L. D., & Lifshitz, E. M. (1965). *Theory of Elasticity*. Oxford: Oxford University Press,.

Laursen, T. A. (2002). Computational Contact and Impact Mechanics. *Fundamentals of Modeling Interfacial Phenomena in Nonlinear Finite Element Analysis.* New York: Springer Verlag.

Lee, E. H. (1962). Viscoelasticity. In W. Flugge (Ed.), Handbook of Engineering Mechanics. New York: McGraw-Hill.

Li. H. (2006). Impact of cohesion forces on practical mixing and segregation. Doctoral dissertation. University of Pittsburgh. Retrieved from http://d-scholarship.pitt.edu/9699/

Makse, H. A., Gland, N., Johnson, D. L., & Schwartz, L. (2004). Granular Packings: Nonlinear elasticity, sound propagation and collective relaxation Dynamics. *Dated: Phys. Rev. E, 70*, 061302.

Menard, K. P. (1999). Dynamic Mechanical Analysis. *A Practical Introduction.* CRC Press LLC.

Meyers, M. A. (1994). Dynamic Behavior of Materials. *J. Wiley.*

Mindlin, R. D. (1949). Compliance of elastic bodies in contact. *Trans. ASME J. Appl. Mech.,* 16, 259-268

Moore, D. F. (1975). *Principles and Application of Tribology.* New York: Pergamon Press, Technology & Engineering.

Moore, D. F. (1978). Principles and Application of Tribology. *Mir, Moscow, Russia.*

Nilsen, L. (1978). Mechanical properties of polymers and polymeric compositions. *Moscow.*

Nilsen, L. E., & Landel, R. F. (1994). Mechanical properties of polymers and polymeric compositions (2nd ed.). Rev., Series: Mechanical Engineering, Marcel Dekker. Ink., New York.

Padovan, J., & Paramadilok, O. (1984). Transient and steady state viscoelastic rolling contact. *Comput Struct, 20*, 545-553.

Persson, B. N. J., Bucher, F., & Chiaia, B. (2002). Elastic contact between randomly rough surfaces. *Comparison of theory with numerical results, Physical Review B, 65*(18), 184106.

Persson, B. N. J. (2010). Rolling friction for hard cylinder and sphere on viscoelastic solid. *Eur. Phys. J. E, 33*, 327-333. Retrieved from http://link.springer.com/article/10.1140%2Fepje%2Fi2010-10678-y

Radok, J. R. M. (1957). Viscoelastic stress analysis. *Q. App. Math, 15.*

Popov. V. L. (2010). *Contact mechanics and friction: Physical principals and application.* Springer/

Popov. V. L., & Hess, M. (2015). *Method of dimensionality reduction in contact mechanics and friction.* Springer. Retrieved from http://www.twirpx.com/file/1756685/

Ramírez, R., Poeschel T., Brilliantov, N. V., & Schwager, T. (1999). Coefficient of restitution of colliding viscoelastic spheres. *Phys. Rev. E, 60*. Retrieved from http://www.ncbi.nlm.nih.gov/pubmed/11970301

Roylance, D. (2001). Engineering viscoelasticity. *Department of Materials Science and Engineering, Massachusetts Institute of Technology Cambridge, MA 02139.* Retrieved from http://ocw.mit.edu/courses/materials-science-and-engineering/3-11-mechanics-of-materials-fall-1999/modules/visco.pdf

Schafer, J., Dippel, S., & Wolf, D. E. (1996). Force schemes in simulations of granular materials. *Journal de Physique I, 6*, 5-20. Retrieved from https://hal.archives-ouvertes.fr/jpa-00247176/document

Schwager, T., & Poschel, T. (2007). Coefficient of restitution and linear-dashpot model revisited. *Granular Matter,* 9, 465-469. Retrieved from http://link.springer.com/article/10.1007%2Fs10035-007-0065-z

Schwager, T., & Poschel, T. (2008). Coefficient of restitution for viscoelastic spheres: The effect of delayed recovery. *Phys. Rev. E, 78*, 051304. Retrieved from http://journals.aps.org/pre/abstract/10.1103/PhysRevE.78.051304

Simon, R. (1967). *Development of a Mathematical Tool for Evaluating Golf Club Performance*. ASME Design Engineering Congress, New York.

Sneddon, I. N. (1965). The Relation between Load and Penetration in the Axisymmetric Boussinesq Problem for a Punch of Arbitrary Profile. *Int. J. Eng. Sci., 3*, 47–57.

Stronge, W. J. (2000). *Impact Mechanics*. Cambridge: Cambridge University Press.

Tabor, D. (1977). Surface forces and surface interactions. *Journal of Colloid and Interface Science, 58*(1), 2-13.

Tabor, D. (1977). The hardness of solids. *Journal Colloid Interface Sci., 58*, 145-179.

Thornton, C. (2009). Special Issue on the 4th. Int. Conf. on Discrete Element Methods. *Powder Technol., 193*, 216-336.

Thornton, C., Cummins, S. J., & Cleary, P. W. (2012). An investigation of the comparative behaviour of alternative contact force models during inelastic collisions. *Powder Technol., 233*, 30-46.

Timoshenko, S., & Goodier, J. N. (1951). Theory of Elasticity. N.Y.: MCGRAW – Hill.

Ting T. (1966). The contact stresses between a rigid indenter and a viscoelastic half space. *Journal of Applied Mechanics, 33*, 845.

Van Krevelen, D. W. (1972). *Properties of polymers correlations with chemical structure.* Amsterdam-London-New York: Esliier Publishing.

Van Zeebbroeck, M. (2005). The Discrete Element Method (DEM) to Simulate Fruit Impact Damage during Transport and Handling. DISSERTATIONES DE AGRICULTURA. *Proefschrift voorgedragen tot het behalen van de graad van Doctor in de Toegepaste Biologische Wetenschappen door.* Retrieved from https://lirias.kuleuven.be/bitstream/1979/46/2/doctoraatmichaelvanzeebroeck.pdf

Webster M. N., & Sayles, R. S. (1986). A numerical model for the elastic frictionless contact of real rough surfaces. *ASME, 108*, 314-320.

16

Beam Dynamics Design Using Analytical Methods for Optimizing Heat Transfer Platens Dynamics Behavior

Heikki Martikka[1] & Ilkka Pöllänen[2]

[1] Himtech Oy, Engineering, Ollintie 4, FIN-54100 Joutseno, Finland

[2] SAV Oy, Savonkatu 21, FIN-45100 Kouvola, Finland

Correspondence: Heikki Martikka, Himtech Oy, Engineering, Ollintie 4, FIN-54100 Joutseno, Finland. E-mail: heikki.martikka@pp.inet.fi

Abstract

The paper presents results of using continuous beam analytical differential equation method to dynamics behavior design of complex heat transfer orthotropic platens which are used in the recovery boilers. The goal is to obtain a predictive model for increasing the reliable useful operation time. The platens are supported vertically with fixed ends. They are loaded by hot flue gas and by accumulating mass loads due to operation conditions. The loads cause alternating stresses fracturing tube branch connections. The beam deflection and bending stress histories at the junction of tube and chamber are obtained using an analytical beam model. The model gives the lowest eigenfrequency close to the measured value which is somewhat larger than that obtained by FEM model. The beam model can be used for further optimal redesign of the tube branch connections and the mean time to failure of the process equipment.

Keywords: beam theory, dynamics, fatigue design, FEM, heat transfer platens, recovery boilers

1. Introduction

Beams are one important basic component of the technological infrastructures. They are used extensively in static and mobile structures. Beams are mostly used as prismatic forms, solid, hollow, composite or profiles. In slender beams the transverse dynamic vibrations are excited even with moderate loads acting too close to resonances.

The object of study is heat transfer platen and other components of a typical recovery boiler.

They have three main functions. One is recycling and regeneration of cooking chemicals. The second is environmental burning of organic materials. The third is heat recovery and energy generation.

Their heat transfer platens are orthotropic and complex in geometry. They are laborious to calculate in detail even with advanced FEM (NX Nastran for FEMAP solver. Version 11.0.0. Serial number: 980-4H-3398-004F. Copyright ©2012 Siemens Product Lifecycle Management Software Inc.) as shown by Martikka, Pöllänen, and Simonen (2006). Therefore cost-effective surrogate models are needed which are much faster in calculations and accurate enough. Several surrogate models can be used to achieve this goal, like matrix, composite structure and continuous beam based methods. The aim is to construct a surrogate beam model for the platen which gives an optimal trade off between simplicity and accuracy. A general trend is to utilize surrogate models in all sciences. Generally surrogate models are combined with some optimization algorithm for maximal utility. The present beam surrogate modeling of vibrating platens is analogous with a study of optimal design of turbine discs as by Huang, Wang, Chen, and Tian (2011) using kriging surrogate models. The utility of combining surrogate models with optimization approach are discussed by Loshchilov, Schoenauer, and Sebag (2010). Jin (2011) discusses surrogate-assisted evolutionary computation combination of surrogate optimization of a swarm type structures. The collective of platens behave like a dynamic interacting swarm. These new approaches offer a promising combination of kriging surrogate models with fuzzy optimum design for future work.

Now the beam models are applied to formulate the dynamic behavior of orthotropic industrial heat transfer platens which are supported vertically with fixed ends. Their endurance is important for the whole process as discussed by Jameel, Araned, and Luks (1998). The actual aim is to maximize profit by keeping the soot layer thickness to a minimum. Beam dynamic theory is discussed by Rao (2007), Hollburg (2007), and by

Dimarogonas and Haddad (1992). This theory is needed to build up surrogate basic models. Many recovery boilers are now required to withstand assigned seismic loads.

The optimum fuzzy design approach is discussed by Martikka and Pöllänen (2009). This fuzzy design method is based on results of Diaz (1988). This method is applicable to the present task as well. Machine design concepts are generated based on design variables ranges. The optimal one is best for the customer. At each beam there are crucial safety critical locations. Structural analysis by many researchers can be used in several calculations (Boresi, 1993; Case, Chilver, & Ross, 1993; Ventsel & Krauthammer, 2001). In order to perform these calculations reliable engineering design data is needed and made available by Beitz and Kuttner (Eds.) (1994). The reliability based design is important to check the reliability of this kind of extremely loaded structure as discussed by Dhillon and Singh (1981), and by Leitch (1988). One goal is to have a reliable, safe and cost-effective operation of the machines. The traditional approach does not consider the dynamics and endurance behavior of the platens. The process control is made by using measured temperatures and pressure histories to give some guidelines for activating the cleansing lance pressure and flow rates. Optimal control is important since they consume 5–12% of the superheated steam produced by the boiler. Results show that intensive cleansing may damage tubes and cause a risk of large annual loss. Subsequently, reducing the pressure by one third requires 20% increase in steam flow to maintain the same cleansing efficiency. This method is a guess and the safe optimum efficiency is not found.

In this work a simplified surrogate modeling is used. The long slender orthotropic heat transfer platens behave at the lowest bending modes as transversely vibrating beams. In this study the analytical beam models are based on differential equations approach. Literature survey of application of beam dynamics in recovery boilers showed that there are few such applications in boilers. Flow induced vibration and flutter and transverse forces can occur in the boiler between parallel platens. Unsteady flow induced forces on heat transfer tubes are discussed by Majander and Siikonen (2002). In their study the stresses and deflections are not considered.

First the internal dynamic load and masses on the platens are measured using an innovative technique.

Secondly predictive models and measurements are needed to locate acute pluggages and optimal intensity of cleansing. Analytical results are verified by using detailed model made by FEM program (NX Nastran).

The third goal is to use the beam model for further optimal redesign of the tube branch connections and the mean time to failure of the process equipment. Using the method of analytical prediction and real time measurements savings may by even 10% of the present level.

2. Analytic Beam Dynamics Theory Is Applied Using the Classical Approach

The objective in this section is to present an overview of the structure and design methodology and detailed geometry and loads.

2.1 Overview of the Studied Structure

The structure sketch is shown in Figure 1 (a). The platens are modeled as beams. The continuous beam theory is applied as discussed by Rao. The goal is to find the dynamical stress history at the critical location and to obtain satisfactory safety factors against fatigue.

2.2 Overview of the Design Methodology

The principle of sequential optimum fuzzy design is illustrated in Figure 1 (b). This approach can be applied in the next stage of this study (Martikka & Pöllänen, 2009; Martikka & Taitokari, 2011, 2012). The principle of using an overall surrogate model for the $P(s)$ function is shown in Figure 1 (c).

The structures are defined by their geometry, materials and function, loads and constraints.

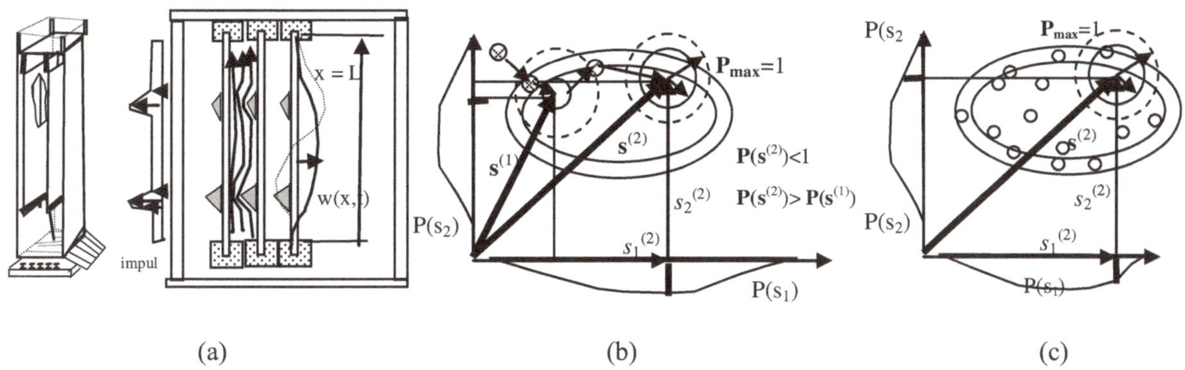

Figure 1. Recovery boiler. (a) Sketch of main structure of a recovery boiler; (b) Principle of sequential fuzzy design is shown on a two dimensional plot. The design progress is shown starting from one initial concept to final optimum. They are located with decision variable vectors. Total satisfactions are shown on the contour plot and also as a circle at tip. The maximal satisfaction is shown as a circle with unit radius; (c) Principle of using an overall surrogate model for the $P(s)$ function.

The design variables are expressed as a vector **x** = (load functions, geometry, materials). From these the decision variable vector **s** = (cost, factors of safety...) = **s**(**x**) for each concept is formulated as intersection of decision variables s_k

$$s^c = [s_1 \cap s_2 \cap s_3 \cap s_4 \cap s_5 \cap s_6]^{concept\ c} \quad (1)$$

The design goal is maximization of the total satisfaction of the customer on the product (Diaz, 1988).

$$P(s) \Rightarrow P(s) = P(s_1) \cdot P(s_2) \cdot \ldots\ldots \cdot P(s_n), Q = \max P \quad (2)$$

In this fuzzy approach the final optimum is found by exhaustive sequential search. Principle of using an overall surrogate model for the $P(s)$ function is shown in Figure 1 (c). The kriging type approach may be applied to randomly generated concepts in the design space and get the intermediate options.

2.3 Detailed Structure of the Object of Study

In Figure 2 some structural details are shown. Figure 2 (a) shows laboratory testing, Figure 2 (b) shows typical pluggages and Figure 2 (c) shows tube fatigue cracks.

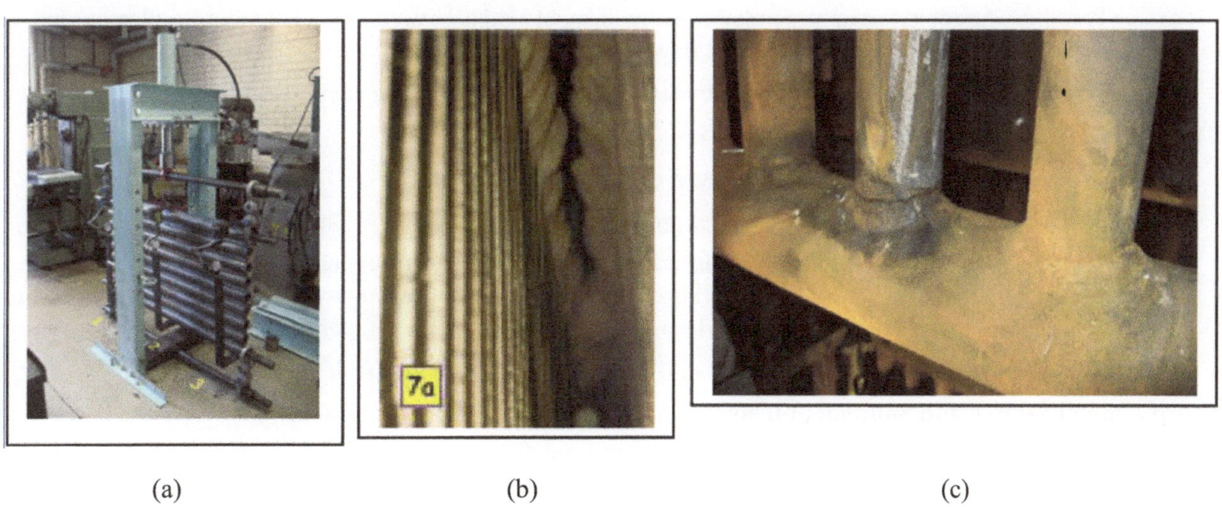

Figure 2. Orthotropic platens of the heat surfaces: (a) Laboratory testing (Talvio, 2009); (b) Typical pluggage prior to tempering by the steam; (c) Tube fatigue cracks

This picture was taken during a chill and blow of the boiler. The pluggage is much localized and shows a nearly closed the gap between the front and back tube bundles (Jameel et al., 1998).

Geometry of the cross section of the orthotropic platen is shown in Figure 3.

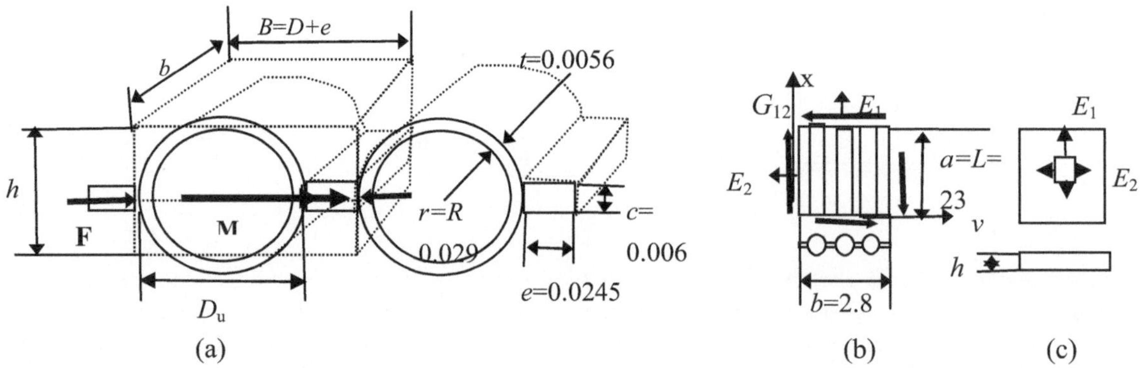

Figure 3. Orthotropic plate definitions. (a) Geometry of the cross section; (b) Orthotropic platen; (c) Surrogate plate model

The dimensions of the cross-section are

$$D_u = 0.0635,\ t = 0.0056,\ D = D_u - t = 0.0579,\ e = 0.0245,\ c = 0.006$$

Here D_u is the outer diameter of the tube, t is wall thickness of the tube D is mean diameter and e and c are the length and thickness of the connection flange plate.

The bending stiffness of the cross-section of transverse length $B=D+e$ is

$$EI = E \cdot \left[\tfrac{\pi}{8} D^3 t + \frac{ec^3}{12} \right] \tag{3}$$

Cross-section area A_b of the basic unit is sum of the tube area and the flange areas. In derivations other notations for it are used also.

$$A_b = \pi Dt + ec, \quad A_b = A(x) = A \tag{4}$$

2.4 Loads on the Heat Transfer Platens

Several loads affect the heat transfer platens.

1) Flow induced vibrations are excited by the pluggages between the platens.

- Pressure variations occur depending on the channel width between two neighboring platens due to flow pressure of gas with density ρ and flow velocity v, as $p(x,t) = \frac{1}{2}\rho v^2$.
- These loads cause variable bending stress under aggressive environmental load.

2) Soot blowing causes also tube fracture if it is not checked

- Consequences of tube fracture have been explosions in the factory.

Importance of soot layer control for the end user profit has been discussed by Wilhelmsen (2012) as:

- "If deposits are allowed to form on heat exchangers, the loss of efficiency can be directly related to extra fuel consumption." Simple heat transfer calculations show that a 1 mm deposit is approximately equivalent to a 10% efficiency loss, and a 3 mm deposit can reduce efficiency by up to 50%.

3) Thermal and chemical loads reduce strength, wall thickness and accelerate crack growth

3. Equation of Motion and Solution of Beam Models for the Heat Transfer Platens

The objective in this section is to present an overview of beam dynamic theory by which the stress history results of this work are derived.

3.1 Beam Under Distributed Transverse Forces

Beam models are shown in Figure 4.

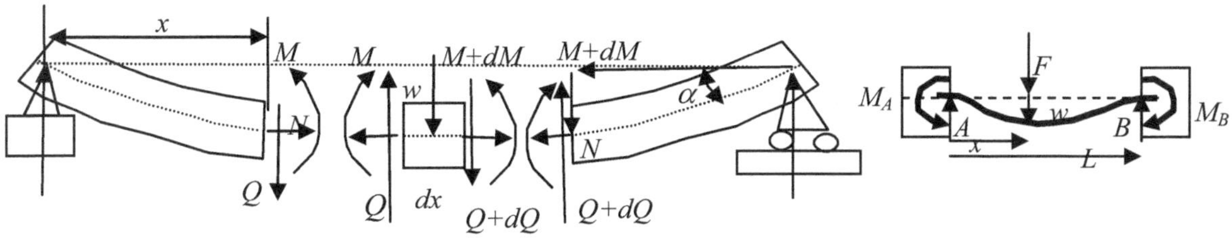

(a) (b)

Figure 4. Beam basics: (a) Beam with free-free supports, Beam deflection w and internal force resultants; (b) Beam with stiff-stiff supports and load force F

The linear equation of motion of the beam may be written as by Rao for forced vibrations with load $f(x,t)$

$$\frac{\partial^2}{\partial x^2}\left[EI\frac{\partial^2 w(x,t)}{\partial x^2}\right]+\rho A(x)\frac{\partial^2 w(x,t)}{\partial t^2}=f(x,t) \tag{5}$$

Here $w(x,t)$ is the transverse deflection of the beam at location x and at time t, ρ is mass density of the beam. A is the cross section area of the prismatic beam, Equation (4), E is the steel elastic modulus and I is second moment of area and EI is bending stiffness, $f(x,t)$ is the load force acting at x and at time t.

Now the range of variation of the co-ordinate is $x = 0$ to L or the length of the beam.

Time variation is from an initial time $t = 0$ to some seconds. The deflection at free vibrations with $f = 0$ is a linear combination of all natural motions of the beam. The assumed solution is

$$w(x,t)=\sum_{1} W_i(x)q_i(t) \tag{6}$$

Here W_i is the ith natural mode shape and q_i is a time dependent function.

They are determined by substituting the trial solution to the free vibration Equation (5). Here temporarily for simplicity the indices are dropped $W= W_j$ for convenience. Now EI is assumed constant. Separation of variables gives two equations

$$W_{xxxx}-\beta^4 W=0 \tag{7}$$

$$\ddot{q}+\omega^2 q=0 \tag{8}$$

The deflection of the beam at location x and at time t by Rao is

$$w(x,t)=\sum_{1}^{\infty} w_n(x,t)=\sum_{1}^{\infty} W_n(x)\cdot(A_n\cos\omega_n t+B_n\sin\omega_n t) \tag{9}$$

Here the eigenfrequency of eigenmode n is

$$\omega_n=(\beta_n L)^2\sqrt{\frac{EI}{\rho A L^4}},\quad \beta_n^4=\frac{\rho A\omega_n^2}{EI} \tag{10}$$

In order to simplify notations one may set for short $\beta_n = \beta$, $\omega_n = \omega$.

The four integration constants are determined with four boundary conditions

$$\begin{aligned}W(x)=&C_1(\cos\beta x+\cosh\beta x)+C_2(\cos\beta x-\cosh\beta x)+\\&+C_3(\sin\beta x+\sinh\beta x)+C_4(\sin\beta x-\sinh\beta x)\end{aligned} \tag{11}$$

3.2 Beam Stiffly Supported at Both Ends

Boundary conditions for the beam deflection are satisfied by the shape function W. The left and right ends of the beam are stiffly supported as illustrated in Figure 4 (b) with zero displacements and zero slopes.

$$\begin{aligned} W(0)=0, \quad W'(0)=0 \\ W(L)=0, \quad W'(L)=0 \end{aligned} \tag{12}$$

The first two conditions give $C_1 = C_3 = 0$. The second two conditions give two homogeneous equations from which the ratio of the two constants can be obtained.

$$\begin{bmatrix} \cos\beta L - \cosh\beta L & \sin\beta L - \sinh\beta L \\ -(\sin\beta L + \sinh\beta L) & \cos\beta L - \cosh\beta L \end{bmatrix} \begin{bmatrix} C_2 \\ C_4 \end{bmatrix} = 0 \tag{13}$$

Some lowest roots are obtained as giving the lowest eigenfrequencies

$$\begin{gathered} \cos\beta L \cdot \cosh\beta L = 1 \\ \beta_1 L = 4.73004, \ \beta_2 L = 7.853, \ \beta_3 L = 10.99, \end{gathered} \tag{14}$$

Thus

$$\omega_n = (\beta_n L)^2 \sqrt{\frac{EI}{\rho A L^4}} = \left(4.73^2, 7.853^2 ....\right) \sqrt{\frac{EI}{\rho A L^4}} \tag{15}$$

The two constants are interdependent. The one independent constant is determined by imposing the conditions of othonormality.

$$C_4 = -K C_2, K = \frac{\cos\beta L - \cosh\beta L}{\sin\beta L - \sinh\beta L}, C_2 = C(m,n) \tag{16}$$

3.3 Orthogonality and Orthonormality of Beam Shape Functions

In the general Equation (5) *EI* was variable. Now the goal is to obtain the modal modes and eigenfrequencies by dynamic equations. The following approach is used by Rao. He substitutes first a harmonic modal time function in Equation (5) and also sets $f = 0$ to get free modal vibrations. Now in order to consider separate modes the indices need to be added. The result is that the equation for the shape functions for each mode i $W_i(x)$ is obtained as

$$\frac{\partial^2}{\partial x^2}\left[EI \frac{\partial^2 W_i(x)}{\partial x^2}\right] = \rho A(x) \omega_i^2 W_i(x) \tag{17}$$

Then following Rao it may be substituted back to Equation (5). Next nonzero load is assumed to act. It does not affect the modal vibrations. The following equation is obtained with nonzero load

$$\rho A(x) \omega_i^2 W_i(x) \cdot q_i(t) + \rho A(x) W_i(x) \frac{\partial^2 q_i(t)}{\partial t^2} = f(x,t) \tag{18}$$

This modal equation may be solved using the orthogonality of the eigenfunctions

(1) First, multiplying Equation (17) by W_j on the left one obtains

$$q_i(t)\left[\rho A(x) \omega_i^2 W_j(x) W_i(x)\right] + \frac{\partial^2 q_i(t)}{\partial t^2}\left[\rho A(x) W_j(x) W_i(x)\right] = W_j(x) f(x,t) \tag{19}$$

(2) Then integrating this over the length of the beam from $x = 0$ to $x = L$ yields

$$q_i(t)\left[\omega_i^2 \cdot \int_0^L \rho A(x) W_j(x) W_i(x) dx\right] + \frac{\partial^2 q_i(t)}{\partial t^2}\left[\int_0^L \rho A(x) W_j(x) W_i(x) dx\right] = \int_0^L W_j(x) f(x,t) dx \tag{20}$$

(3) The next step is to require that the shape functions of different eigenmodes are mutually orthogonal.

$$q_i(t)\left[\omega_i^2 \cdot \delta_{ji}\right] + \frac{\partial^2 q_i(t)}{\partial t^2} \cdot \delta_{ji} = Q_j \tag{21}$$

(4) The normal mode shape functions are made orthonormal using the property of delta function

$$\delta_{ji} = \int_0^L \rho A(x) W_j(x) W_i(x) dx = 1, \text{ if } i = j, \qquad (22)$$
$$= 0, \text{ if } i \neq j$$

(5) The orthonormalization condition gives equations for the constants $C(m, n)$.

The following model is assumed for W_n

$$W_n(x) = C(m,n) \cdot \left[(\cos\beta_n x - \cosh\beta_n x) - K(\sin\beta_n x - \sinh\beta_n x) \right] = C(m,n) \cdot H_n(x) \qquad (23)$$

Thus the orthonormalization constant is obtained depending on indices m, n as

$$C(m,n)^2 \rho A \int_0^L H_n(\beta_n x) H_n(\beta_n x) dx = 1 \Rightarrow C(m,n) = \frac{1}{\left[\rho A \int_0^L H_n H_n dx \right]^{\frac{1}{2}}} \qquad (24)$$

3.4 Solution of Beam Dynamical Deflections With Various Loads

The objective in this section is to obtain the beam deflection as superposition of modal solutions depending on the measured load history.

3.4.1 Modal Equations

The equation of motion of the mode i is obtained from Equation (21)

$$\ddot{q}_i + \omega_i^2 q_i = Q_i(t) \qquad (25)$$

Here the generalized force is obtained from Equation (20)

$$Q_i(t) = \int_0^t W_i(x) f(x,t) dx, \; i = 1,2... \qquad (26)$$

Complete solution is obtained as sum of free and forced solutions

$$q = q_{\text{free}} + q_{\text{forced}} \qquad (27)$$

The forced solution can be expressed with the Duhamel's integral as

$$q_i(t) = A_i \cos\omega_i t + B_i \sin\omega_i t + \frac{1}{\omega_i} \int_0^t Q_i(\tau) \sin\omega_i (t-\tau) d\tau \qquad (28)$$

Substitution of this solution to Equation (5) to the trial solution of Equation (6) gives the deflection

$$w(x,t) = \sum_1^\infty W_i(x) \left[A_i \cos\omega_i t + B_i \sin\omega_i t + \frac{1}{\omega_i} \int_0^t Q_i(\tau) \sin\omega_i (t-\tau) d\tau \right] \qquad (29)$$

3.4.2 Initial and Boundary Conditions

Boundary conditions of stiff supports are satisfied at the ends of the beam $x = 0$ and $x = L$ by the previously obtained shape function $W(x)$.

$$w(x,t) = \sum_1^\infty W_i(x) q_i(\omega_i t, A_i, B_i)$$
$$w(0,t) = \sum_1^\infty W_i(0) q_i(\omega_i t, A_i, B_i) = 0, \quad w'(0,t) = \sum_1^\infty W'_i(0) q_i(\omega_i t, A_i, B_i) = 0 \qquad (30)$$
$$w(L,t) = \sum_1^\infty W_i(L) q_i(\omega_i t, A_i, B_i) = 0, \quad w'(L,t) = \sum_1^\infty W'_i(L) q_i(\omega_i t, A_i, B_i) = 0$$

Initial time conditions can be given for a mode i as initial shape and initial velocity at time $t = 0$

$$w(x,0)=\sum_{1}^{\infty}W_i(x)q_i(\omega_i 0,A_i,B_i)=w_0(x)$$
$$\dot{w}(x,0)=\sum_{1}^{\infty}W_i(x)q_i(\omega_i 0,A_i,B_i)=\dot{w}_0(x) \tag{31}$$

From these two equations for each mode *i* the two mode constants can be solved. Now all initial conditions are set to zero and the solution of free vibration is zero by setting $A_i = 0$ and $B_i = 0$

3.4.3 Concentrated Harmonic Forces Acting on the Beam

Using this model many load forces F_k, k=1,2··· may be set to act at random locations x_k and at random time t_k. When excitation starts at time $\tau = 0$ one obtains for the ith time function as the forced solution only

$$q_i(t)=0+\frac{1}{\omega_i}\int_0^t Q_i(\tau)\sin\omega_i(t-\tau)d\tau \Rightarrow q_{i,forced} \tag{32}$$

Here the generalized force is by Equation (26)

$$Q_i(\tau)=\int_0^L W_i(x)f(x,\tau)dx,\ i=1,2... \tag{33}$$

Possible definitions of the load function are shown in Figure 5.

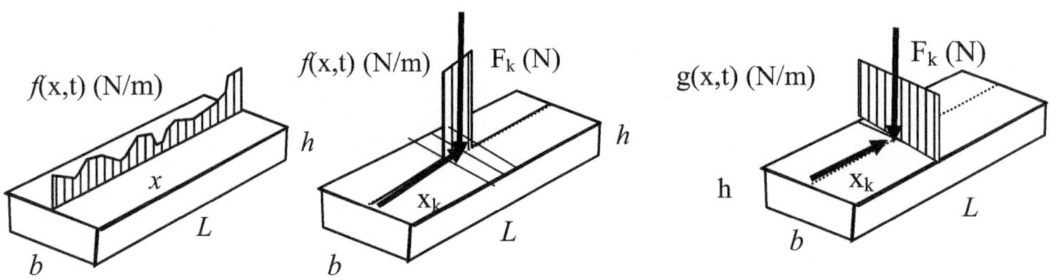

(a) (b) (c)

Figure 5. Definition of the load forces on beam: (a) Axial line load; (b) Axial block load; (c) Transverse line load

In Figure 5 (a) $f(x)$ is axial line load. It is a distributed load along a line upon the beam. In Figure 5 (b) an axially narrow block load model is used. The concentrated force F_k acts at location x_k.

In Figure 5 (c) a transverse line load is modeled. The concentrated force acts as transverse line force

$$f(x,t)=\frac{F_k(x_k,t)}{2\Delta x},\ g(x,t)=\frac{F_k(x_k,t)}{b} \tag{34}$$

A load force with only one harmonic term is described as

$$f(x,\tau)=F_1\sin\Omega_1\tau\cdot\delta(x-x_1) \tag{35}$$

This force acts at location $x = x_1$, starts at time $t = 0$ and then acts harmonically with amplitude F_1 and frequency Ω_1.Using this option one obtains for the modal load force

$$Q_i(\tau)=\int_0^L W_i(x)\cdot F_1\sin\Omega_1\tau\cdot\delta(x-x_1)\cdot dx,\ i=1,2... \tag{36}$$

The property of the delta function is useful to localize the force

$$\delta(x-x_1)=1,\ if, x=x_1 \tag{37}$$

The modal force is obtained as

$$Q_i(\tau) = F_1 \cdot W_i(x_1) \cdot \sin \Omega_1 \tau \tag{38}$$

This modal force is next substituted into the equation for the time function

$$q_i(t) = \frac{1}{\omega_i} \int_0^t Q_i(\tau) \sin \omega_i (t - \tau) d\tau \tag{39}$$

The result is a time function due to one harmonic localized force excitation on mode i

$$q_i(t) = \frac{F_1}{\omega_i} W_i(x_1) \int_0^t \sin \Omega_1 \tau \cdot \sin \omega_i (t - \tau) d\tau \tag{40}$$

Substituting this forced solution to Equation (29) and $A_i = 0$, $B_i = 0$ gives as by Rao for the beam deflection

$$w(x,t) = \sum_1^{\infty} W_i(x) \left[A_i \cos \omega_i t + B_i \sin \omega_i t + \frac{F_1}{\omega_i} W_i(x_1) \int_0^t \sin \Omega_1 \tau \cdot \sin \omega_i (t - \tau) d\tau \right] \tag{41}$$

3.4.4 Deflection Response of Beam to Two Harmonic Forces at Different Locations

Now the start times of the load forces are set to zero. The solution is by Rao. This is used for bending stress calculations.

$$w(x,t) = \sum_{n=1} W_n(x) \cdot \frac{F_k W_n(x_k)}{\omega_n^2 - \Omega_n^2} \left(\sin \Omega_n t - \frac{\Omega_k}{\omega_n} \sin \omega_n t \right)$$
$$w(x,t) = \sum_{n=1} W_n(x) \cdot q_n(t) \tag{42}$$

3.4.5 Impact Force on the Beam

One block force

This method is illustrated by using first one block force. The narrow time range of the activation of the force is expressed using a block function B such that $B = 1$ within time range $t_a < t < t_b$, and 0 outside. Thus

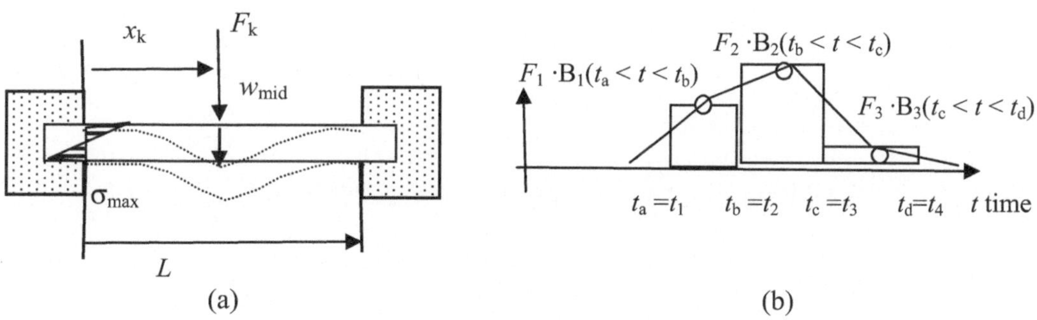

Figure 6. Impact force on beam: (a) Co-ordinate space; (b) Time space

$$f(x,\tau) = F_1 \sin \Omega_1(\tau) \cdot B(t_a, t_b) \cdot \delta(x - x_1), \quad \delta(x - x_1) = 1, \ if, x = x_1 \tag{43}$$

Now the force starts at time 0. Using this one obtains

$$Q_i(\tau) = \int_0^L W_i(x) \cdot F_1 \sin \Omega_1(\tau) \cdot B(t_a, t_b) \cdot \delta(x - x_1) \cdot dx, i = 1, 2... \tag{44}$$

Thus

$$Q_i(\tau) = F_1 \cdot W_i(x_1) \cdot \sin \Omega_1(\tau) \cdot B(t_a, t_b) \tag{45}$$

This may be substituted to the forced time function

$$q_i(t) = 0 + \frac{1}{\omega_i}\int_0^t Q_i(\tau)\sin\omega_i(t-\tau)d\tau \tag{46}$$

Rearrangement gives

$$q_i(t) = \frac{F_1}{\omega_i}W_i(x_1)\int_0^t \sin\Omega_1(\tau)\cdot B(t_a,t_b)\cdot\sin\omega_i(t-\tau)d\tau \tag{47}$$

Now for simplicity it may be assumed that within the block there is no harmonic time dependency

$$q_i(t) = \frac{F_1}{\omega_i}W_i(x_1)\left[\int_{\tau=t_a}^{\tau=t_b} 1\cdot B(t_a,t_b)\cdot\sin\omega_i(t-\tau)d\tau\right] \tag{48}$$

Three block forces are used to model a simple force impulse on the beam

$$q_i(t) = \frac{1}{\omega_i}W_i(x_1)\left[\int_{\tau=0}^{\tau=t} 1\cdot\sum_{k=1}^{k=3=Nk} F_k B_k(t_s(k),t_e(k))\cdot\sin\omega_i(t-\tau)d\tau\right] \tag{49}$$

here for block k: $B_k\ F_k = 1\cdot F_k$ within the interval $t_s(k) < t < t_e(k)$, s = start, e = end

By continuing the algorithm any measured load force spectrum can be applied to the beam and the stress history can be obtained. Performing the integration gives

$$q_i(t) = \sum_{k=1}^{k=Nk}\left\{\frac{F_k}{\omega_i}W_i(x_k)\left[\left(-\frac{1}{\omega_i}\right)\cos\omega_i(t-\tau)\right]_{t_s(k)}^{t_e(k)}\right\} \tag{50}$$

Substituting here the time limits gives

$$q_i(t) = -\sum_{k=1}^{k=Nk}\left\{\frac{F_k}{\omega_i^2}W_i(x_k)\left[\cos\omega_i(t-t_e(k)) - \cos\omega_i(t-t_s\{k\})\right]\right\} \tag{51}$$

Expanding the sum gives three time functions

$$q_i(t)_{F1(x1)} = -\frac{F_1}{\omega_i^2}W_i(x_1)\left[\cos\omega_i(t-t_2) - \cos\omega_i(t-t_1)\right] \tag{52}$$

$$q_i(t)_{F2(x2)} = -\frac{F_1}{\omega_i^2}W_i(x_2)\left[\cos\omega_i(t-t_3) - \cos\omega_i(t-t_2)\right] \tag{53}$$

$$q_i(t)_{F3(x3)} = -\frac{F_3}{\omega_i^2}W_i(x_3)\left[\cos\omega_i(t-t_4) - \cos\omega_i(t-t_3)\right] \tag{54}$$

The total time function of the ith mode due to three excitation forces is a superposition

$$q_i(t)_{\Sigma F} = q_i(t)_{F1(x1)} + q_i(t)_{F2(x2)} + q_i(t)_{F3(x3)} \tag{55}$$

The deflection response is weighted sum for each mode shape function at location x and time function due to excitation forces $F_k(x_k)$ at locations x_1, x_2 and x_3

$$w(x,t) = \sum_1^{\infty} W_i(x)q_i(t)_{\Sigma F} \tag{56}$$

4. Derivation of the Model for Stress Calculation at the Stiff Joint of the Beam

The objective in this section is to derive the stress history modeled from the deflection model.

Industrial experience has shown the need to re-design optimally the tube joints. To reach this goal the load and stress histories are needed. Dominant stress is dynamic bending at the two lowest modes. But torsional stresses are dominant at the third mode. Now the lowest frequency is considered as most dominant. When the deflection equation is known the stresses may be calculated from the second derivative of the deflection curve

$$M = EIw_{xx},$$
$$\sigma = \frac{M}{I}z = \frac{M}{W} = \frac{EIw_{xx}}{I}z = Ew_{xx}z = E\frac{z}{R} \tag{57}$$

The maximum stress is

$$\sigma_{\max}(x=0,t) = E\tfrac{1}{2}D \cdot w_{xx}(0,t) \tag{58}$$

Here

$$w_{xx}(x,t) = \sum_{1}^{\infty} W_n''(x) q_n(t) \tag{59}$$

The second derivative is needed

$$W_k(x) = C \cdot \left[(\cos\beta_k x - \cosh\beta_k x) - K(\sin\beta_k x - \sinh\beta_k x) \right]$$
$$W_k(x) = C \cdot H_k(x) \tag{60}$$
$$W_k''(x) = C \cdot H''_k(x)$$

Using here the function

$$H_n(x) = (\cos\beta_n x - \cosh\beta_n x) - K(\sin\beta_n x - \sinh\beta_n x) \tag{61}$$

One obtains the second derivative of the shape function as

$$W_n''(x) = C \cdot \beta_n^2 \left[(-\cos\beta_n x - \cosh\beta_n x) - K(-\sin\beta_n x - \sinh\beta_n x) \right]$$
$$W_n''(x) = C \cdot \beta_n^2 U(\beta_n x) \tag{62}$$

Simplification gives

$$W_n''(x) = C \cdot \beta_n^2 U(\beta_n x), \; \beta_n^2 = \omega_n \sqrt{\frac{\rho A}{EI}}$$
$$W_n''(x) = C \cdot \omega_n \sqrt{\frac{\rho A}{EI}} \cdot U(\beta_n x) \tag{63}$$

At the stiff joint at $x = 0$ the second derivative of the mode form function is

$$W_n''(0) = C \cdot \beta_n^2 U(\beta_n \cdot 0)$$
$$W_n''(0) = C \cdot \beta_n^2 \cdot \left[(-\cos\beta_n 0 - \cosh\beta_n 0) - K(-\sin\beta_n 0 - \sinh\beta_n 0) \right] \tag{64}$$

Thus

$$W_n''(0) = C \cdot \beta_n^2 \cdot (-2) \tag{65}$$

Here the orthonormalization constant C is given in Equation (24). Numerical solution gives $C = 0.0739$. Here the eigenfrequencies are

$$\omega_n = (\beta_n L)^2 \sqrt{\frac{EI}{\rho A L^4}}, \quad (\beta_n L) = 4,73,7.853..... \tag{66}$$

The bending stress is obtained as

$$\sigma_{\max}(x=0,t) = E\tfrac{1}{2}D \cdot \sum_{1}^{\infty} W_n''(0) q_n(t) \tag{67}$$

$$\sigma_{\max}(x=0,t) = ED\cdot\sum_{1}^{\infty}\{C\cdot\beta_n^{\,2}\}q_n(t) \tag{68}$$

5. Results of Stress Calculations and Fatigue Design

The objective in this section is to present results eigenfrequency calculations and dynamic responses and fatigue design.

5.1 Eigenfrequencies

In the case of the heat transfer platen of length L = 23 m the beam model gave the lowest eigenfrequency as

$$f_1 = \tfrac{1}{2\pi}\omega_1 = 4.73^2\sqrt{\frac{EI}{\rho A}}\frac{1}{L^2} = \tfrac{1}{2\pi}4.73^2\sqrt{\frac{2.1\cdot10^{11}\cdot4.273\cdot10^{-7}}{7850\cdot1.1656\cdot10^{-3}}}\frac{1}{L^2} = \tfrac{1}{2\pi}\frac{22373}{L^2} \tag{69}$$
$$f_1 = 0.67Hz,\quad L = 23m$$

This value 0.67 Hz is close to the measured frequency f_{meas} = 0.63 Hz. Using FEM and a longer actual model L = 23 m and same cross-section the eigenfrequency is obtained as f_{FEM} = 0.545 Hz. The results agree well.

5.2 The Dynamic Stress Response to a Dynamic Load

Now a beam of length L = 20 m (not 23) is considered. One force F_k = 100 N is applied at $x_k = \frac{1}{2}L$ = 10 m. The excitation frequency is set to a typical possible value of half of the resonance value

$$f(x,t) = F_k\sin\Omega_k t\cdot\delta(x-x_k),\; F_k = 100,\; \Omega_k = 0.5\cdot\omega_1 = 0.5\cdot5.56\tfrac{rad}{s} \tag{70}$$

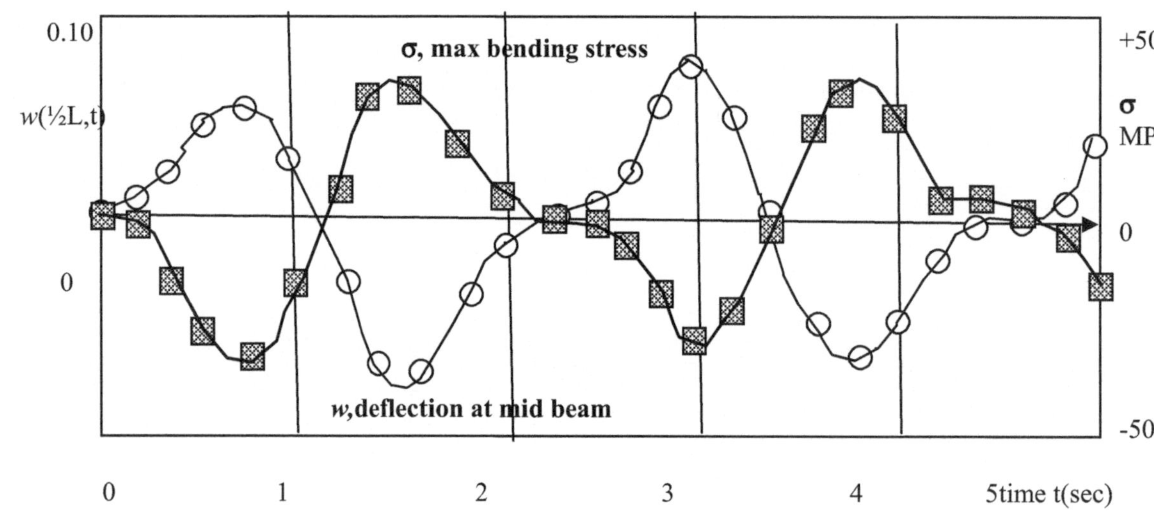

(a)

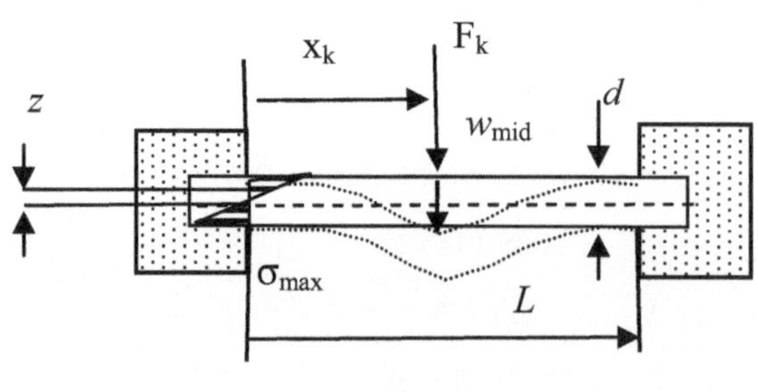

(b)

Figure 7. Results of dynamical simulations: (a) Maximum bending stress ; (b) Geometry

5.3 Fatigue Design

There are several methods for fatigue dimensioning. One method is to use fatigue diagrams based on assumption that fatigue cracks have not yet initiated (Meyer, 1985).

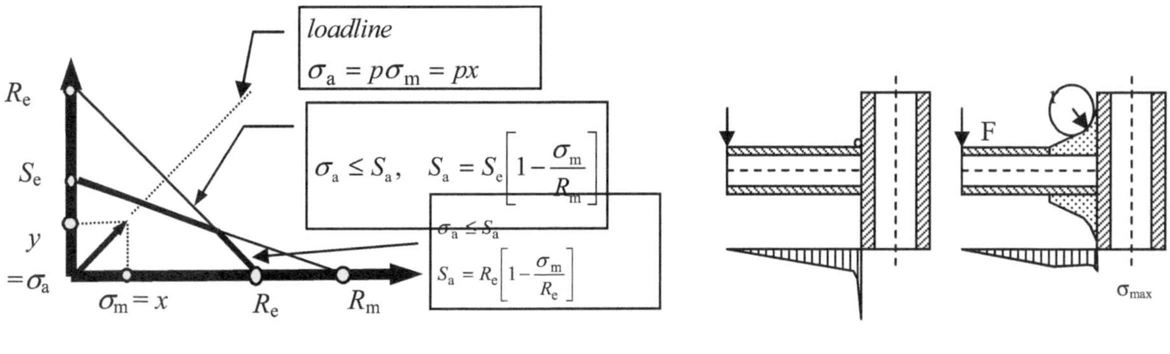

Figure 8. Fatigue modeling. (a) Fatigue diagram; (b) Conventional tube to tube joint with large stress at the joints; (c) Improved joint with small stress at the joint and even stress distribution giving even reliability

The mean stress is due to self weight of a heat transfer platen of length $L = 23$ m.

$$\sigma_m = \rho g L \tag{71}$$

The stress amplitude is the maximum bending stress

$$\sigma_a = \frac{M}{I}z = \frac{EIw_{xx}}{I}z = Ew_{xx}z > \sigma_m, \quad p = \frac{\sigma_a}{\sigma_m} >> 1 \tag{72}$$

Thus the stress ratio p is large and the bending stress is the dominant stress.

The endurance condition is

$$R(1) = S_a - \sigma_a \ge 0 \quad \rightarrow \quad S_e = C \cdot 0.5R_m = 0.4 \cdot 0.5 \cdot 400 = 40MPa \tag{73}$$

Dynamic simulation gave the stress amplitude as 30 MPa.

$$S_a = \frac{1}{N}\left[\frac{pS_e}{\frac{S_e}{R_m}+p}\right] \approx \frac{1}{N}S_e \ \ \ge \ \ \sigma_a \Rightarrow \frac{1}{N}40 > 30 \Rightarrow \max.N = 1.33 \tag{74}$$

The fatigue life of tube joints can be maximized using shape optimization and protective coatings.

Figure 8 (b) shows two joints types.

- The old conventional joint type has high nominal stress and high stress concentration due to small notch radius. The fatigue lives been about a year in many cases.
- The new joint type is obtained by applying preliminary shape optimization to increase fatigue life. It has an intermediate conical tube part and large radius of curvature giving low stress concentration.
- The ratio of estimates for fatigue lives of the two models is obtained as

$$\frac{N_{new}}{N_{old}} = \left(\frac{\sigma_{new}}{\sigma_{old}}\right)^{-3} = \left(\frac{\{K\sigma_{nom}\}_{new}}{\{K\sigma_{nom}\}_{old}}\right)^{-3} = \left(\frac{1.5 \cdot 20}{3 \cdot 40}\right)^{-3} = 4^3 = 256 \tag{75}$$

With the new design the fatigue life at critical joint may be increased by decades.

6. FEM Models

Some FEM results for one heat transfer platen are shown is Figure 9. They are obtained using (NX Nastran) FEM model with 1.15 million elements. The goal was to get high accuracy in details and also in total behavior.

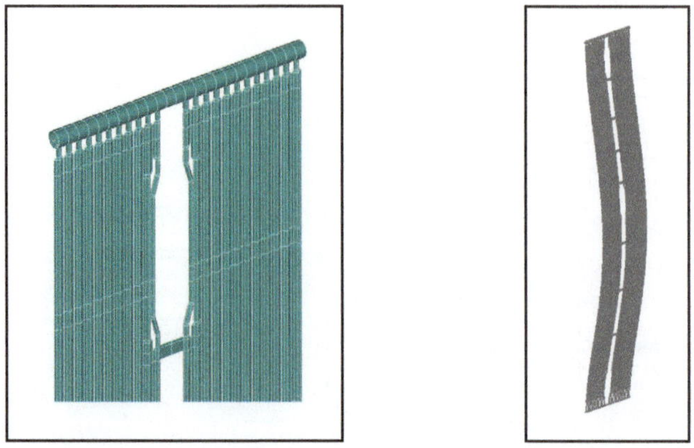

(a) (b)

Figure 9. FEM model results: (a) Upper part. The length L= 23 and total width 2.8 m; (b) The first modal form with eigenfrequency 0.545 Hz. The analytic beam model gave 0.67 Hz and the experimental measurement gave 0.63 Hz

7. Conclusions

The main conclusions of the present study can be summarized as follows.

(1) *Dynamic behavior is modeled reliably with the surrogate beam model*

The continuous beam transverse dynamics model is accurate enough for describing the behavior of actual platens. Lowest bending eigenfrequencies and modes are obtained which are close to the accurate results of large FEM beam bending model. The torsional analytical dynamics is studied later. The third eigenfrequency is torsional.

(2) *Fatigue design is based on reliable load and beam dynamics interaction*

The beam dynamic models are useful to get reasonable dynamic bending stress history under realistic loads.

Loads on the beams are soot mass and flow induced impacts and also cleansing flow forces. By optimally targeted cleansing the annual heat transfer net profit can be increased notably with minimal maintenance and cleansing costs. The advantage of the present general dynamic beam model is that the load history can be taken into account in the fatiguing load history at the critical welds.

The large detailed FEM model gave accurate detailed stresses which are needed in detailed fatigue life analysis. These are important for optimizing the fatigue endurance of tube welds at the joints. The factor of safety against initiation of fatigue failure was obtained with a reasonably conservative fatigue diagram.

(3) *Tube joint optimum design can be done using the innovative concept, fuzzy design and FEM*

There are many available methods for improving notably the tube weld fatigue life or the mean time to failure. Optimal redesign by FEM of the joint geometry can be used to minimize stresses. Ceramic coating may be applied to protect the surface against corrosion and oxidation and concomitant cracks.

(4) *The present approach focuses directly on the task of removing the soot with minimal damage*

Nearly all conventional approaches measure process variables like temperatures, pressure and steam flow rates.

But this information is not useful for revealing the soot pluggage locations. The present method is based on measurements and their data analysis and location predictions based on dynamic behavior simulations. It can give accurate enough coordinates of soot pluggages for cleansing robots to focus on. Unnecessary cleansing and forces are avoided.

Acknowledgements

This research is financed by Himtech Oy and SAV Oy.

References

Boresi, A., Schmidt, R., & Sidebottom, O. (1993). *Advanced mechanics of materials*. USA: John Wiley.

Case, J., Chilver, A. H., & Ross, C. F. T. (1993). *Strength of materials and structures with an introduction to finite element methods*. London: Edward Arnold.

Dhillon, B. S., & Singh, C. (1981). *Engineering reliability*. Wiley-Interscience.

Diaz, A. (1988). Goal aggregation in design optimization. *Engineering Optimization, 13*, 257-273. http://dx.doi.org/10.1080/03052158808940958

Dimarogonas, A. D., & Haddad, S. (1992). *Vibration for engineers*. Prentice Hall.

Beitz, W. & Kuttner, K. H. (Eds.). (1994). *Dubbel Handbook of Mechanical Engineering*. Springer Verlag.

Hollburg, U. (2007). *Maschinendynamik*. Oldenburg.

Jameel, M. I., Araned, H., & Luks, G. (1998). Extending recovery boiler runtime through the temperation of steam at the soot blower. *TAPPI Engineering Conference*.

Jin, Y. (2011). Surrogate-assisted evolutionary computation: Recent advances and future challenges. *Swarm and Evolutionary Computation, 1*(2), 61-70. http://dx.doi.org/10.1016/j.swevo.2011.05.001

Leitch, R. D. (1988). *Basic reliability engineering analysis*. London: Butterworths.

Loshchilov, I. M., Schoenauer, & Sebag, M. (2010). Comparison-Based Optimizers Need Comparison-Based Surrogates. *Parallel Problem Solving from Nature* (*PPSN XI*) (pp. 364-373). Springer.

Majander, P., & Siikonen, T. (2002). A comparison of time integration methods in an unsteady low-Reynolds-number flow. *International Journal for Numerical Methods in Fluids, 39*, 361-390. http://dx.doi.org/10.1002/fld.286

Martikka, H., Pöllänen, I., & Simonen, J. (2006). *Design of optimally safe recovery boilers against occurrence and consequences of internal explosions* (pp. 227-236). Southampton: Wessex Institute of Technology.

Martikka, H. I., & Pöllänen, I. (2009). Multi-objective optimisation by technical laws and heuristics. *Memetic computing, 1*(3), 229-238. http://dx.doi.org/10.1007/s12293-009-0020-0

Martikka, H., & Taitokari, E. (2011). Optimum Fuzzy Design of Ecological Pressurised Containers. *Advances in Fuzzy Systems, 2011*. http://dx.doi.org/10.1155/2011/791687

Martikka, H., & Taitokari, E. (2012). Design of perforated shell dryings drums. *Mechanical Engineering Research, 2*(2), 31-44. http://dx.doi.org/10.5539/mer.v2n2p31

Meyer, J. A. (1985). Finite life under combined stress. *Machine Design*, August, 83-84.

Rao, S. R. (2007). *Vibration of Continuous systems*. USA: John Wileys.

Talvio, T. (2009). *Testing of tube platens*. KyAMK Osk.

Ventsel, E., & Krauthammer, T. (2001). *Thin plates and shells*. New York, NY: Marcel Dekker. http://dx.doi.org/10.1201/9780203908723

Wilhelmsen. (2012). Retrieved from http://www.wilhelmsen.com/services/maritime/companies/buss/products/MarineChemicals/Chemicals/FuelOil/Pages/571240.aspx

Huang, Z. J., Wang, C. G., Chen, J., & Tian, H. (2011). Optimal design of aeroengine turbine disc based on kriging surrogate models. *Computers & Structures, 89*(1-2), 27-37. http://dx.doi.org/10.1016/j.compstruc.2010.07.010

17

Recycling of DLC-Coated Tools for Dry Machining of Aluminum Alloys via Oxygen Plasma Ashing

Tatsuhiko Aizawa[1], Eijirou Masaki[1], Eiji Morimoto[2] & Yoshio Sugita[3]

[1] Department of Engineering and Design, Shibaura Institute of Technology, Tokyo, Japan

[2] Green Tool Co. Ltd., Okayama, Japan

[3] YS-Electric Industry, Co. Ltd., Kofu, Japan

Correspondence: Tatsuhiko Aizawa, Department of Engineering and Design, Shibaura Institute of Technology, Tokyo 108-8548, 3-9-14 Shibaura, Japan. E-mail: taizawa@sic.shibaura-it.ac.jp

Abstract

DLC coated tools have been widely utilized in the dry milling and cutting of aluminum alloy parts and components for automotive industries and cellular phone makers. Most of these tools are inevitably exchanged with new ones to preserve high quality of machined products. Then, the used DLC coating must be removed or ashed away before grinding and polishing. Complete removal of used DLC coating plays a key to make DLC-recoating onto the ashed surface of tools for their reuse. High density oxygen plasma ashing provides a solution to make perfect ashing of DLC coating together with metallic interlayers before recoating. In the present paper, this plasma system with quantitative diagnosis instrumentation is first stated to describe the effect of processing conditions on the ashing behavior. Two types of end-milling tools were prepared to prove that the whole DLC-coating should be completely removed with fast rate. This ashed tool substrate was ground and DLC-recoated to demonstrate that the reused tools have the same machinability in dry as the original virgin tools through dry milling experiments.

Keywords: DLC-coated tools, oxygen plasma ashing, perfect removal of DLC coating, recoating, dry machinability tests, aluminum alloys, surface quality

1. Introduction

The aluminum alloys have been widely utilized as a light-weight structural material with higher specific strength in the aircraft components and automotive parts for further reduction of CO_2 emission in transportation (Cole, 1997). In recent years, mobile phone cover cases are fabricated by directly machining the aluminum alloys with high cost competitiveness (Samsung, 2013). In the above mass-productive machining, tool life extension becomes the first necessary condition to shorten the leading time with low production cost and to save materials and resource. Environmentally friendly machining under MQL (minimum quantity lubrication) as well as dry machining becomes the second necessary condition to be responsible for environmentally benign manufacturing. In particular, little or no need for post-cleansing of dry- or semi-dry machined aluminum alloy parts is attractive to industries (Gangopadhyay, Acharya, Chattopadhyay, & Sargade, 2010).

DLC (Diamond-Like Carbon) coating is high-lighted as a promising method to prolong the tool life in practice. As had been reported in Jones (2012) and Zolgharni, Jones, Bulpett, Anson, and Franks (2008), adhesion of aluminum alloys to cutting tools was significantly reduced by DLC coating. In case of bare tools, the wrought aluminum alloy was easy to glue into teeth of tools even at the beginning stage of machining. On the other hand, DLC-coated cutting tools were free from adhesion of work materials. However, the DLC film thickness was inevitably reduced by abrasive wear during dry milling or cutting. This suggests that the used DLC coating should be removed once for recoating the original tool substrates to reuse them. This recycling of tool substrate materials is indispensable to save the tool materials cost such as WC (Co) or super-alloys but also to preserve the product surface quality with proof of originally tailored tool surfaces.

The authors (Aizawa, 2011; Aizawa & Sugita, 2011; Aizawa & Fukuda, 2012, 2013; Aizawa, Mizushima, Redationo, & Yang, 2013) have developed the high density oxygen plasma ashing process to make fast-rate removal of DLC coating on the molds and dies without deterioration of their surface quality. In fact, this approach was successfully applied to reuse of WC (Co) die for mold-stamping the oxide-glass lens by ashing the used DLC coating and by recoating DLC onto the recycled die. At the presence of DLC coating in mold-stamping, adhesion of oxide glass melts onto the mold-die surface was much reduced even in the mold-stamping at 943 K in the inert gas atmosphere. In every 300 to 400 shots in stamping, the used DLC coating was ashed away for recoating. This recycling process resulted in much reduction of production cost with preservation of the original surface quality of mold-die. That is, complete ashing method of DLC coating established the recycling process of mold-die material without quality loss of mold surfaces. Furthermore, the authors (Aizawa & Fukuda, 2013; Aizawa, Masaki, & Sugita, 2014) discussed the possibility on the complete ashing of DLC coating from cutting tool surfaces.

The present paper is concerning with the suitable design of oxygen plasma ashing process for complete removal of used DLC coating from the end-milling tools with complex shaped teeth on them. First, the high density plasma ashing system is introduced to explain how to remove the DLC coating via the quantitative plasma diagnosis. High electron and ion density state is preferable to the fast rate ashing. In second, the DLC-coated end-milling tools are prepared as a specimen to demonstrate that DLC-coating on the tooth surfaces could be ashed away to leave the original tool material surfaces without its deterioration. In third, the ashed end-milling tools are DLC-recoated to make dry machinability testing. Direct comparison of recoated tool surfaces after dry machining reveals that the DLC-recoated tools are proven to have the equivalent quality in dry machinability to the virgin DLC coated tools.

2. Experimental Procedure

2.1 High Density Oxygen Plasma Ashing System

High density oxygen plasma ashing system was illustrated in Figure 1. In the conventional DC- or RF-plasma generators, the plasmas are ignited and generated in the frequency of 13.56 MHz or its multiples. This high density oxygen plasma ashing system has no mechanical matching box. Input and out powers are automatically matched by frequency adjustment around 2 MHz. This difference in power matching reflects on the response time to temporally varying plasma states. The conventional mechanical matching requires for long response time in the order of 1 s to 10 s to adjust the RF-power. While, in the present system, it is shortened down to less than 1 ms. There is no time delay in power control to drive the ashing process. In addition, the vacuum chamber is electrically neutral. Hence, RF-voltage and DC-bias are controlled independently from each other. RF-voltage was controllable up to 250 V, while DC bias, 0 V to −600 V. A dipole electrode is utilized to generate RF-plasma, and, DC bias is directly applied to the specimens.

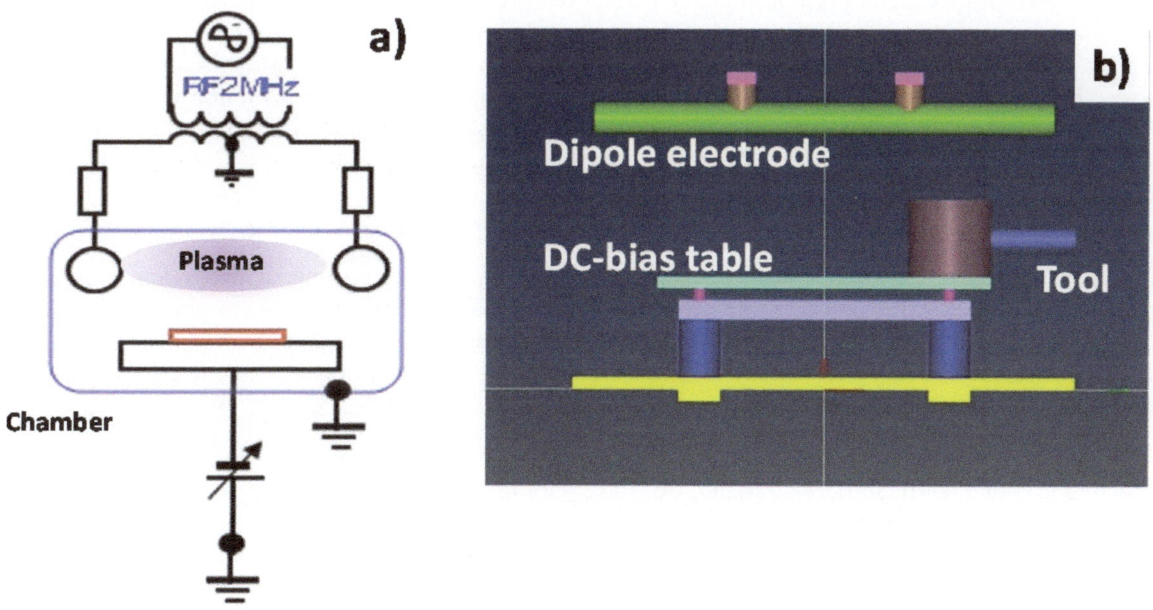

Figure 1. High density RF-DC oxygen ashing system. a) Illustration on the ignition of plasmas, and, b) Location of specimen holder on the cathode table

Spectroscopic analyzer of emissive light from plasmas is instrumented in this system. The detector of emissive light is placed onto a silica window of chamber to transfer this light to the analyzer via the optical fibers. In addition, this spectroscopic analysis is performed to make on-line monitoring of plasmas. Through this analysis, each activated species in the plasmas is identified by its corresponding peak at the specified wave length. The Langmuir probe is also equipped to the present ashing system. Both the electron density and temperature are also on-line monitored during this ashing process. Both the density and temperature for electrons, ions and radicals are important parameters to define the plasma state for ashing process and to describe the effect of plasma processing parameters on the ashing behavior.

In the following ashing experiments, the specimens were located on the cathode table before evacuation down to the base pressure of 0.1 Pa. Then, a carrier gas was introduced into the chamber to attain the specified pressure. Both oxygen and argon gases were available in this system besides the nitrogen gas for bent. With use of magnetic lens, the ignited RF-DC oxygen plasmas were focused onto the surface of specimens during ashing. Table 1 lists the standard processing conditions for ashing of DLC coating.

Table 1. Standard processing conditions for ashing of DLC coating

Processing Item	**Parameters**
RF-Voltage	250 V
DC-Bias	−600 V
Carrier Gas	O_2 Only
Partial Pressure	15 Pa
Mass flow rate	65 ml/min
Ashing duration time	7.2 ks

2.2 Specimens

Two types of specimen were prepared for ashing experiments. DLC-coated WC (Co) plate specimen, as shown in Figure 2a), is employed to describe the ashing behavior and to measure the removal rate of DLC coating. No metallic interlayers are present between the main DLC-film and the WC (Co) substrate. Two end-milling tools are employed as a specimen to demonstrate the perfect removal of DLC coatings in practice. One specimen for long distance cutting test is a DLC-coated end-milling tool. It has three teeth with the skew angle of 45°, its diameter, 10 mm, and, its total length, 100.15 mm as depicted in Figure 2b). The other for large volume milling test also has three teeth. Its diameter is 10 mm, and, its total length, 100 mm as shown in Figure 2c). Each tooth has a skew angle of 45° and a length of 15 mm. The thickness of DLC film is constant at 3.0 μm for both end-milling tools. DLC coating was performed by the ion-plating system in the Japan Coating Center, Co. Ltd. (JCC, 2014).

2.3 Observation and Analysis

Laser microscopy was employed to measure the thickness of masked DLC coating. Optical microscope and SEM (Scanning Electron Microscope) were utilized for surface observation of DLC-coated specimens before and after ashing. Raman spectroscopy was also used to detect the residual DLC fragments on the ashed tools. EDX (Electron Diffraction X-ray) was utilized to detect the chemical component distribution at the specified area of tool surface after ashing process.

2.4 Dry End-Milling Test

A machining center, DMU50 (Deckel Maho, Co. Ltd.) with the spindle of BT 40 was commonly employed for the following two dry end-milling tests. Table 2 summarizes the dry milling conditions for the long distance cutting test up to 200 m with use of DLC-recoated end-milling tools as shown in Figure 2b). As listed in Table 3, both the feed per tooth and the depth of cut are forwarded in case of the large volume milling up to 240 cm^3 with use of DLC-recoated tools as depicted in Figure 2c). Type A5052 aluminum alloy is used as a work material.

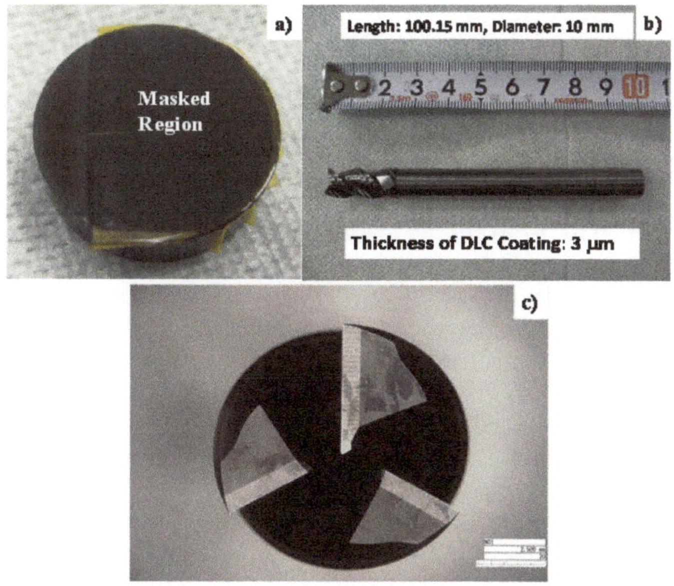

Figure 2. DLC-coated WC (Co) specimens. a) Plate specimen, b) End-milling tool specimen for long distance cutting, and c) End-milling tool for large volume machining

Table 2. Dry end-milling test condition for cutting in the distance of 200 m

Processing Items	Parameters
Spindle speed	6350 min^{-1}
Cutting speed	200 m/min
Feed rate	2250 mm/min
Feed per tooth	0.12 mm/tooth
Axial depth of cut	10 mm
Radial depth of cut	1 mm
Overhang	30 mm
Cutting direction	Down-cutting
Actual cutting speed	200 m/min

Table 3. Dry end-milling test condition for cutting up to the removed volume of 240 cm^3

Processing Items	Parameters
Spindle speed	6350 min^{-1}
Cutting speed	200 m/min
Feed rate	2250 mm/min
Feed per tooth	0.15 mm/tooth
Axial depth of cut	12 mm
Radial depth of cut	5 mm
Perpendicular depth of cut	5 mm
Overhang	30 mm
Cutting direction	Down-cutting
Actual cutting speed	200 m/min

3. Experimental Results

3.1 Plasma Assisted Oxidation of DLC Coating

With aid of quantitative plasma diagnosis in Saloum, Naddaf, and Alkhaled (2010) and Suenaga, Yunata, and Aizawa (2013), the activated plasma state is described to understand the reactivity between the oxygen plasmas and the DLC coating. Through the spectroscopic analysis of emissive light from oxygen plasmas, the state of activated species is analyzed to optimize the plasma processing parameters. The activated oxygen species are classified into O_2^+, O* and O (j) for j = II, III, and IV (or O^+, O^{2+}, and O^{3+}). These population distributions are strongly dependent on the RF/DC conditions and plasma pressure. Activation into oxygen ions and radicals is accompanied with increase of electron density. Figure 3a) shows a typical emissive light spectrum of the fully developed plasma state. Besides a single, significant peak of O_2^+, almost all the peaks come from the activated oxygen atoms and ions. This is because of cascading electron attachment and detachment reactions in series where $O_2^+ + e \rightarrow 2O^*$, $O^* \rightarrow O^+ + e$, $O^+ \rightarrow O^{2+} + e$, and, $O^{2+} \rightarrow O^{3+} + e$. Hence, huge amount of electrons is expected to be present with these activated species of O* and $\{O^+, O^{2+}, O^{3+}\}$. The Langmuir probe method is employed to measure the electron density and temperature in the present plasma state with comparison to the reference data for ICP (Inductively Coupled Plasma) in Cvelbar, Krstulovi, Miloševi, and Mozeti (2007). Table 4 assures that higher electron density is attained by the present method than ICP, intensified plasma state in the narrowed region. This higher population of O* and O^+ is mainly responsible for reaction of oxygen atoms and ions with the carbon in DLC.

Figure 3b) depicts the emissive light spectrum from reactive oxygen plasma during ashing through narrow scanning in the spectroscopy. Significant peaks corresponding to CO (Aizawa et al., 2013a; Aizawa & Fukuda, 2013; Cuesta, Mart'ınez-Alonso, & Tasco'n, 2001) are detected together with activated oxygen atom and ion peaks. This proves that DLC coating is ashed away by the chemical reaction of C (in DLC) + O (in plasmas) → CO.

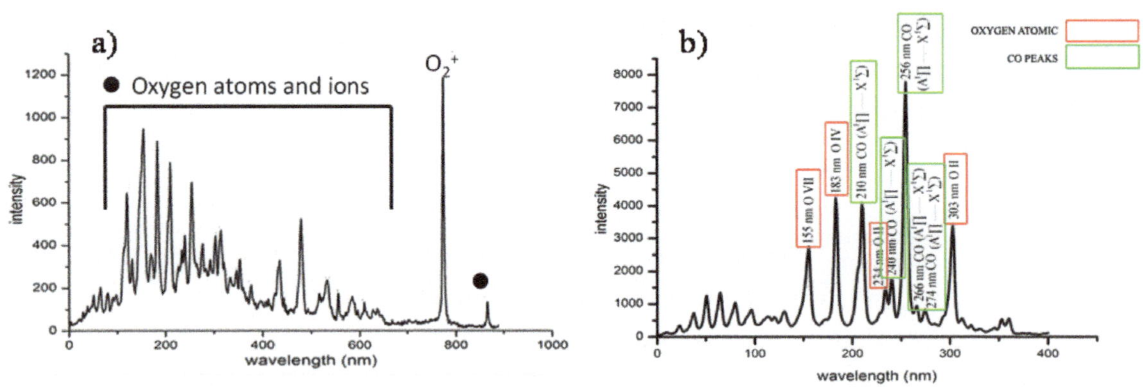

Figure 3. Spectrum of activated oxygen atoms, ions and resultants in the plasmas. a) measured spectrum of activated oxygen species in the plasmas, and b) detection of resultant, CO, from direct reaction of carbon in DLC with oxygen atoms in plasmas

Table 4. Comparison of oxygen plasma parameters between the present plasmas and ICP

Process Parameters	Inductively Coupled Plasma (Cvelbar, 2007)	Present Plasma
Pressure	75 Pa	75 Pa
Plasma Configuration	27.12 MHz	2 MHz
	300 W	< 300 W
Diameter of Chamber	40 mm	800 mm
Length of Chamber	600 mm	1,000 mm
Electron Density	1×10^{16} m^{-3}	4.8×10^{16} m^{-3}
Electron Temperature	5 eV	3.9 eV

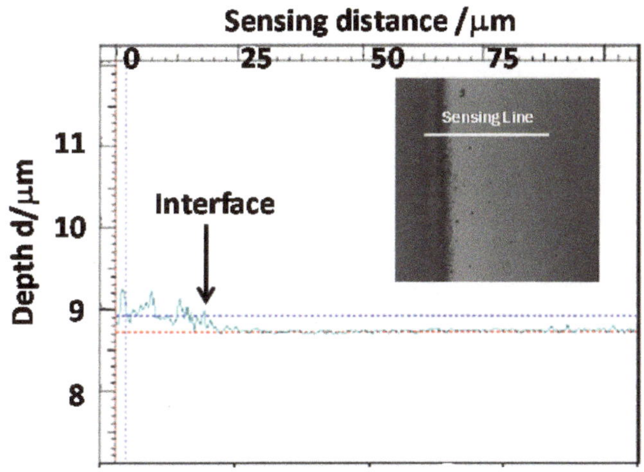

Figure 4. Depth profile of ashed DLC coating on the WC (Co) plate specimen with use of masking, analyzed by the laser-microscopy

3.2 Ashing Performance of DLC-Coated WC (Co) Plates

DLC-coated WC (Co) plate specimens are utilized to describe the plasma ashing behavior and to measure the ashing rate of DLC coating without any interlayers between the WC (Co) substrate and DLC coating. A half of DLC film is masked by polyimide tape. Figure 4 depicts the depth profile of ashed DLC coating on the WC (Co) plate specimen after 100 s. Across the boundary of masking, the film thickness abruptly reduces from the original thickness of 200 nm to zero. Since no residual DLC films are left on the unmasked half of specimen, the original black colored surface of DLC film changes to be in metallic shining. The measured ashing rate is 2 nm/s or 7.2 μm/H.

3.3 Ashing Performance of DLC-Coated WC (Co) End-Milling Tools

The DLC-coated end-milling tool specimen, as shown in Figure 2b), is employed in the ashing experiment to prove that DLC coating is completely ashed away by the present process. Figure 5a) depicts the ashed specimen after two hours or 7.2 ks with use of plasma ashing conditions in Table 1. Black-colored tool surface comes back to an original metallic shining. SEM image around the tooth is also shown in Figure 5b). No residuals of DLC film are seen even at the vicinity of tooth. The machined traces by grinding before DLC coating are only seen on the skewed surface of tooth.

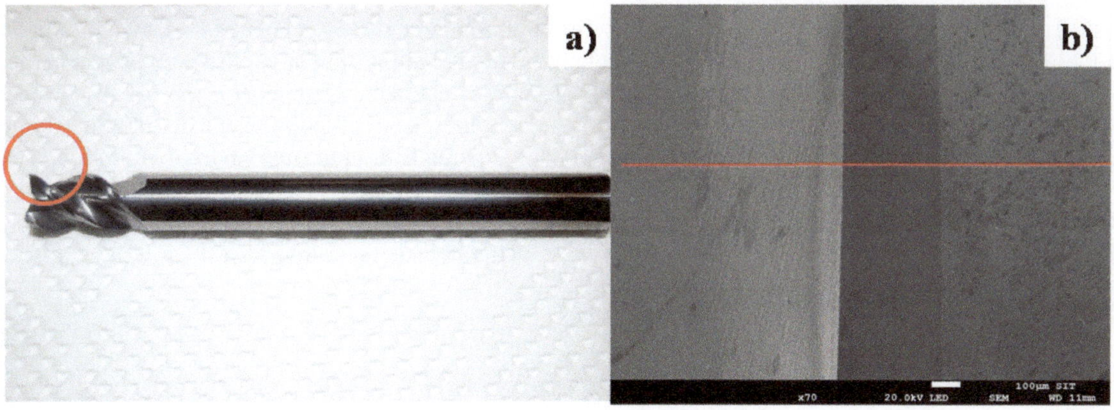

Figure 5. Observation of DLC-coated WC (Co) end-milling tool after ashing. a) Outlook of ashed WC (Co) tool, and b) SEM image of ashed WC (Co) tool at the vicinity of the first tooth

In general, the amorphous carbon film is characterized by the Raman spectroscopy to have a pair of two broad peaks around 1320 cm^{-1} and 1600 cm^{-1} respectively in the measured spectrum. Local tetragonal carbon structure or sp3 bonding state corresponds to the former D-peak; while the local planar structure or sp2 bonding state, to the latter G-peak. In addition, the area intensity ratio, or, I(D)/I(G) of these sp3- and sp2-peaks has correlation with hardness. That is, hardness is expected to be higher with increasing this I(D)/I(G) ratio. Figure 6 a) depicts the Raman spectrum of original DLC coating before ashing. Higher area intensity ratio of two D and G-peaks proves that this DLC coating has high hardness enough to a protective coating for cutting tools. With reduction of film thickness, these peak intensities also reduce themselves; the similar spectra might be detected even for very thin DLC coating.

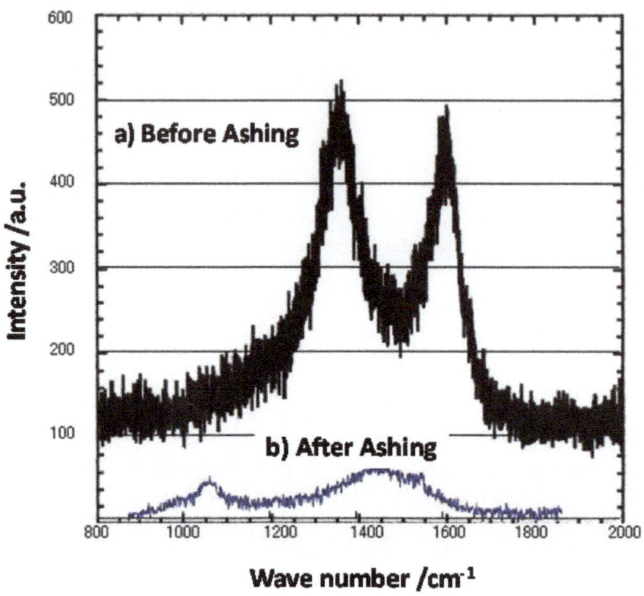

Figure 6. Comparison of Raman spectra for DLC-coated WC (Co) end-milling tool before and after ashing. a) Before ashing, and b) After ashing

Figure 6b) shows the measured Raman spectrum after ashing. No peaks are detected at the above specified wave numbers. Much broader spectra with weak peak intensities are only seen in this Figure 6b). This detection reveals that residuals on the tool surface could be only dirt or surfactant if they were. That is, the whole DLC coating layer is ashed away by this process for 7.2 ks.

Different from the plate specimen in Figure 2a), the chromium undercoat was deposited as a metallic interlayer on the tool surface before DLC coating. Aiming at reliable DLC-recoating, this metallic interlayer must be also removed together with DLC coating. Then, once the chromium could be detected on the ashed surface, the metallic interlayer must be left as a residual. Figure 7 depicts the chemical component distribution along the scanning line in Figure 5b). Since the tool material is WC (Co), both the tungsten and cobalt are detected by EDX analysis. In particular, the average content of Co was 10 at %; this is just corresponding to the original cobalt content of WC (Co). In comparison to these data, little chromium is detected in Figure 5b). Its average of line-detected content is only 0.3 at % or less than by EDX. This reveals that almost all the interlayer should be ashed away.

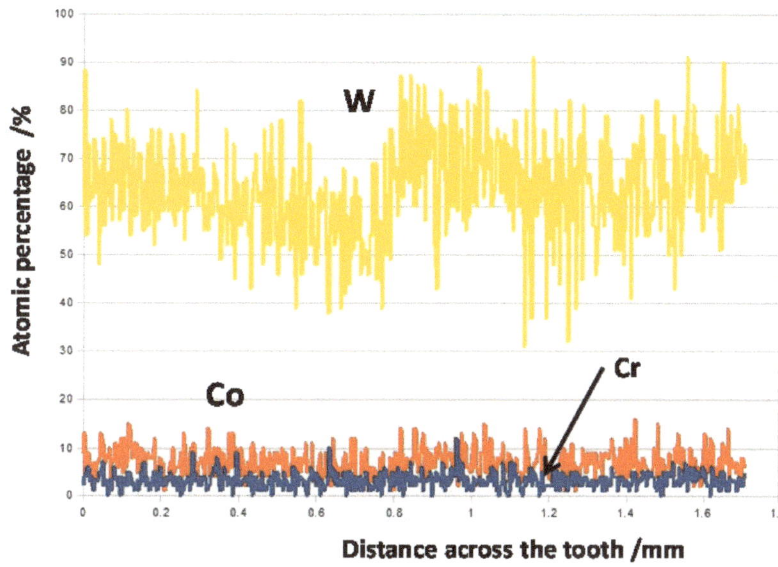

Figure 7. Distribution of chemical components on the ashed tool surface around the first tooth analyzed by EDX

3.4 Dry End-Milling Test of DLC-Recoated WC (Co) Tools up to 200 m in Distance

After complete ashing of used DLC coating as shown in Figure 5a), the original end-milling substrate was ground and polished before DLC-recoating. The same ion-plating system was also used for this recoating process. The DLC-recoated end-milling tool was fixed into a holder in DMU50 with BT40 spindle. Dry machining test was continued under the milling conditions in Table 2. Figure 8 depicts the optical microscopic images of the first tooth in the DLC-recoated end-milling tool after dry machining for 200 m. No chipping or tiny defects are seen both at the tooth edge and on the skew tooth surface. This proves that the quality of original tool material and its tooth configuration are preserved even by DLC-ashing and by DLC-recoating, respectively

3.5 Dry End-Milling Test of DLC-Recoated WC (Co) Tools up to 240 cm³ in Removed Volume

Different from the milling of steels, a bulk part is often directly produced by machining the aluminum alloy block. Hence, dry machinability in three dimensional milling has to be evaluated by DLC-recoated tools. The objective volume of aluminum alloys to be removed was determined to be 240 cm^3; the three-tooth cutting tool as shown in Figure 2c) was employed for this dry machining test.

Figure 9 shows the tooth configuration for three teeth after dry milling to remove 240 cm^3. No chipping or damages are seen in each tooth surface even in the heavy machining conditions as listed in Table 3. In particular, no deterioration on the skew tooth surfaces and edges demonstrates that bulk aluminum alloy parts could be machined out in dry from material block even by using the DLC-recoated tools.

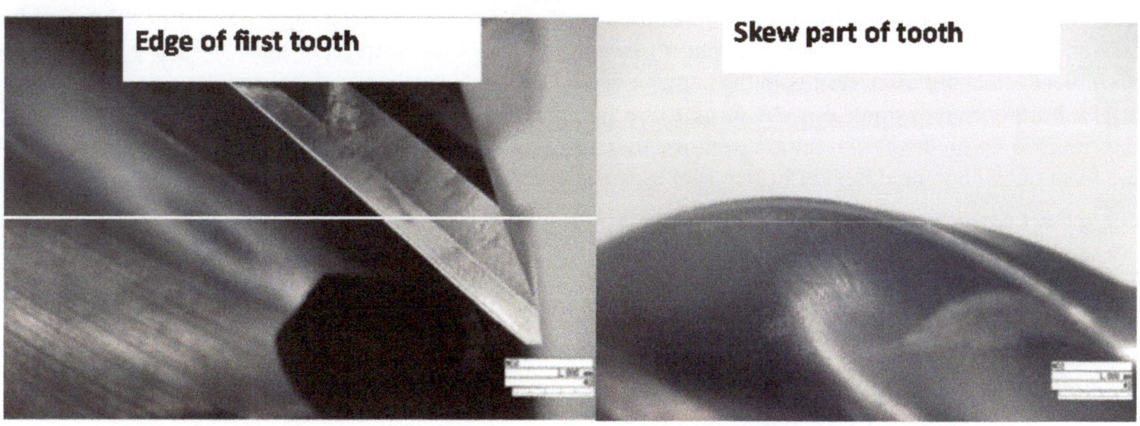

Figure 8. Tooth surfaces of DLC-recoated end-milling tools after dry cutting in 200 m

Figure 9. Tooth surfaces of DLC-recoated end-milling tools after dry machining to remove 240 cm^3

4. Discussion

4.1 Ashing Rate of DLC-Coated Tools

The effective ashing rate of DLC coating from the tool substrate is determined by three processes: i.e. two ashing processes of DLC layer and metallic interlayers and a homogeneous ashing process of the whole tool surfaces. As reported in (Okimoto, Kumakiri, & Tamagaki, 2002), the oxygen ashing system in commercial succeeded in removal of DLC coating. The hollow-cathode discharging method removed the DLC coating by the rate of 0.5 μm/H. In the present system, the ashing rate of bare DLC coating is 7.2 μm/H, or, faster than the above standard approach by fourteen times. This difference comes from higher population of activated oxygen ions and radicals accompanied with high electron density.

In usual DLC-coating of tools, metallic interlayers such as chromium or tungsten are also coated onto the bare tool surfaces. As the second process, these metallic layers must be also removed in fast rate. Here the high population of oxygen ions and radicals is needed to make ashing the metallic layer. In fact, no metallic layers could be removed by the conventional oxygen plasma processing in (Okimoto, Kumakiri, & Tamagaki, 2002). On the other hand, chromium interlayer is completely removed together with main DLC films by the present approach. This might be because metallic components (M) could react with oxygen flux of O^* or O^+ by $M + O^* \rightarrow MO$ or $M + O^+ + e \rightarrow MO$ and the resultant metallic monoxide (MO), be ejected from the system. More ashing time is needed for removal of metallic layers; e.g. the ashing rate for DLC-coated end-milling tools is 1.5 μm/H, or, around 1/5 of the ashing rate for bare DLC coating.

Homogeneous ashing process is controlled by the electro-magnetic field in the chamber. When the specimen holder is located at the center of cathode table, the DLC film thickness at the root of teeth was reduced but not completely removed. On the other hand, no DLC films were left on every part of tooth surfaces in Figure 2 when the specimen holder was located on the edge of cathode table as illustrated in Figure 1b). This difference might come from the strength of electro-magnetic field to keep plasma immersion state around the specimen. Hence, a cathode design is of much importance to drive the homogeneous ashing.

4.2 Ashing Behavior of DLC-Coated Tools

In general, two processes work in oxygen plasma ashing of the used DLC coatings: i.e. physical bombardment process and chemical reaction process. In the former, the higher bias voltage in plasma ashing stimulates this bombardment effect but often results in the damage of tool substrate surface quality or in the deterioration of tool materials. Furthermore, ashing process turns to be inhomogeneous by the shadow effect, where the activated species attack only on the partial region of tools. In the present study, DC bias was limited at most by −600 V to be free from heavy bombardment effects during the ashing process.

Among the activated species, both O^* and O^+ are responsible for chemical reaction to carbon in DLC by $C + O^* \rightarrow CO$ and $O^+ + e + C \rightarrow CO$. Relatively higher intensity of CO detected in Figure 3b) proves that high ion and electron density state should be necessary to promote the above chemical reaction. Once the carbon-base coatings were under this high density plasma immersion state, it must be ashed away in fast rate. As depicted in Figure 3a), oxygen ions with higher covalent number or $\{O^{2+}, O^{3+}, \ldots\}$ were present together with O^* and O^+ in the activated plasmas. Their role in the chemical reaction might be also considered to make quantitative description on the ashing process.

4.3 Characteristics for Green Manufacturing

The present recycling of end-milling tools provides a solution to requirement for green manufacturing by dry machining of aluminum alloy parts. First, a large amount of lubricating oils is saved by this technology to be in the minimum level. After the report on the impact of dry forming in (Aizawa & Morita, 2012), the production cost might be reduced in 30% besides the post-treatment cost of used lubricating oils. In second, the originally designed tool substrate is repeatedly reused by complete ashing and recoating the DLC coating. Although light grinding and polishing processes might be necessary to adjust the tooth configuration, the tooling cost must be much reduced without deterioration of accuracy in machining. More resource for tool substrate materials is further saved by this reuse of tools; after (Allwood & Cullen, 2012), this resource material saving becomes a key as a sustainable manufacturing. In third, the chemical cleansing process for machined products is never necessary to ship the machined parts and components after mechanical polishing. This reduction of environmental burdens is also essential in the sustainable manufacturing.

5. Conclusion

High density oxygen ashing system was developed and applied to make removal of DLC coating from DLC-coated milling tools for dry cutting of aluminum alloys. Perfect ashing of DLC-coating as well as removal of metallic interlayers with high ashing rate is preferable to recycling the used tool substrate. The DLC-recoated tools after the present ashing process have the same machinability as the virgin DLC-coated tools. This demonstrates that the tool substrates are reused only by post-grinding and recoating after the present ashing process.

Plasma immersion state might be necessary to preserve the fast-rate ashing process even for DLC-coated tools with complexity in geometry. In particular, high electron and ion density distribution is required for complete ashing of the carbon base coatings such as DLC or CVD diamond coatings to reuse the WC (Co) tools. In the latter case, the present recycling method is indispensable to repeatedly reuse the WC (Co) tool substrate materials toward the dry machining and milling of CFRP (Carbon Fiber Reinforced Plastics) or CFRTP (Carbon Fiber Reinforced Thermoplastic Plastics) plates and components for aircraft and automobiles. In addition, a massive ashing of used tool stock in the batch mode requires for fine control of oxygen atom and ion flux even in the oxygen plasma immersion state.

Acknowledgements

The authors would like to express their gratitude to Dr. H. Matsuoka (Matsuoka Technology Research, Co. Ltd.) for his advice in experiments. This study was financially supported in part by MEXT-project with the contract of #411419, and, by SIT-project with the contract of #411501, respectively.

References

Aizawa, T. (2011). Micro-patterning onto diamond like carbon coating via RF-DC oxygen plasma etching. *Proc. 5th SEATUC Conference* (Vietnam, Hanoi), 425-428.

Aizawa, T., & Fukuda, T. (2012). Micro-patterning onto carbon-based coatings via oxygen plasma etching. *Res. Rep. Shibaura Institute of Technology, 56*(2), 47-56.

Aizawa, T., & Fukuda, T. (2013). Oxygen plasma etching of diamond-like carbon coated mold-die for micro-texturing. *Surface and Coatings Technology, 215*, 364-368. http://dx.doi.org/10.1016/j.surfcoat.2012.07.095

Aizawa, T., & Morita, H. (2012). Semi-dry transfer stamping by nano-laminated DLC coated dies and tools. *Res. Rep. Shibaura Institute of Technology, 56*(1), 27-36.

Aizawa, T., & Sugita, Y. (2011). High density plasma technology for etching and ashing of carbon materials. *Res. Rep. Shibaura Institute of Technology, 55*(1), 13-22.

Aizawa, T., Masaki, E., & Sugita, Y. (2013). Oxygen plasma ashing of used DLC coating for reuse of milling and cutting tools. *Proc. 10th International Conference on Materials Processing Technology* (BKK, Thai), 15-20.

Aizawa, T., Masaki, E., & Sugita, Y. (2014). Complete ashing of used DLC coating for reuse of the end-milling tools. *Manufacturing Letters, 2*(1) 1-3. http://dx.doi.org/10.1016/j.mfglet.2013.10.006

Aizawa, T., Mizushima, K., Redationo, T. N., & Yang, M. (2013). Micro-imprinting onto DLC and CNT coatings via high density oxygen plasma etching. *Proc. 8th ICOMM Conference* (Canada, Victoria), 459-466.

Allwood, J. M., & Cullen, J. M. (2012). *Sustainable materials*. UIT.

Cole, G., Glove, A., Jeryan, R., & Davies, G. (1997). *Steel World, 2*, 75-83.

Cuesta, A., Mart'ınez-Alonso, A., & Tasco'n, J. M. D. (2001). Carbon reactivity in an oxygen plasma: a comparison with reactivity in molecular oxygen. *Carbon, 39*, 1135-1146.

http://dx.doi.org/10.1016/S0008-6223(00)00235-9

Cvelbar, U., Krstulovi, N., Miloševi, S., & Mozeti, M. (2007). Inductively coupled RF oxygen plasma characterization by optical emission spectroscopy. *Vacuum, 82*(2), 224-227. http://dx.doi.org/10.1016/j.vacuum.2007.07.016

Gangopadhyay, S., Acharya, R., Chattopadhyay, A. K., & Sargade, V. G. (2010). Effect of cutting speed and surface chemistry of cutting tools on the formation of bul or bue and surface quality of the generated surface in dry turning of AA6005 aluminum alloy. *Machining Science Technology, 14*(2), 208-223. http://dx.doi.org/10.1080/10910344.2010.500961

Japan Coating Center. (2014). Retrieved 11th January, 2014 from http://www.jcc-coating.co.jp

Jones, B. J. (2012). Substrate and material transfer effects on the surface chemistry and texture of diamond-like carbon deposited by plasma-enhanced chemical vapor deposition. *Surface and Interface Analysis, 26*, 431-436. http://dx.doi.org/10.1002/sia.4871

Okimoto, T., Kumakiri, T., & Tamagaki, H. (2002). Application of hollow cathode plasmas to SiO_x coating. *Future: Advance thin film technology in Research Report of KOBELCO 52*(2), 2002, 13-18.

Saloum, S., Naddaf, M., & Alkhaled, B. (2010) Active species characterization in RF remote oxygen plasma using actinometry OES and electrical probes. *Vacuum, 85*, 439-442. http://dx.doi.org/10.1016/j.vacuum.2010.08.007

Samsong. (2013). Retrieved 19th April, 2013 from http://au.ibtimes.com/articles/457503/20130416/iphone-5s-samsung-galaxy-s4-note-3.thm

Suenaga, R., Yunata, E. E., & Aizawa, T. (2013). Quantitative plasma diagnosis on high density RF-DC plasmas for surface processing. *Proc. 7th SEATUC Conference. OS-6-19* (Bandung), 1-4.

Zolgharni, M., Jones, B. J., Bulpett, R., Anson, A. W., & Franks, J. (2008). Energy efficiency improvements in dry drilling with optimized diamond-like carbon coatings. *Diamond and Related Materials, 17*(7-10), 1733-1737. http://dx.doi.org/10.1016/j.diamond.2007.11.012

CNC Machining Path Planning Optimization for Circular Hole Patterns via a Hybrid Ant Colony Optimization Approach

Adel T. Abbas[1], Karim Hamza[2] & Mohamed F. Aly[3]

[1] Department of Mechanical Engineering, King Saud University, Riyadh, Saudi Arabia

[2] Department of Mechanical Engineering, University of Michigan, Ann Arbor, USA

[3] Department of Mechanical Design, American University in Cairo, New Cairo, Egypt

Correspondence: Adel T. Abbas, Department of Mechanical Engineering, King Saud University, Saudi Arabia, 11421, P.O.Box 800. E-mail: aabbas@ksu.edu.sa

Abstract

This paper presents a path-planning optimization study for a Computer Numerical Control (CNC) machining center tasked with machining jobs involving a large number of holes to drill that are mostly arranged in concentric circular patterns. Benefits of this research may contribute to shortening the machining time in certain components used in heat exchangers, boilers, condensers, trammel screens and food separators. Optimization of tool travel distance and machining cost are typically overlooked aspects when generating tool paths and CNC codes from commercially available CAD software packages. Tool path travel distance minimization can be modelled Travelling Salesman Problem (TSP). Optimization algorithms have been heavily applied in the literature to the TSP with varying levels of success. Ant Colony Optimization (ACO) is one of the most prominent approaches that mimics the natural behavior of ant colonies. The research in this paper proposes a hybrid ACO that has a biasing mechanic designed to take advantage of the geometric hole-pattern arrangement, as well as a local search. Simulation examples show the proposed approach exhibiting superior performance compared to the classic ACO approach, a genetic algorithm (GA) approach, as well as the simple spiral path generated via commercial CAD software. The proposed approach is then applied to the drilling path planning of a two-thousand-hole food-industry separator plate.

Keywords: CNC path planning, automated drilling operations, ant colony optimization, genetic algorithm

1. Introduction

Total machining time and cost of production in Computer Numerical Control (CNC) drilling operations of large number of holes arrangements is of a significant concern to manufacturing companies. Associated time investment in process optimization is often well justified. The fraction of machining time spent on drilling holes can vary from an average of 40% for a typical product (Shunmugam et al., 2000) to more than 80% for products that incorporate dense matrices of holes such as heat exchangers, food-processing separators (Figure 1), as well as drum and trammel screens. And as such, cost optimization of CNC drilling operations has been an important field of study in the literature (Ke et al., 2006; Zhu & Zhang, 2008; Satishkumar & Asokan, 2008). Along with the opportunity of reducing the production via small improvements in the cutting conditions, the total travelled distance of the tool between drilling locations can be reduced through optimum path planning. Such reduction can account for up to 20% of the total machining time of the part, which in turn can have a significant impact on cost.

Path planning to minimize travel time between drilling locations can be formulated as a Travelling Salesman Problem (TSP), which is classified as NP-hard, well known and difficult to solve in integer programming (Wolsey & Nemhauser, 199). Several approaches have been applied to TSP including: Dynamic Programming (Denardo, 2004), Genetic Algorithms (Goldberg, 1989; Michalewiz, 1992; Anand et al., 1999), Simulated Annealing (Aarts, 1989; Van Laarhoven & Aarts, 2009), Particle Swarm Optimization (Zhu, 2006), and Ant Colony Optimization (ACO) (Dorigo, 1996; Stutzle & Dorigo, 2002; Dorigo & Stutzle, 2004; Dorigo et al., 2004; Zhou, 2009; Arnaout et al., 2010), whose run-time modeling (Stutzle & Dorigo, 2002; Dorigo et al., 2004) suggests it as one of the most efficient algorithms to address TSP types of problems. After settling on optimum paths, it is a fairly straight-forward task to automatically generate the G-code for CNC machining centers (Abbas & Megahed, 2005;

Abbas, 2012). Applications of modeling drilling operations as TSP include food-industry separators (Abbas et al., 2011) and printed circuit boards (Montiel-Ross et al., 2012; Saealal et al., 2012).

Figure 1. Food-industry processing separators

This paper extends the approach in (Abbas et al., 2011) with a focus on the optimization of a CNC drilling tool path between holes lying in concentric circular patterns. Simple examples studies as well as a full-scale industrial application of a food processing separator screen are tested. Results of the studies show better performance of the modified ACO algorithm compared to the classic ACO, the modified version in (Abbas et al., 2011), a typical genetic algorithm, as well as establish the scalability to large problems.

The paper starts with a brief introduction and a review of relevant work reported in the literature that highlights the motivation to carry out this research. The formulation of the optimization problem with details of the classic ACO algorithm and the proposed modifications is then presented. Example studies and an industrial application are then presented to demonstrate the algorithms' performance. The paper then concludes with a brief discussion and prospective future extensions.

2. Problem Formulation

Tool path planning for the CNC drilling tool is modelled as a TSP when ignoring the tool dynamics (which is a reasonable assumption when the tool has to dwell a reasonable amount of time in each visited location to be drilled). TSP can be formulated as an integer linear program (Padberg & Sung, 1991) as follows:

$$\text{Minimize:} \quad f(x) = \sum_{i}^{n} \sum_{j}^{n} X_{ij} D_{ij} \tag{1}$$

$$\text{Subject to:} \quad \sum_{i=1}^{n} X_{ij} = 1, j = 1, \ldots, n \tag{2}$$

$$\sum_{j=1}^{n} X_{ij} = 1, i = 1, \ldots, n \tag{3}$$

$$\sum_{i=k}^{k+l-1} \sum_{j=k}^{k+l-1} X_{ij} < l, k = 1, \ldots, n-l+1, l = 1, \ldots, n-1 \tag{4}$$

$$n = n_h + 1 \tag{5}$$

where:

n_h is the total number of holes

n is the total number of visited location by the drilling tool, which is equal to the number of holes to be drilled, plus the home position of the tool

D_{ij} is the travel distance from location i to location j

$X_{ij} \in \{0, 1\}$ are the design variables used to describe the tool path. $X_{ij} = 1$ means that the tool will travel from location i to location j as part of the path that loops on all the holes in the matrix. Otherwise, $X_{ij} = 0$ means that the path from location i to location j is not a part of the tool path.

The objective function is formulated as the summation of all the distances, between holes, travelled by the tool in the chosen path (Equation (1)). A minimization procedure is ran on this objective until an optimized path is selected. The set of constraints described in Eqn. (2), ensure that each hole j is only visited once in the path described by Xij, while the set of constraints described in Equation (3), ensure that the path coming out of every hole i goes to one other hole j. Further discussion of the integer linear programming formulation of TSP may be found in (Padberg & Sung, 1991) .Elements of the distance matrix Dij are calculated as the in-plane distance between the node centers as:

$$D_{ij} = D_{ji} = \sqrt{(x_i - x_j)^2 + (y_i - y_j)^2} \tag{6}$$

where:

x_i is the coordinate position of location i along the X-axis

y_i is the coordinate position of location i along the y-axis

x_j is the coordinate position of location j along the X-axis

y_j is the coordinate position of location j along the y-axis

3. Path Planning Algorithms

CNC drilling tool path planning in this paper is formulated as a TSP in section 2 of this paper. Of the popular algorithms for solving TSP, Ant colony optimization was reported in the literature (Dorigo, 1996; Stutzle & Dorigo, 2002; Dorigo & Stutzle, 2004; Dorigo et al., 2004) as exhibiting superior performance in discovering globally optimal (or close to globally optimal) solutions within reasonable computational time. The work in this paper also introduces a modification to the basic ACO in order to take advantage of the geometrical arrangement of the holes to be drilled. Another popular algorithm used in several continuous and discrete optimization problems (TSP included) is genetic algorithm (GA) Goldberg (1989); Michalewiz (1992); Anand et al. (1999). This section explains the details of the algorithms tested in this paper.

3.1 Basic Ant Colony Optimization Algorithm

The classic Ant-Colony Optimization (ACO) (Dorigo et al., 2004) technique is naturally inspired from ants driven strategy to always find the shortest route from among multiple possible paths between their nest and food location. The basic mechanics behind the achievement of such path is realized through the following steps:

- Different ants start by trying different paths.
- As the ants travel through different paths they deposit a chemical compound known as *pheromone* along each path.
- Subsequent ants traveling along each of those paths have the ability to smell the pheromone and to make decisions on which path to take
- Ants choosing between two paths are *more likely* to follow the one with more pheromone deposit (with some randomness to this decision)
- Deposited pheromone tends to evaporate over time. Eventually the best path will be the most travelled one, and thus richest in pheromone

It should be noted that ACO (and GA, which is later considered in this paper) formulate the design variables $\boldsymbol{X}$ as permutation of a sequence of numbers between 1 and n (corresponding to the order of visiting the path nodes, which implicitly satisfies the constraints in Eqns. 2-4), rather than integer formulation. Main steps of a basic ACO algorithm (Dorigo et al., 2004) are described as:

```
Ant-Colony-TSP(n, D_ij)
    P ← (1/n) × ones(n, n)          ▷ Initial pheromone distribution is uniform
        ▷▷ (equally likely for an ant at any location to choose          ▷▷ to go to
any other location)
    X* ← Generate-Path(n, P)          ▷ An initial solution is recorded
    while Termination-Condition() = FALSE
        do  i ← 1
        while i < nAnts
            do  AntPaths ← ϕ
                AntPaths ← AntPaths ∪ Generate-Path(n, P)
                i ← i + 1
            X° ← Min-Distance(AntPaths)
            if f(X°) < f(X*)
                then X* ← X°
            P ← Update-Pheromone(AntPaths, P)
    return       X*
```

where:

P is the $n \times n$ *Pheromone probability matrix*, for path choices to be made by ants. P_{ij} is an indicator of the probability that an ant currently located at i will choose to go to location j next

`Termination-Condition()` is chosen as a fixed number of iterations, after which, the algorithm returns its best discovered path. Alternative termination criteria are used by other approaches such as the examination of the pheromone matrix P to invoke termination of the search when only one path becomes predominant in its pheromone content. This can only be applicable for problems with no local optima as the termination condition may never be met.

`Generate-Path()` is a sub-function in the ACO algorithm that generates a randomized path that visits all the target locations, with probability to select a path according to the distribution of the pheromone matrix P. Without loss of generality in the interpretation of the TSP path, the path always starts at the node corresponding to the tool home point. The probability of going from node i to node j is equal to:

$$\Pr(i \rightarrow j) = \frac{Q_j}{\sum_{j=1}^{n} Q_j} \tag{7}$$

$$\text{where } Q_j = \begin{cases} 0 & \text{if } j \text{ has already been visited} \\ P_{ij} & \text{otherwise} \end{cases} \tag{8}$$

`Update-Pheromone()` is a sub-function in the ACO algorithm that updates the pheromone matrix P. This incorporates that only ants that discovered the few best- paths will be the ones allowed to deposit pheromone on these paths, hence making it more likely for other ants to follow similar path choices.

The main steps of the function `Generate-Path()` for generating randomized paths are described as:

```
Generate-Path (n, P)
    X ← φ
    X ← X ∪ {1}                 ▷ Always start at first node
    while |X| < n
        do i ← X_|X|
           for j = 1 to n
               do Q_j ← P_ij
           for j = 1 to |X|
               do Q_X_j ← 0
           S ← Sum (Q_j, j = 1 … n)
           for j = 1 to n
               do Q_j ← Q_j / S
           nextNode ← Select (n, Q)   ▷ Selects a random integer between 1 to n (uniform distribution)
                                      ▷▷ with probability of selection Q_j
           X ← X ∪ {nextNode}
    return X
```

The function `Update-Pheromone()` updates the pheromone matrix by calling a `Simple-Update-Pheromone()` sub-function several times but with different values of pheromone deposition of the few current best known paths. The `Simple-Update-Pheromone()` sub-function is used to update the current pheromone matrix via a path $\boldsymbol{X}$ and a deposited pheromone value of σ as follows:

```
Simple-Update-Pheromone (X, P, σ, nBestPaths)
    k ← 1
    i ← X_k
    k ← k + 1
    while   k < nBestPaths       ▷ depositing pheromone
        do j ← X_k
           P_ij ← P_ij + σ
           i ← j
           k ← k + 1
    for i = 1 to n          ▷ normalizing P and diminishing the older pheromone values
        do S ← Sum (P_ij, j = 1 … n)
           for j = 1 to n
               do P_ij ← P_ij / S
    return    P
```

It should be noted that the basic ACO (Dorigo et al., 2004) includes an explicit step for evaporation of existing pheromone before new pheromone is deposited. In this implementation however, the pheromone evaporation happens implicitly via the re-normalization of the pheromone matrix after the new pheromone is deposited.

3.2 Modified Ant Colony Optimization Algorithm

The basic ACO algorithm described in section 3.1 uses a uniform distribution for the initialization of the pheromone matrix P. At the start of the algorithm, an ant at some location i is equally likely to choose to visit any of the other locations j. Taking into consideration the special holes layout (holes with certain geometrical patterns such as concentric circles), it may not make much sense for an ant to travel from one end of a circle to its diametric opposite. As such, a modified initialization of the pheromone matrix (line #1 in the basic ACO algorithm discussed in section 3.1) follows along approaches that use heuristics to bias the probability of the choices made by the ants (Dorigo & Stutzle, 2004). The modification seeks to reduce the probability of selecting a step to the next hole if there are other closer choices:

$$P_{ij}^{N} = \begin{cases} \dfrac{1}{D_{ij}^{p}} & \text{if } D_{ij} \neq 0 \\ 0 & \text{if } D_{ij} = 0 \end{cases} \tag{9}$$

$$P_{ij} = \frac{P_{ij}^{N}}{\sum_{j=1}^{n} P_{ij}^{N}} \tag{10}$$

where:

p is an exponent for tuning the effect of the distance between the locations to be visited on the probability that ants make the choice to travel between them

P_{ij}^{N} is an un-normalized pheromone value that is designed to have higher values when the locations to be travelled between are close to one another. Normalization of P_{ij}^{N} via Eqn. 9 gives the pheromone value to be used in the ACO algorithm

It is noted that the proposed modification is generalization of both the basic ACO and the approach in (Abbas et al., 2011). Setting the exponent p in Eqn. 8 to zero causes the initial pheromone matrix to have equal values for travel between any locations (basic ACO), and setting p to a *very large number* (ACO would only consider immediate neighbors) equates to the approach in (Abbas et al., 2011). In some perspective, instead of using a heuristic matrix to influence the decision making of the ants as in (Dorigo & Stutzle, 2004), the perception that near neighbors are likely better choices is directly implemented in the initial pheromone matrix. As such, the proposed approach is perceived to have the following advantages:

a) No a-priori knowledge of a good path is required
b) Better computational efficiency (fewer terms to compute in the calculation of the probability of choosing a node)
c) Heuristic perception of good pathing automatically fades in favor of learned pathing as the search progresses (with pheromone updating) without the new for an additional tunable control parameter

3.3 Genetic Algorithm

The genetic algorithm tested in this paper is based on one of the implementations specific to TSP described in (Michalewiz, 1992, Chapter 10). Basic steps of a GA (Goldberg, 1989) are:

1. Initialize a random "Population" of solutions
2. Best known solution is recorded separately
3. Generate new solutions from existing population of solutions via: i) Selection, ii) Crossover, and iii) Mutation, and then place the newly generated solutions in a new population
4. When the number of solutions in the new population reaches the same size as current population, the new population replaces the current one, and the algorithm repeats at step 2
5. Steps 2-4 are repeated for a prescribed number of generations, then the best encountered solution is returned

Different implementations of GA will assure the maintenance of the overall algorithm structure, but will differ in the way a solution is represented, or the way new solutions are generated via selection, crossover and mutation. The setting for the GA tested in this paper is as follows:

3.3.1 Solution Representation: Path Representation

As in (Michalewiz, 1992, Chapter 10), the chromosome is encoded as an ordered string of integers that represent the sequence by which the nodes are visited in the TSP. For example, a parent chromosome P1: {1 2 3 4 5 6 7 8 9} means the start is at node 1, then node 2 is visited next, then node 3… until the last node, then a return from node 9 to node 1.

3.3.2 Selection: Binary Tournament

Selection of solutions from existing population to generate new solutions via crossover and mutation depends on the "fitness" of current solutions. In binary tournament selection (Michalewiz, 1992), two solutions are randomly picked from the current population, then the one that has better fitness (less travel distance in TSP) "wins the tournament" and becomes a candidate parent. Another tournament is conducted to select a second parent, and then two child solutions are generated via crossover and mutation from the parent chromosomes.

3.3.3 Crossover: Order of Appearance

Crossover is the main operator that GA uses to generate new solutions from existing ones. The Order of Appearance Crossover (Michalewiz, 1992, Chapter 10) uses a randomly located (with uniform distribution) sub-segment of the chromosome of one parent, and then fills the rest of the newly generated chromosome via nodes chosen from another parent according to their order of appearance.

Two parent chromosomes P1 and P2 with segments to be exchanged are demonstrated as:

P1: {1 2 3 | 4 5 6 7 | 8 9}

P2: {4 5 2 | 1 8 7 6 | 9 3}

First, the segment to be exchanged is taken from first parent chromosome:

C1: {x x x | 4 5 6 7 | x x}

Then, the empty location in the new child chromosome is filled with the nodes that do not exist yet on the chromosome (i.e. nodes: 1, 2, 3, 8, 9) according to their order of appearance in the second chromosome:

C1: {2 1 8 | 4 5 6 7 | 9 3}

3.3.4 Mutation: Reciprocal Exchange

Newly generated child chromosomes have small probabilities of a mutation occurring. In reciprocal exchange (Michalewiz, 1992, Chapter 10), mutation randomly selects (with uniform distribution) two nodes in the chromosome and exchanges their spots.

3.4 Hybridization with Local Search

A known drawback of basic ACO and GA is that the later stages of convergence to an optimum (after a quasi-optimum solution is discovered) may take a long time. One approach discussed in the literature is to conduct local search within the global optimization (Bonabeau et al., 1999), however, in earlier experimentation with study problems, premature convergence of the global search was a concern. Alternatively, another common approach was adopted, which is to perform a local optimization search after ACO or GA is done, while using the solution returned by ACO/GA as a starting point for the location search.. A popular local search for TSP is based on Greedy search. The version adopted in (Abbas et al., 2011) is explained as:

`Local-Search-TSP` (n, X^G, D_{ij})

1 $X^* \leftarrow X^G$ ▷ Initialize best solution with the optimized solution ▷▷ returned from ACO or GA

2 **for** $i = 1$ to n

3 **for** $j = i + 1$ to n

4 $X^t \leftarrow$ `Swap-Order` (X^*, i, j)

5 **if** $f(X^t) < f(X^*)$

6 **then** $X^* \leftarrow X^t$

7 **return** X^*

Where X^G, X^*, X^t are respectively the initial path, final path and a temporary test path. The function `Swap-Order()` in line 4 simply swaps the order of two node locations i and j on the travel path. If the swap is beneficial (less overall travel distance), the swap is retained, otherwise, the swap is discarded.

4. Example Studies

4.1 Problem # 1: Three Complete Circular Patterns

A simple case study of a three concentric circular patterns (Figure 2) is investigated to evaluate the performance of the proposed algorithm. In the upcoming studies, tool velocity transients are not considered, thus the shortest distance tool path is also the one having the minim machining time. In all tests, a local search (as described in section 3.4) is conducted after the completion of each run of the tested algorithm. Comparison is conducted between 10 runs of each of: i) basic ACO ($p = 0$), ii) modified ACO with various settings of the exponent p, and iii) genetic algorithm. The results from various algorithms are also compared verses the typical spiral path (Figure 3) that is generated by most CAD software.

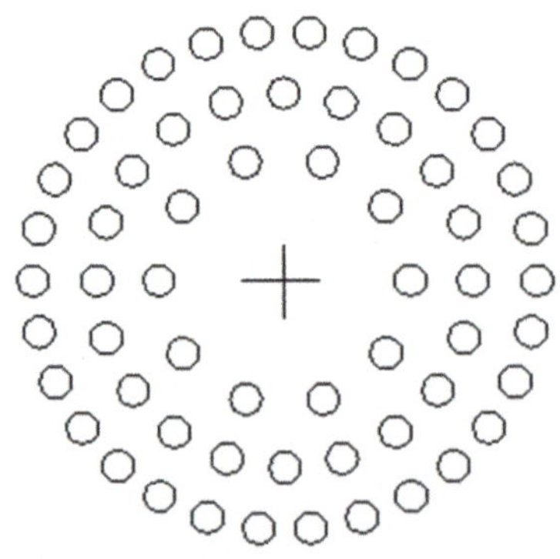

Figure 2. Circular pattern with three concentric circles

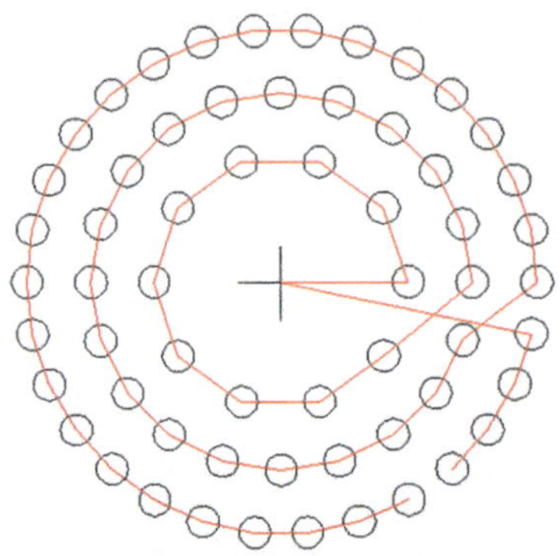

Figure 3. Typical spiral path generated by most CAD software

4.2 Problem # 2: Four Incomplete Circular Patterns

Similar numerical experiments for comparison of the tested algorithms discussed in section 4.1 are conducted for a machining task with more complex geometry (Figure 4), which could represent a more complex design of a food separator screen; one that will have handles at top and bottom portions of the plate and a pipe passing through the right side section.

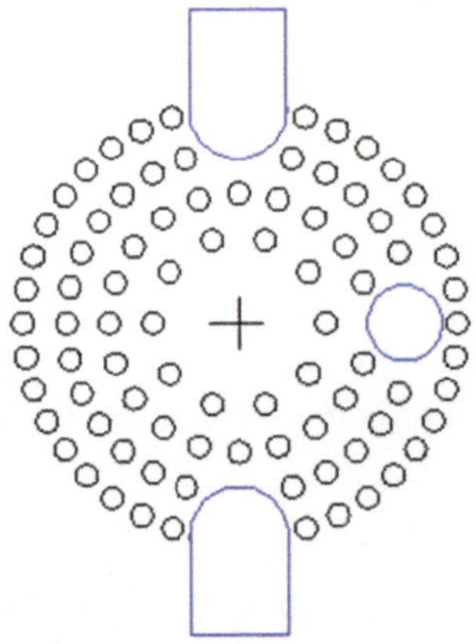

Figure 4. Circular pattern with four concentric circles and three zones vacated for design considerations

4.3 Results and Discussion

Details of the holes geometrical layout for test problems 1 and 2 are listed in Table 1. Tuning parameters for the ACO and GA (based on general recommendations in (Michalewiz, 1992; Dorigo et al., 2004)) are listed in Tables 2 and 3 respectively. While in general, an optimization problem with larger search space requires a larger GA population, it was found in earlier experimentation with problem #2 that using a population size of 100 with 60 generations was producing better results for GA than 120×50, so 100×60 was the chosen setting in order to present the best obtained results from GA.

Table 1. Geometrical layout of the example study problems

Problem #	1	2
Number of Circles	3	4
Inner Pitch Circle Diameter (mm)	40	40
Pitch Circle Diametric Increment (mm)	20	20
Total number of holes	60	86
Length of Spiral Path (mm)	625	944

Table 2. Tuning Parameters for ACO

Problem #	1	2
Number of Ants (*nAnts*)	50	60
Number of iterations until termination	100	100
Number of ants that update the pheromone matrix (*nBestPaths*)	5	5
Pheromone value used in updating the pheromone matrix (σ)	0.1	0.1

Results of 10 runs of the tested algorithms are plotted in Figure 5 and Figure 6 for problems #1 and #2 respectively. Numerical values of the results are also listed in Tables 4 and 5. Travel paths for one of the best obtained solutions for each problem are plotted in Figure 7. A general observation was GA did not exhibit very good performance and was not able to discover better paths than the simple spiral path. This may be due to the size of the TSP problems considered in the example studies; problem #1 has 61 nodes and problem #2 has 87 nodes. Such relatively large number of nodes often makes it difficult for GA to discover the optimum path, even when local search is invoked after the GA runs.

Table 3. Tuning Parameters for GA

Problem #	1	2
Population size	100	100
Number of Generations until termination	50	60
Crossover Probability	0.8	0.8
Mutation Probability	0.1	0.1

Table 4. Results summary for Problem #1

Algorithm	Average Path Length (mm)	Min.	Max.	Standard Deviation
Spiral Path	625	-	-	-
GA	900	828	967	42.3
Basic ACO	711	653	761	29.5
Modified ACO (p = 1.0)	683	646	735	25.0
Modified ACO (p = 3.0)	659	613	656	13.2
Modified ACO (p = 5.0)	622	603	641	14.0
Modified ACO (p = 8.0)	608	594	630	12.4
Modified ACO (p = 12.0)	595	594	608	4.6
Modified ACO (p = 20.0)	594	594	594	0.0

Table 5. Results summary for Problem #2

Algorithm	Average Path Length (mm)	Min.	Max.	Standard Deviation
Spiral Path	944	-	-	-
GA	1434	1330	1509	62.1
Basic ACO	1144	1062	1280	67.2
Modified ACO (p = 1.0)	1021	928	1106	54.0
Modified ACO (p = 3.0)	946	922	980	20.2
Modified ACO (p = 5.0)	904	878	938	18.6
Modified ACO (p = 8.0)	833	813	859	15.6
Modified ACO (p = 12.0)	810	804	820	6.6
Modified ACO (p = 20.0)	804	804	810	2.0

Performance of the basic ACO was observed to be better than GA in both problems #1 and #2, however neither algorithm were capable of discovering paths with shorter travel distance than the simple spiral that would be generated via a typical CAD system (Figure 5, 6). On the other hand, the modified ACO with $p \geq 5$ was consistently capable of discovering paths that were better than the simple spiral. The best performance for the modified ACO was realized at $p = 20$, where the consistency in solution quality was very good (small value for standard deviation amongst the conducted runs) and the discovered paths (Figure 7) are deemed among the best-known for the problem. The fact that large values of p produce favorable results implies that "nearest neighbor" strategies are indeed favorable for this type of problem. However, due to the layout of the holes (many holes have comparable neighbor distances), favoring nearest-neighbor isn't realized except at larger values of p. It may be also worth noting the problems do have multiple equally-good solutions, but there is no mechanism to control which solution the ACO converges to. Plots in Figure 7 are simply those first discovered in the 10 runs.

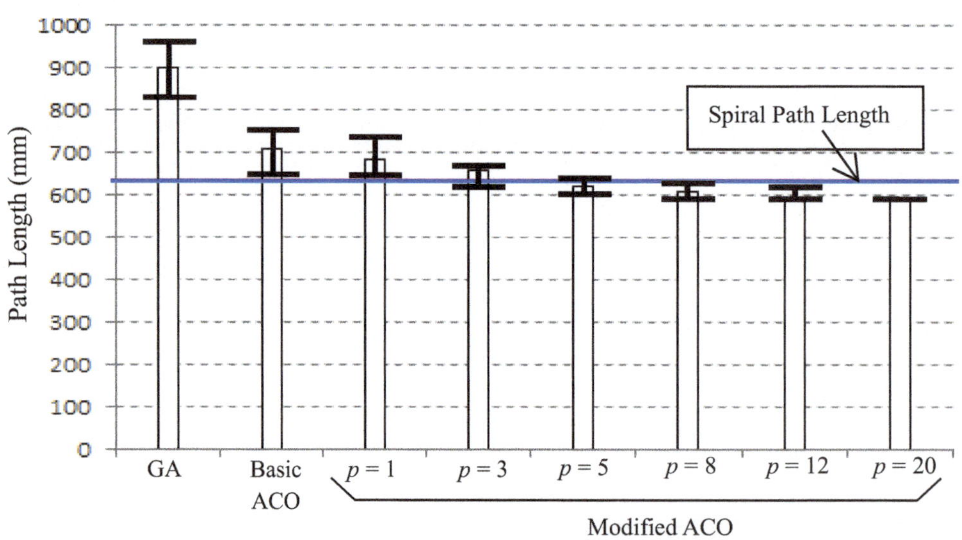

Figure 5. Results summary for problem #1

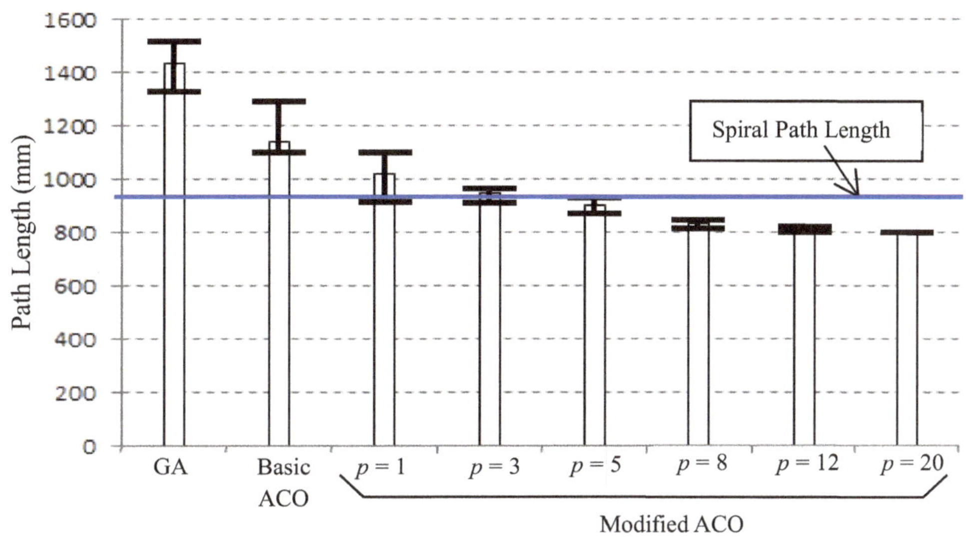

Figure 6. Results summary for problem #2

It is also worth noting that the solutions discovered via the modified ACO have 5% and 15% reduction in travel distance compared to the spiral path for problems #1 and #2 respectively. The travel distance savings via the proposed approach in problem #2 is observed to be more significant. This likely due to the increase in complexity of the geometric layout of the holes to be drilled, as opposed to problem #1 where the spiral path is already close to being optimal. This demonstrates both the capability and context of the proposed algorithm, where it becomes most beneficial when manufacturing similar complex products.

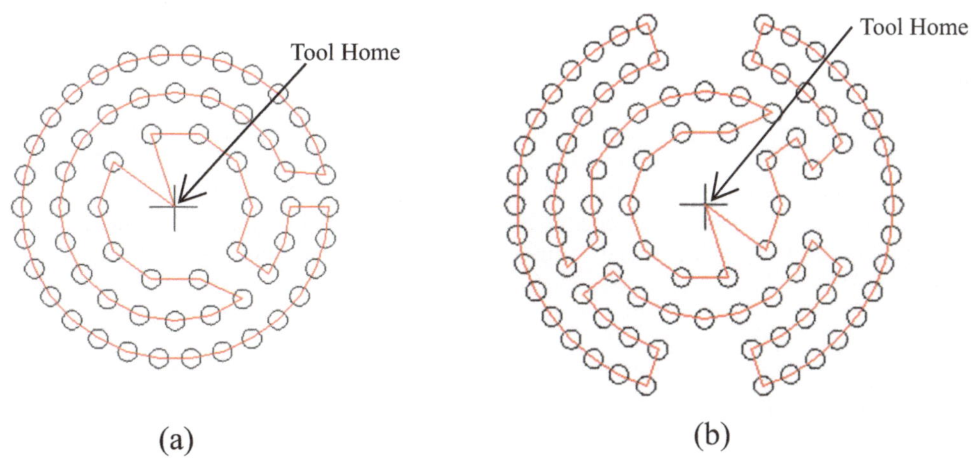

Figure 7. Selected best found paths for a) Problem #1, and b) Problem #2

5. Application Study

A full-scale food processing separator screen is chosen as an applications study to demonstrate the scalability of the proposed approach (Figure 1). A CAD model of the separator screen is shown in Figure 8 and its main geometrical parameters are listed in Table 6. The separator screen has a total of 2100 holes to be drilled (2101 nodes in the TSP), which is a very large sized problem from a TSP perspective. However, approaches derived from ACO generally scale well in terms of run-time. In this study, 5 runs of the modified ACO, with p = 20 were conducted on an Intel i3-2310M CPU @ 2.1 GHz, with all the algorithms implemented via in-house code on Matlab (MathWorks, 2010). The average time per run of the algorithm was approximately 90 min, which is a fairly reasonable time-investment when planning a machining job. Summary of results of the conducted runs are listed in Table 7 and plots of the obtained paths are provided in Figure 9.

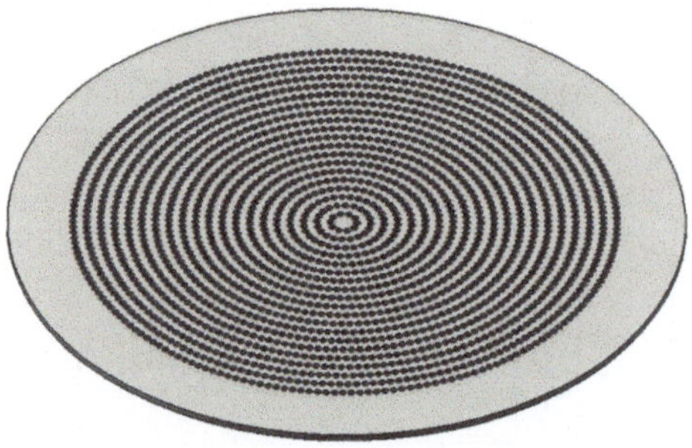

Figure 8. CAD model of separator screen

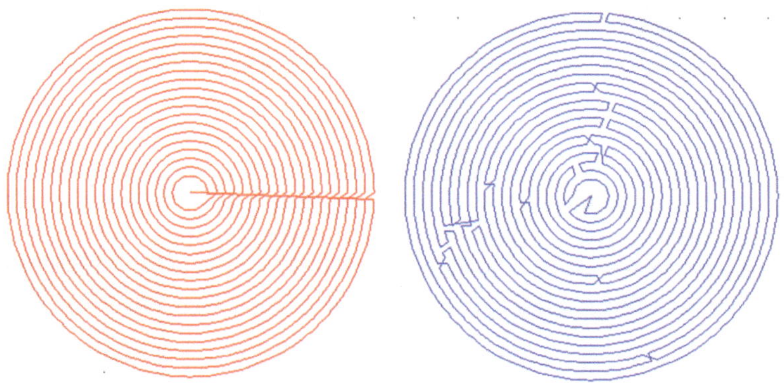

Figure 9. Tool paths for manufacturing the separator screen: a) Spiral path, and b) best found

Examining the best found tool-path (Figure 9.b), it is apparent that there are still some folds and crossovers that may be possible to improve upon. It is however noted that all five runs using the proposed approach (Table 7) were capable of discovering better solutions than the simple spiral path (99% confidence level using statistical *t*-test). This is despite the large size of the associated TSP and the fact that the spiral path is generally accepted as "close-to-optimum" by CAD systems and commercial tools for generating CNC G-code.

Table 6. Geometrical parameters of the Processing Separator

Parameter Description	Value
Hole diameter (mm)	5
Hole depth (mm)	10
Diameter of first pitch circle (mm)	40
Increment in pitch circle diameters (mm)	20
Number of holes in the first pitch circle	10
Increment in number of holes per pitch circle	10
Number of concentric pitch circles	20
Total number of holes	2100

Table 7. Results summary for industrial case study

Algorithm	Average Path Length (mm)	Min.	Max.	Standard Deviation
Spiral Path	14773	-	-	-
Modified ACO (p = 20.0)	14687	14655	14707	23.0

6. Conclusion

This paper presented a study on optimum path planning for drilling operation for products with large number of holes arranged in concentric circular patterns. The path planning is formulated as a traveling salesman problem. Solution of the TSP is tested via basic ant colony optimization and genetic algorithm, as well as a modified ACO where initialization of the pheromone matrix is biased towards traveling between nearby locations via a penalty function on the travel distance. Numerical simulations of example studies showed superior performance of the modified ACO algorithm compared to genetic algorithm and the basic ACO, with up to 15% reduction in total travel distance from the default spiral path in complex hole layouts. A model of a full-scale food separator screen with 2100 holes was used to demonstrate the scalability of the proposed approach.

Future extensions of this research may include application of the modified ACO approach to path planning in three-dimensional CNC tool-paths with axes re-orientation and/or tool changes.

Acknowledgments

This Project was supported by King Saud University, Deanship of Scientific Research, College of Engineering Research Center.

References

Aarts, E., & Korst, J. (1989). *Simulated annealing and Boltzmann machines: a stochastic approach to combinatorial optimization and neural computing*. New York: John Wiley & Sons Inc.

Abbas, A. T. (2012). Enhanced CNC machines capabilities by adding circular patterns cycle. *International Journal Precision Engineering and Manufacturing, 13*(10), 1753-1758. http://dx.doi.org/10.1007/s12541-012-0230-0

Abbas, A. T., & Megahed, S. M. (2005). A general algorithm for drilling holes lying in a matrix. *Robotics and Computer Integrated Manufacturing, 21*(3), 235-239. http://dx.doi.org/10.1016/j.rcim.2004.08.001

Abbas, A. T., Aly, M. F., & Hamza, K. T. (2011). Optimum drilling path planning for a rectangular matrix holes using ant colony optimization. *International Journal of Production Research, 49*(19), 5877-5891. http://dx.doi.org/10.1080/00207543.2010.507608

Anand, S., McCord, C., & Sharma, R. (1999). An integrated machine vision based system for solving the nonconvex cuttings stock problem using genetic algorithm. *Journal of Manufacturing Systems, 8*(6), 114-119

Arnaout, J. P., Rabadi, G, & Musa, R. (2010). A two-stage ant colony optimization algorithm to minimize the makespan on unrelated parallel machines with sequence-dependent setup times. *Journal of Intelligent Manufacturing, 21*, 693–701. http://dx.doi.org/10.1007/s10845-009-0246-1

Bonabeau, E., Dorigo M., & Theraulaz, G. (1999). *Swarm Intelligence: From Natural to Artificial Systems*. New York: Oxford University Press.

Denardo, E. V. (2004). Dynamic Programming: Models and Applications, Dover Publications

Dorigo, M. (1996). Ant system: optimization by a colony of cooperating agents. *IEEE Transactions on Systems, Man and Cybernetics-Part B Cybernetics, 26*(1), 29-41. http://dx.doi.org/10.1109/3477.484436

Dorigo, M., & Stützle, T. (2004). Ant colony optimization. Cambridge, MA: MIT Press.

Dorigo, M., Birattari, M., Blum, C., Gambardella, L. M., Mondada, F., & Stuetzle, T. (2004). *Ant colony optimization and swarm intelligence*. 4th International Workshop, Springer-Verlag, Berlin Heidelberg

Goldberg, D. (1989). *Genetic Algorithms in Search, Optimization and Machine Learning*. New York: Addison-Wesley Inc.

Ke, F., Ni, J., & Stephenson, D. A. (2006). Chip thickening in deep-hole drilling. *International Journal of Machine Tools and Manufacture, 46*(12-13), 1500-1507. http://dx.doi.org/10.1016/j.ijmachtools.2005.09.022

MathWorks. (2010). MATLAB Release 2010a. Natick, Massachusetts, United States: The MathWorks, Inc.

Michalewiz, Z. (1992). *Genetic algorithms + data structures = evolution programs* (2nd ed.). New York: Springer-Verlag.

Montiel-Ross, O., Medina-Rodríguez, N., Sepúlveda, R., & Melin, P. (2012). Methodology to Optimize Manufacturing Time for a CNC using a High Performance Implementation of ACO. *International Journal of Advanced Robotic Systems*. http://dx.doi.org/10.5772/50527

Padberg, M., & Sung T. (1991). An analytical comparison of different formulations of the travelling salesman problem. *Mathematical Programming, 52*, 315-357. http://dx.doi.org/10.1007/BF01582894

Saealal, M., Abidin, A., Adam, A., Mukred, J., Khalil, K., Yusof, Z., Ibrahim, Z., & Nordin, N. (2012). An ant colony system for routing in PCB holes drilling process. *International Journal of Innovative Management, Information & Production, 3*(1), 50-56

Satishkumar, S., & Asokan, P. (2008). Selection of optimal conditions for CNC multitool drilling system using non-traditional techniques. *International Journal of Machining and Machinability of Materials, 3*(1-2), 190-207. http://dx.doi.org/10.1504/IJMMM.2008.017633

Shunmugam, M. S., Bhaskara reddy, S. V., & Narendran, T. T. (2000). Optimal selection of parameters in multi tool drilling. *International Journal of Material Processing Technology, 103*, 318–323. http://dx.doi.org/10.1016/S0924-0136(00)00500-8

Stutzle, T., & Dorigo, M. (2002). A short convergence proof for a class of ant colony optimization algorithms. *IEEE Transactions on Evolutionary Computations, 6*(4), 358-365. http://dx.doi.org/10.1109/TEVC.2002.802444

Van Laarhoven, P., & Aarts, E. (2009). *Simulated annealing: theory and applications*. Berlin Heidelberg: Springer-Verlag.

Wolsey, L. A., & Nemhauser, G. L. (1999). *Integer and combinatorial optimization*. New York: John Wiley & Sons Inc.

Zhou, Y. (2009). Runtime analysis of an ant colony optimization algorithm for TSP Instances. *IEEE Transactions on Evolutionary Computations, 13*(5), 1083-1092. http://dx.doi.org/10.1109/TEVC.2009.2016570

Zhu, G. Y. (2006). Drilling path optimization based on swarm intelligent algorithm. *IEEE International Conference on Robotics and Biomimetics* (PP. 193-196). Kunming, China. http://dx.doi.org/10.1109/ROBIO.2006.340357

Zhu, G. Y., & Zhang, W. B. (2008). Drilling path optimization by the particle swarm optimization algorithm with global convergence characteristics. *International Journal of Production Research, 46*(8), 2299-2311. http://dx.doi.org/10.1080/00207540601042480

19

Increase in Visibility of Light Signals

Pablo Ixtaina[1], Matías Presso[2] & Gustavo Marín[2]

[1] Laboratorio de Acústica y Luminotecnia de la Comisión de Investigaciones, Científicas de la Provincia de Buenos Aires LAL CIC, Argentina

[2] Instituto de Investigación en Tecnología Informática Avanzada INTIA – UNCPBA, Argentina

Correspondence: Pablo Ixtaina, Laboratorio de Acústica y Luminotecnia de la Comisión de Investigaciones Científicas, Camino Centenario y 506, Gonnet, Argentina. E-mail: pixtaina@yahoo.com

Abstract

This work describes a study of perception of light signals used in aeronautics. Devices known as "obstruction light" (L-810 type, according to the Organization of International Civil Aviation OACI and the Federal Aviation Administration FAA) with LED technology were fed with square pulses of voltage, attaining flashes with imperceptible rise times. The lights were presented to observers, simulating the habitual vision conditions for these devices.

With this diagram, perceptive comparisons for different flash frequencies were carried out, exploring especially the transition zone between flickering and steady vision of light. The test showed that an important number of people experienced a perceptive increase in such transition zone. The phenomenon, which can be related with to Broca-Sulzer effect, could be used in order to improve the signal visibility.

Keywords: visual perception, pulsing light, leds, signal, sensation, intensity

1. Introduction

When the frequency of a periodic luminous stimulus is lower than certain value, the visual system perceives the successive "offs", producing the sensation of flickering light. The limit or transition frequency is known as critical frequency of fusion (FCF). Above this frequency, the luminance variation stops being seen and the result is the sensation of "steady light".

In this last condition, that is, frequencies higher than FCF, the sensation produced by pulsing lights can be estimated by means of the mathematical relation known as Talbot's law (Moon, 1936; Talbot, 1834), which assigns a temporal integrating function to the visual system. Thus, if an intensity light $L_l(t)$, flickering periodically, is presented to an observer, the visual sensation L is adjusted to the equation:

$$L = \frac{1}{T}\int L_l(t)dt \qquad (1)$$

In the Equation (1), L is the average of the train of light pulses and T is de pulse period.

If the flickering frequency is kept above the critical frequency, the light will be undistinguishable from another steady of equal value to that arising from Talbot's law. The critical frequency is around 30~50 Hz and depends on the stimulus intensity and the observer's age, among other factors.

Below the critical frequency, the light pulses will be perceived individually as a kind of flashes. The apparent brightness of the flashes is in agreement with their duration. For pulses lower than 0.2 s, Bloch's law (Moon, 1936) establishes the persistence in the illumination product in the retina by duration of the "flash". Therefore, the sensation produced by a light pulse of short duration is lower than that generated by the stimulus if it was permanent. There are several mathematic models that try to quantify the phenomenon (Ohno, 2002), being the model suggested by Blondel and Rey (1911) and Illuminating Engineering (1964) one of the most accepted ones at the beginning of the last century. In this model, the Effective Intensity I_{eff} of a pulsing light is expressed by the following equation:

$$I_{eff} = \frac{\int_{t_1}^{t_2} I(t)dt}{a+(t_2-t_1)} \quad (2)$$

Where $I(t)$ is the instant intensity of the luminous pulse and "a" a constant (Blondel-Rey constant, $a = 0.2$ s). t_1 and t_2 are the times (in seconds) of the beginning and end of that part of the the instant intensity $I(t)$ when its value exceeds Ieff, with de condition of the choice of t_1 and t_2 maximizes the value of I_{eff}.

For pulses between 50 and 100 ms, of constant luminous intensity, it is observed an increase in the apparent brightness of flashes. This phenomenon is known as Broca – Sulzer (Brown, 1965; Hart Jr, 1987) effect and it mainly depends on the initial retinal illumination and on the pulse width.

2. The Study

The purpose of this work was to study the visual perception of the aeronautical light signal stimulated by pulses with frequencies close to the critical frequency at which flicker appears. The light signals, with LED technology, were fed with square pulses, verifying that the produced light also responded to this shape, with rise times practically negligible in relation to the pulse duration.

Thirty-five people, with normal vision and ages between 30 and 45 years old participated in the experience. From a fixed observation position, farther than 12 m from the source, the observers indicated their visual sensation for the presented alternatives. Figure 1 shows aspects of the experience.

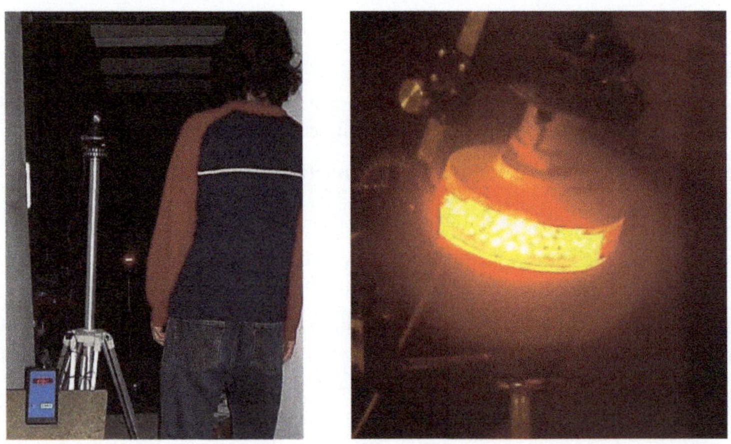

Figure 1. Observer in position. Obstruction light used in the experience

The experience was developed in two stages. In the first one, variations of the permanent light intensity (not pulsing) that an observer is able to distinguish around a continuous intensity of reference were determined. For this, the observer was shown alternatively a same beacon, fed with a continuous reference voltage first, and then with a variable continuous voltage. Figure 2 shows a diagram of the test.

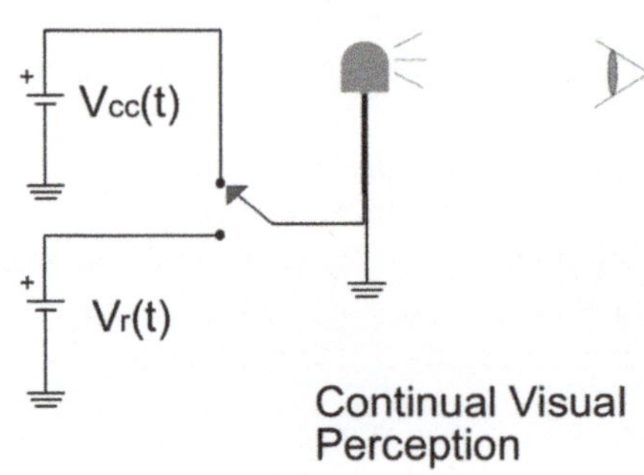

Figure 2. Circuit of visual perception test of continuous light

The variable source ($V_{cc}(t)$) took different values, slightly higher and lower than that of the reference ($V_r(t)$). The light was presented to the observer, who should indicate if he perceived it with higher, equal or lower brightness than the reference condition. With this test it was determined the range of intensities where the observers could not distinguish differences between both luminous intensities (perception threshold). Figure 3 shows voltage, current (excitations and reference) and perception graphs.

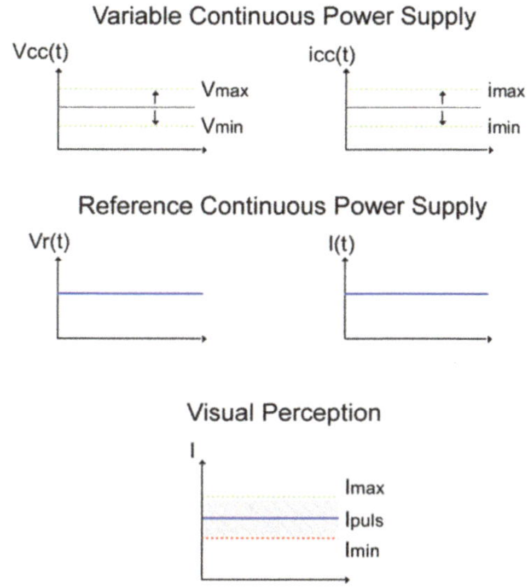

Figure 3. Graphs of voltage, current (excitations and reference) and perception

The second test consisted in exciting the light signal (beacon) by alternating either a square pulsing voltage of reference or a variable continuous voltage.

The frequency of voltage pulses was slightly increased in order to reach the critical frequency of each observer. Once the frequency was adjusted, the levels of voltage and current of the pulsing signal ($V_{puls}(t)$) were fixed. The experience began by presenting alternatively the beacon fed with this pulsed voltage and the variable continuous voltage to the observer, while being asked about his perception. Again, the observer had to indicate whether he perceived the beacon with higher, equal or less brightness than the condition of reference. Figures 4 and 5 show this second experience.

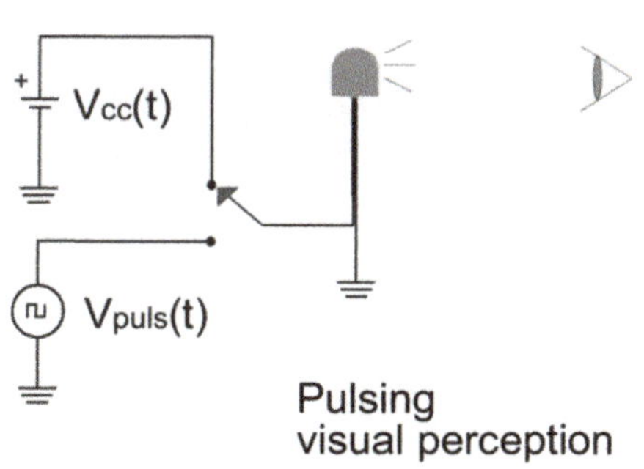

Figure 4. Circuit of visual perception test of continuous/pulsed light

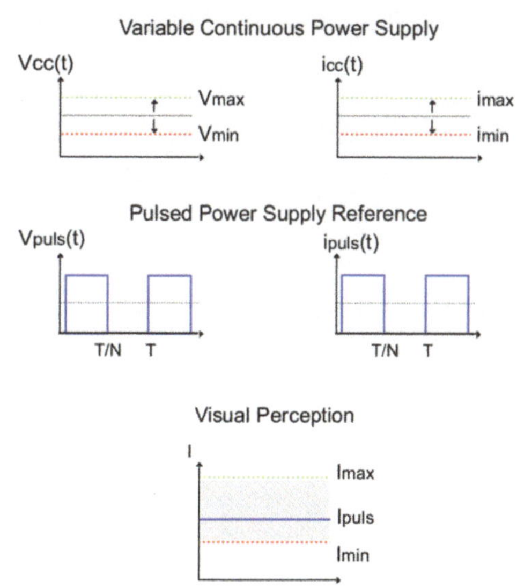

Figure 5. Graphs of voltage, current (excitations and reference) and perception

In both experiences, the switching of power source was carried out at random, so that the observer could not suspect which power source was being assessed. The signal was presented during the time necessary for the observer to give his opinion (higher, equal or lower than the previous vision), usually between 10 and 15 s. The switching to the following presentation was fast, tenths of seconds, in order to place a noticeable darkness between both images. The experience ended when the observer did not detect differences between successive presentations of the beacon.

During the tests, the light intensity I was measured permanently with a photo detector oriented in line with the observer's vision.

2.1 Voltage, Current and Supply Power

Voltages and currents of reference, either in the case of continuous supply or pulsed supply, were regulated in such a way that the consumed power per beacon was the same in both cases in order to be able to carry out a comparative analysis.

For the continuous supply, the power was:

$$P_r = I_{rcc} \cdot V_{rcc} \tag{3}$$

For the supply of the pulsed signal, the power was the product of effective values:

$$P_{efpuls} = I_{efpuls} \cdot V_{efpuls} \tag{4}$$

For a square wave:

$$I_{efpuls} = \frac{I_{p\max}}{\sqrt{N}} \tag{5}$$

Where 1/N is the fraction of the period where the square signal is not null.

Equally, for the effective voltage

$$V_{efpuls} = \frac{V_{p\max}}{\sqrt{N}} \tag{6}$$

The power was:

$$P_{efpuls} = \frac{I_{p\max} \cdot V_{p\max}}{N} \tag{7}$$

For the carried out tests the cycle of pulses was 50%, that is N=2.

As the beacon is basically a set of diodes, the total operation voltage is the sum of polarization voltages of diodes and is kept approximately constant for a range of current values. The ratio between power and current is approximately linear with a slope given by the total operation voltage of the beacon. This indicates that in a certain range, the beacon power is regulated proportionally with the current (green line in Figure 6).

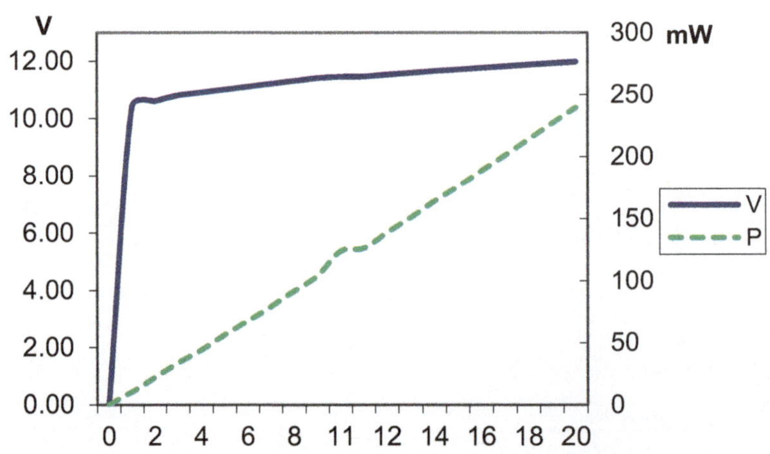

Figure 6. Voltage Curves – Current, Power – Beacon Current

For studying comparatively the visual sensation, the pulsing and continuous powers of reference were equaled:

$$P_r = P_{efpuls} \tag{8}$$

As N=2 and $V_{p\max} \approx V_{rcc}$:

$$I_{p\max} \approx 2I_{rcc} \tag{9}$$

Such ratio was used for adjusting the reference sources.

3. Results

3.1 Continuous - Continuous Supply Comparison

Figure 7 shows the results obtained in this experience. There the intensity values regarding the reference versus each observer's appreciation were yielded.

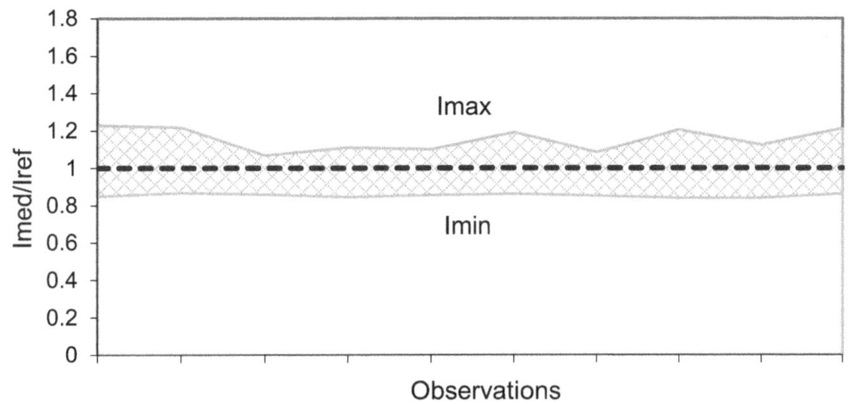

Figure 7. Light intensities and observer's appreciation between two continuous stimuli

The dots in the lower curve (I_{min}) determine the limit where each observer began to notice less light intensity from the variable source than from the reference. The dots of the upper curve (I_{max}) determine the limit where the observers noticed that the variable source was higher than the reference.

The marked zone (between I_{min} and I_{max} curves) corresponds to ratios for which the observers did not distinguish differences as regards the reference and it determines the perception threshold. For the experience which compares continuous supply versus continuous, this perception threshold was roughly ±15% around the reference, verifying that on average, any variation in intensity within that region was not perceived by the observers.

3.2 Continuous Supply - Pulsing Supply Comparison

First, a scanning in frequency was carried out in order to locate the critical frequency for each observer. Then, the source was adjusted close to this value with the condition that flashes could not be distinguished. Next, the steps described for the comparison were followed. The obtained results are indicated in Figure 8.

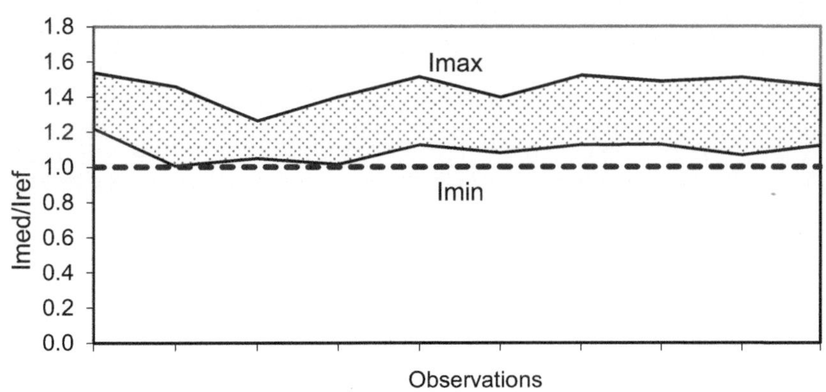

Figure 8. Light intensities and appreciation for a continuous stimulus and another pulsed

Just like in Figure 7, three regions are determined. Above curve I_{max} and below I_{min}, the observers detected differences (higher and lower than the reference, respectively). Between both curves (marked zone) the observers did not distinguish differences with respect to the reference. The lowest and highest average limit values were 1.09 and 1.45 respectively, and the range of difference was 0.362.

The same experience carried out at a higher frequency than the critical one yielded results similar to those in Figure 7, in some way verifying Talbot's law (Figure 9).

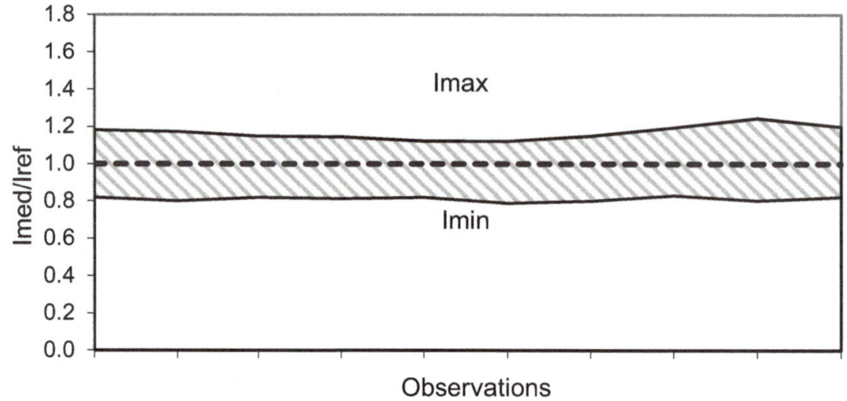

Figure 9. Light intensities and appreciation for a continuous stimulus and another pulsed (f>FCF)

4. Conclusions and Subsequent Analysis

The group of observers that participated in the tests did not distinguish variations of ±15% in the successive presentations of the beacon with continuous light. Taking into account the approximate symmetry around the reference, the range where existed observers' uncertainty was around 30%. The same tendency was verified by comparing continuous supply with pulsed supply, at higher frequencies than FCF. As expected, equivalence between supply power with continuous voltage and supply with pulses was evidenced.

Instead, the comparison carried out against the light pulsed at a frequency very close to FCF, just in the limit of non-flickering vision indicated that the observers noticed the difference when the continuous intensity was higher than 45% of the reference (as higher limit) or when the steady intensity was lower than 1.09 times the reference (as lower limit). In other words, the sensation produced in the observers by the beacon generating light pulses at the indicated frequency was the same as that produced by a continuous light with intensity between 9%-45% higher.

Translated to supply power (Figure 10) we can see a clear gain with respect to the pulsed light: it is needed higher continuous power (marked stripe) in order to obtain the same perceptive effect as with the pulsed supply (dotted line, taken as reference in the experience).

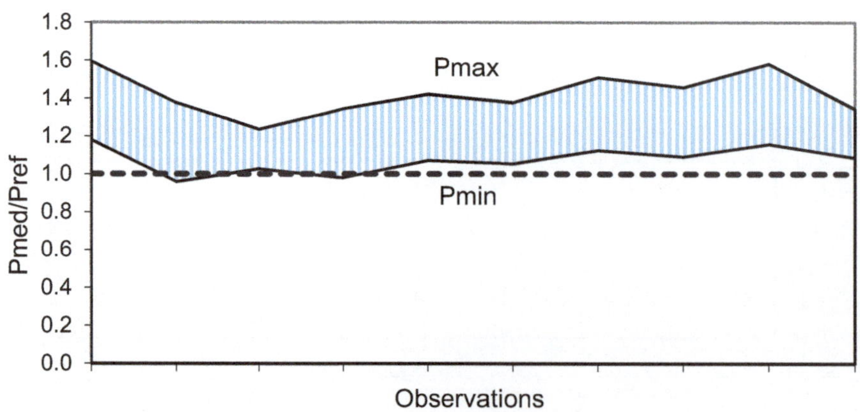

Figure 10. Power associated to light stimuli

Finally, unlike the isolated pulses that have a slighter effect than its real intensity (Ieff, of Blondel - Rey), multiple pulses, in the limit of continuous perception (critical frequency), would produce certain visual gain. This effect, which could be related to Broca – Sultzer, showed in average levels of perception increase of 36% and may well be used to improve the visibility of signals and beaconing devices

References

Blondel, A., & Rey, J. (1911). Sur la perception des lumières brèves à la limite de leur portée. *J. Phys. Theor. Appl.*, *1*(1), 530-550. http://dx.doi.org/10.1051/jphystap:0191100107053000

Brown, J. L. (1965). Flicker an Intermittent Stimulation. In C. H. Graham (Ed.), *Vision and Visual Perception*. New York: John Wiley and Sons, Inc.

Hart Jr, W. M. (1987). The temporal responsiveness of vision. In R. A. Moses, & W. M. Hart (Eds.), *Adler's Physiology of the eye, Clinical Application*. St. Louis: The C. V. Mosby Company.

Illuminating Engineering. (1964). *IES Guide for Calculating the Effective Intensity of Flashing Signal Lights* (pp. 747-753).

Moon, P. (1936). The Scientific Basis of Illuminating Engineering. McGraw Hill Company, London.

Ohno, Y., (2002). Physical Measurement of Flashing Lights – Now and Then. *Proc. CIE Expert Symposium on Temporal and Spatial Aspects of Light and Colour Perception and Measurements*.

Talbot, H. (1834). Experiments on Light. *Phil Mag, 5*.

20

Influence of Thermal Shrinkage on Protective Clothing Performance during Fire Exposure: Numerical Investigation

Ahmed Ghazy[1]

[1] Faculty of Engineering, Helwan University, Cairo, Egypt

Correspondence: Ahmed Ghazy, Faculty of Engineering, Ibrahim Abel Razik Street, Ain Shams E., 11718, Cairo, Egypt. E-mail: ahmed.ghazy@usask.ca

Abstract

The thermal shrinkage of protective clothing during fire exposure plays a crucial rule in reducing the clothing protective performance. The transversal reduction in the fabric perimeter around the body due to the fabric thermal shrinkage causes a dynamic reduction in the air gap between the clothing and the body. This leads to a dynamic change in the heat transfer modes within the gap. Despite of its influential effect on the clothing performance, the thermal shrinkage of protective clothing during fire exposure has not been yet addressed in the literature. This can be attributed to the absence of a gap model that can capture the reciprocal change in heat transfer modes within the gap due to clothing shrinkage. This paper develops a finite volume model to investigate the influence of the fabric thermal shrinkage on protective clothing performance. A special attention was drawn to the model of the air gap between the clothing and skin as it responds directly to the clothing thermal shrinkage. The influence of a variation in the fabric shrinkage rate and the overall reduction in the fabric dimensions was investigated. The paper demonstrates that the clothing protective performance continuously decreases with the reduction in the fabric dimensions while the decay in the clothing protective performance is limited to small shrinkage rates of the fabric. Moreover, this decay in the clothing performance vanishes at high shrinkage rates of the fabric.

Keywords: fabric shrinkage, fire exposure, conduction-radiation, thermal radiation, finite volume method.

1. Introduction

Protective clothing is widely used in many industries and applications such as petroleum and petrochemical industries and municipal firefighting to seek protection from thermal and fire exposures. The thermal protective performance (TPP) of the clothing is determined by estimating heat transfer from the thermal source to the skin through clothing, which causes skin burns as a result. Standard bench top tests (ISO 9151, 1995, ASTM D 4108, 1987, ASTM F, 1999, and NFPA, 2007) are used to evaluate the TPP of fabric specimens while manikin test (ASTM F, 2000) is used to evaluate the TPP of the whole garment at different locations of the body.

Modeling the thermal performance of protective clothing has been extensively reported in the literature during the past decade. Torvi (1997) and Torvi and Dale (1999) modeled heat transfer in Kevlar®/PBI and Nomex® flame-resistant fabric during a flame TPP test. Mell and Lawson (2000) modeled heat transfer in multiple layers protective garment during radiant exposure. Tan, Crown and Capjack (1998) studied the design of flightsuit protective garment for optimum protection to flight personnel. Chitrphiromsri and Kuznetsov (2005) and Song, Chitrphiromsri and Ding (2008) modeled heat and moisture transfer through firefighters' clothing in fire exposure and local flame test, respectively. Zhu and Zhang (2009) considered the fabric thermal degradation at high temperature during radiant exposure. Mercer and Sidhu (2008, 2009) investigated the performance of protective clothing with embedded phase change material.

The air gap between the fabric and skin plays an essential role in determining the performance of protective clothing during fire exposure. This role was acknowledged in the literature in several studies. For example, Torvi, Dale and Faulkner (1999) investigated the effect of the gap width on the protective performance of a flame-resistant fabric during a flame TPP test. Sawcyn and Torvi (2005) and Talukdar, Torvi, Simonson and Sawcyn (2010) attempted to improve the modeling of the air gap in bench top tests of protective fabrics. The 3-D body scanning technology was used (Song, Barker, Hamouda, Kuznetsov, Chitrphiromsri, & Grimes, 2004; Kim, Lee, Li, Corner, & Paquette, 2002; Mah & Song, 2010a; Mah & Song, 2010b) to determine the widths and distribution of air gaps between flame-resistant garment and manikin body. Ghazy and Bergstrom (2010) developed a numerical model for single layer protective clothing that considers the combined conduction-radiation heat transfer between the fabric and the skin. Then, Ghazy and Bergstrom (2011) further investigated the influence of the conduction-radiation in the gap between protective clothing and the skin on the overall performance of the clothing. Ghazy and Bergstrom (2012) also developed a model for heat transfer in multiple layers firefighters' clothing that considers the combined conduction-radiation heat transfer within the air gaps between clothing layers. Ghazy (2013) developed a novel air gap model that stands middle way between the conduction-radiation model introduced in Ghazy (2011) and the approximate air gap model exists elsewhere in the literature.

The thermal shrinkage of protective clothing in thermal or fire exposures significantly affects the clothing performance. The transversal reduction in the fabric perimeter around the body due to fabric thermal shrinkage causes a dynamic reduction in the air gap between the clothing and the body. This leads to a corresponding variation in the total heat transfer through the gap and an interchanging variation in its modes. In addition, the reduction in the gap width caused by thermal shrinkage and hence the overall protective performance of the clothing depends on the total reduction in fabric dimensions and the shrinkage rate of the fabric.

The influence of the thermal shrinkage on the performance of protective clothing during fire exposure has not been yet addressed in the literature. This is because the lack of knowledge about shrinkage rates of fire-resistant fabrics during fire exposure. In addition, the approximate analysis of the air gap that most of the models in the literature adopted is not capable of considering the dynamic variation in the air gap between the clothing and the skin due to fabric thermal shrinkage. This paper numerically investigates the effect of the fabric's thermal shrinkage during fire exposure on the overall performance of protective clothing. A special attention was drawn to modeling heat transfer through the gap since it responds directly to the fabric thermal shrinkage. The temperature dependence of the thermophysical properties of the air gap and the fabric was accounted for. The influence of a variation in the fabric shrinkage rate and the reduction in the fabric dimensions on the clothing protective performance was studied to capture different forms of fabric thermal shrinkage.

2. Problem Description

A typical protective clothing system is shown in Figure 1. The clothing system comprises a fire-resistant fabric that is exposed to a heat flux of about 80 kW/m^2 from a lab burner, the human skin that consists of epidermis, dermis and subcutaneous layers and an air gap enclosed between the fabric and the skin. The energy equations for the clothing elements are expressed as follows.

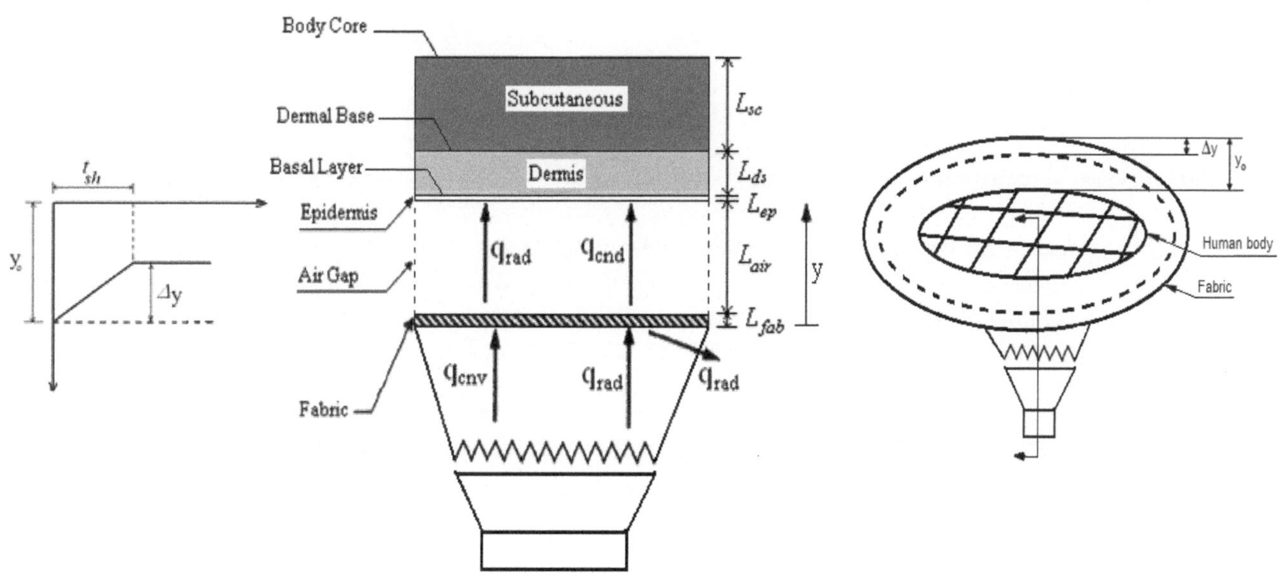

Figure 1. A schematic diagram of protective clothing

2.1 Heat Transfer in Fire-Resistant Fabrics

The transient energy equation for the Kevlar®/PBI fabric was introduced by Torvi (1997) and Torvi and Threlfall (2006) as

$$\rho C^A(T)\frac{\partial T}{\partial t}=\frac{\partial}{\partial y}\left(k(T)\frac{\partial T}{\partial y}\right)+\gamma\sigma\varepsilon_g T_g^4\exp(-\gamma y) \qquad 0<t\le t_{\exp} \qquad (1a)$$

$$\rho c_P(T)\frac{\partial T}{\partial t}=\frac{\partial}{\partial y}\left(k(T)\frac{\partial T}{\partial y}\right) \qquad t>t_{\exp} \qquad (1b)$$

where ρ is the fabric density, C^A is the apparent heat capacity of the fabric, k is the fabric thermal conductivity, σ is Stefan-Boltzmann constant, T_g is hot gases temperature, ε_g is hot gases emissivity, γ is the extinction coefficient of the fabric and $t_{\exp}$ is the exposure duration.

The boundary conditions of the fabric are

$$-k_{fab}(T)\frac{\partial T_{fab}}{\partial y}\bigg|_{y=0}=h_{fl}(T_g-T_{fab(y=0)})-\sigma\varepsilon_{fab1}(1-\varepsilon_g)(T^4_{fab(y=0)}-T^4_{amb}) \qquad 0<t<t_{\exp} \qquad (2a)$$

$$-k_{fab}(T)\frac{\partial T_{fab}}{\partial y}\bigg|_{y=0}=h_{cnv}(T_{fab(y=0)}-T_{amb})+\sigma\varepsilon_{fab1}(T^4_{fab(y=0)}-T^4_{amb}) \qquad t>t_{\exp} \qquad (2b)$$

$$-k_{fab}(T)\frac{\partial T_{fab}}{\partial y}\bigg|_{y=L_{fab}}=q''_y(\vec{r})\Big|_{y=L_{fab}}-k_{air}(T)\frac{\partial T_{air}}{\partial y}\bigg|_{y=L_{fab}} \qquad t>0 \qquad (3)$$

where h_{fl} is the flame convection heat transfer coefficient, h_{cnv} is the convection heat transfer coefficient from the clothing to the ambient surroundings, ε_{fab1} is the emissivity of the fabric exposed surface, T_{amb} is ambient temperature and $q''_y(\vec{r})\big|_{y=L_{fab}}$ is the emitted radiation from the fabric backside surface, which is discussed in section 2.2.

The initial condition of the fabric is

$$T_{fab}(y,t=0)=T_{amb} \qquad (4)$$

2.2 Heat Transfer in the Air Gap

The transient heat transfer in the gap is written as

$$\rho(T)c_P(T)\frac{\partial T}{\partial t}=\frac{\partial}{\partial y}\left(k(T)\frac{\partial T}{\partial y}\right)-\frac{\partial q''_R}{\partial y} \qquad (5)$$

where ρ, c_P, and k are the density, specific heat, and thermal conductivity of the air gap, respectively and $\frac{\partial q''_R}{\partial y}$ is the divergence of the radiative heat flux through the air gap. Note that for gap widths of 1/4 in. (6.4 mm) or less, the computed Rayleigh number within the gap is less than the critical Rayleigh number for nature convection heat transfer. That makes radiation and conduction are the dominant modes of heat transfer within the gap.

The width of the gap between the fabric and the skin due to fabric shrinkage is

$$y=y_o-\left(\frac{\Delta y}{t_{shk}}\times t\right) \qquad t<t_{shk} \qquad (6)$$

$$y=y_o-\Delta y \qquad t>t_{shk} \qquad (7)$$

where y is the air gap width at any time t, y_o is the nominal air gap width, Δy is the reduction in the gap width due to the fabric thermal shrinkage and t_{shk} is the time over which the fabric thermal shrinkage takes place, as shown in Figure 1.

The Radiative Transfer Equation (RTE) of the air gap (Modest, 2003) is written as

$$\frac{dI(\vec{r},\hat{s})}{ds} = -\kappa(\vec{r})I(\vec{r},\hat{s}) + \kappa(\vec{r})I_b(\vec{r}) \tag{8}$$

where I is the radiation intensity, s is the geometric distance, $\vec{r}$ is the spatial position, $\hat{s}$ is the angular direction and κ is the gap absorption coefficient.
The unit direction $\hat{s}$ is defined in y-direction as

$$\hat{s} = (\sin\theta\sin\phi)\hat{e}_y \tag{9}$$

where θ is polar angle, ϕ is azimuthal angle and $\hat{e}_y$ is unit vector in y-direction.

The black body intensity, I_b, is defined as

$$I_b = \frac{\sigma T^4}{\pi} \tag{10}$$

where T is the medium absolute temperature and σ is the Stephan-Boltzmann constant.

The boundary conditions for the RTE of the air gap (Equation 8) are written as

$$I_{fab}(\vec{r},\hat{s}) = \varepsilon_{fab2}(\vec{r})I_{b,fab}(\vec{r}) + \frac{\rho_{fab2}(\vec{r})}{\pi}\int_{\hat{s}'\cdot\hat{n}<0} I(\vec{r},\hat{s})|\hat{s}'\cdot\hat{n}|d\Omega' \quad \text{at } y = L_{fab} \tag{11}$$

$$I_{ep}(\vec{r},\hat{s}) = \varepsilon_{ep}(\vec{r})I_{b,ep}(\vec{r}) + \frac{\rho_{ep}(\vec{r})}{\pi}\int_{\hat{s}'\cdot\hat{n}<0} I(\vec{r},\hat{s})|\hat{s}'\cdot\hat{n}|d\Omega' \quad \text{at } y = L_{fab} + L_{air} \tag{12}$$

where ε_{fab2} and ρ_{fab2} are the emissivity and reflectivity of the fabric backside, ε_{ep} and ρ_{ep} are the emissivity and reflectivity of the epidermis surface, $\hat{s}'$ is reflected ray unit direction, $\hat{n}$ is unit normal to the surface and $d\Omega'$ is solid angle containing the reflected ray.

The divergence of radiative heat flux in the air gap energy equation (Equation 5) is calculated as

$$\frac{\partial q''_R}{\partial y} = \kappa(4\pi I_b(\vec{r}) - G(\vec{r})) \tag{13}$$

where the incident radiation, $G(\vec{r})$, is defined as

$$G(\vec{r}) = \int_{4\pi} I(\vec{r},\hat{s})d\Omega \tag{14}$$

The radiation heat flux emitted from the fabric and that is incident on the skin surface are calculated as

$$q''_y(\vec{r})\Big|_{y=L_{fab}} = \int_{4\pi} I(\vec{r},\hat{s})(\hat{s}\cdot\hat{e}_y)d\Omega\Big|_{y=L_{fab}} \tag{15}$$

$$q''_y(\vec{r})\Big|_{y=L_{fab}+L_{air}} = \int_{4\pi} I(\vec{r},\hat{s})(\hat{s}\cdot\hat{e}_y)d\Omega\Big|_{y=L_{fab}+L_{air}} \tag{16}$$

The RTE (Equation 8) is solved along with its boundary conditions using the Finite Volume Method (Chai & Patankar, 2000) where the facial intensity was related to the nodal one as follows.

$$a_P^l I_P^l = a_N^l I_N^l + a_S^l I_S^l + b^l \tag{17}$$

where

$$a_N^l = \max(-A_n D_{cn}^l, 0), \quad a_S^l = \max(-A_s D_{cs}^l, 0) \tag{18}$$

$$a_P^l = \max(A_n D_{cn}^l, 0) + \max(A_s D_{cs}^l, 0) + \kappa_P \Delta V_P \Delta\Omega^l \tag{19}$$

$$b^l = S_P^l \Delta V_P \Delta\Omega^l \tag{20}$$

$$D_{cn}^l = \int_{\Delta\Omega^l} (\hat{s}^l\cdot\hat{e}_y)d\Omega, \quad D_{cs}^l = -D_{cn}^l \tag{21}$$

$$\Delta\Omega^l = \int_{\Delta\Omega^l} d\Omega \tag{22}$$

$$S_P^l = \kappa_P I_{b,P} \tag{23}$$

The boundary conditions for the air gap energy equation (Equation 5) are as follows.

$$T_{air}\big|_{y=L_{fab}} = T_{fab}\big|_{y=L_{fab}} \qquad t>0 \tag{24}$$

$$T_{air}\big|_{y=L_{fab}+L_{air}} = T_{ep}\big|_{y=L_{fab}+L_{air}} \qquad t>0 \tag{25}$$

where T_{ep} is the epidermis surface temperature and L_{air} is the air gap width.

The initial condition of the air gap is

$$T_{air}(y,t=0) = T_{amb} \tag{26}$$

2.3 Heat Transfer in the Human Skin

Heat transfer in the human skin is modeled by the bioheat transfer equation developed by Pennes (1948). The energy equations for the epidermis, dermis and subcutaneous layers of the skin are written as

$$(\rho c_P)_{ep}\frac{\partial T}{\partial t} = \frac{\partial}{\partial y}\left(k_{ep}\frac{\partial T}{\partial y}\right) \tag{27}$$

$$(\rho c_P)_{ds}\frac{\partial T}{\partial t} = \frac{\partial}{\partial y}\left(k_{ds}\frac{\partial T}{\partial y}\right) + (\rho c_P)_b \omega_b (T_{cr} - T) \tag{28}$$

$$(\rho c_P)_{sc}\frac{\partial T}{\partial t} = \frac{\partial}{\partial y}\left(k_{sc}\frac{\partial T}{\partial y}\right) + (\rho c_P)_b \omega_b (T_{cr} - T) \tag{29}$$

where ω_b is the blood perfusion rate, T_{cr} is the core body temperature.

The boundary conditions of the skin are

$$-k_{ep}(T)\frac{\partial T_{ep}}{\partial y}\bigg|_{y=L_{fab}+L_{air}} = q''_y(\vec{r})\big|_{y=L_{fab}+L_{air}} - k_{air}(T)\frac{\partial T_{air}}{\partial y}\bigg|_{y=L_{fab}+L_{air}} \qquad t>0 \tag{30}$$

$$T_{sc}\big|_{y=L_{fab}+L_{air}+L_{ep}+L_{ds}+L_{sc}} = T_{cr} \qquad t>0 \tag{31}$$

where $q''_y(\vec{r})\big|_{y=L_{fab}+L_{air}}$ is the incident radiation heat flux on the skin (Equation 16) and L_{ep}, L_{ds} and L_{sc} are the thicknesses of the epidermis, dermis, and subcutaneous layers, respectively.

The initial conditions of the skin are represented by a linear temperature distribution from the epidermis surface (32.5ºC) to the subcutaneous base (37ºC). Skin burn injury takes place when the basal layer (the base of the epidermis layer) temperature reaches 44ºC. Henriques' integral (Henriques & Moritz, 1947) is employed to predict times for the skin to receive burn injuries as follows.

$$\varphi = \int_0^t P\exp\left(-\frac{\Delta E}{RT}\right)dt \tag{32}$$

where the values for the activation energy ΔE of the skin and the pre-exponential factor P were determined by Weaver and Stoll (1996) for second–degree burns and by Takata, Rouse and Stanley (1973) for third–degree burns. The basal layer temperature is employed in the aforementioned integral to predict times to first– and second–degree burns. First– and second–degree burns take place when φ reaches 0.53 and 1, respectively. Whilst the dermal base (the base of the dermis layer) temperature is employed in the integral to estimate times to third–degree burns, which occur when φ reaches 1.

3. Numerical Solution

The fabric, air gap and the skin (epidermis, dermis, and subcutaneous) energy equations were solved along with their boundary conditions using the finite volume method (Patankar, 1980) using the Gauss-Seidel point-by-point iterative scheme. The solution proceeds as follows. Within each time step, temperatures calculated in the previous time step are used as initial guess for the iteration loop. The air gap width is updated according to Equations 6 and 7. A uniform reduction in the air gap control volumes sizes is assumed whereas the air properties in each control volume do not change. Within the iteration loop, temperatures of the air gap, fabric backside and epidermis surface are used to solve the RTE of the gap. Next, the radiation heat flux at the boundaries of the gap and the divergence of radiative heat flux are estimated to be used in the boundary conditions of the fabric and the skin and in the energy equation of the gap. The source terms and properties of the fabric and air gap are updated in accordance with the most recent temperatures within the loop. Then, new temperatures are calculated by solving the discrete equations from the fabric exposed surface to the subcutaneous base. When temperature convergence is reached for a time step, temperatures of the basal layer and dermal base are used in Henriques' integral (Equation 32) to predict times to skin burns.

4. Results and Discussion

Simulations were carried out for an exposure period of 10 seconds followed by a cool down period. The parameters used in the simulation are listed in Table 1 whilst the thermophysical properties of the human skin are listed in Table 2. The dynamic reduction in the gap width between the fabric and the skin was introduced by the percentage reduction in the gap width and the rate at which this reduction takes place. These two parameters are representatives for the percentage reduction in the fabric perimeter around the body due to fabric shrinkage and the shrinkage rate of the fabric, respectively.

Table 1. Simulation parameters

Property		Value
ρ_{fab}	Fabric density	323 kg/m^3
γ_{fab}	Fabric extinction factor	0.01
ε_{fab}	Fabric emissivity	0.9
τ_{fab}	Fabric transmissivity	0.01
L_{fab}	Fabric thickness	0.6 mm
T_g	Hot gases temperature	2000 K
ε_g	Hot gases emissivity	0.02
h_{fl}	Flame convective heat transfer coefficient	40 W/m^{2}K

Table 2. Thermophysical properties of the human skin

Property	Epidermis	Dermis	Subcutaneous	Blood
Density (Kg/m^3)	1200	1200	1000	1060
Specific Heat (J/Kg°C)	3598	3222	2760	3770
Thermal Conductivity (W/m°C)	0.255	0.523	0.167	-
Thickness (m)	8×10^{-5}	2×10^{-3}	1×10^{-2}	-
Blood Perfusion Rate (m^3/s)/m^3 tissue	-	-	-	1.25×10^{-3}

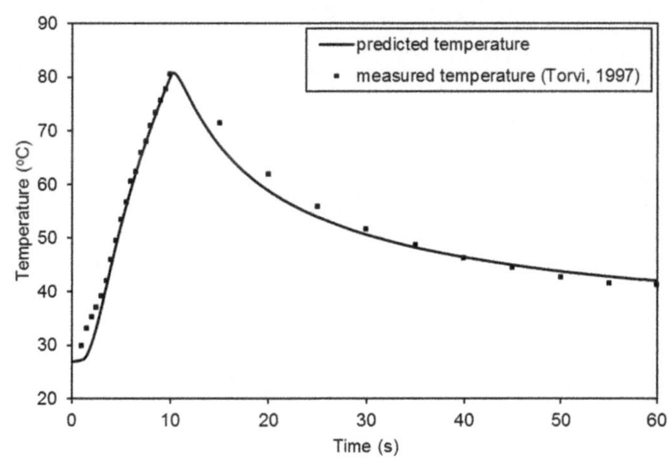

Figure 2. Comparison between measured and predicted temperatures of the skin simulant surface

The model validation has been done by simulating the open flame bench top test performed by Torvi (1997) using skin simulant gauge and a constant air gap width of 6.4 mm (1/4 in.). A comparison between the predicted and measured temperatures of the skin simulant surface during the exposure and cool down periods are held in Figure 2. The model successfully predicted the skin simulant surface temperature during both the exposure and cool down periods. The small deviation between the prediction and measurements can be attributed to other factors that are not considered in the model such as the fabric thermal pyrolysis. However, identifying the source of the small discrepancy noted in Figure 2 may require extensive experimental data.

The effect of a variation in the percentage reduction in the gap width on heat transfer within the gap from the fabric to the skin is compared to the case of no fabric shrinkage and illustrated by Figure 3. The conduction heat flux through the air gap increases as the percentage reduction increases, Figure 3(a), while the radiation heat flux decreases with the increase in the percentage reduction, Figure 3(b). Nevertheless, the variation in the conduction heat flux with a variation in the percentage reduction is bigger than that in the radiation heat flux. In addition, the divergence in the conduction heat flux with a variation in the percentage reduction starts earlier than that of the radiation heat flux. Moreover, the divergence in the radiation heat flux drops quickly during the cool down period compared to the conduction heat flux. In addition, the divergence in the conduction and radiation heat fluxes within the gap is not linearly proportional to the variation in the percentage reduction. For example, the divergence in the energy transfer (conduction and/or radiation) within the gap due to a variation in the percentage reduction from 10% to 30% is less than that due to a variation from 30% to 50%. Furthermore, the remarkable deviation in the conduction and radiation heat fluxes within the gap from the case of no fabric shrinkage proves the influential effect of the fabric thermal shrinkage on the heat transfer to the skin and hence on the clothing protective performance.

(a)

(b)

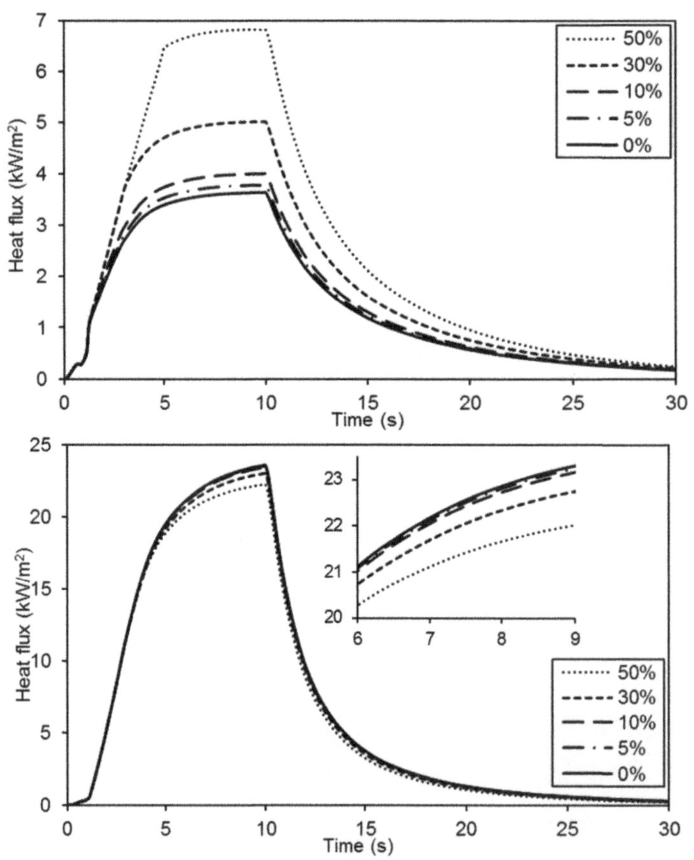

Figure 3. Effect of a reduction in the gap width on heat transfer through the gap: (a) conduction heat flux from the fabric, (b) emitted radiation from the fabric

(a)
(b)
(c)

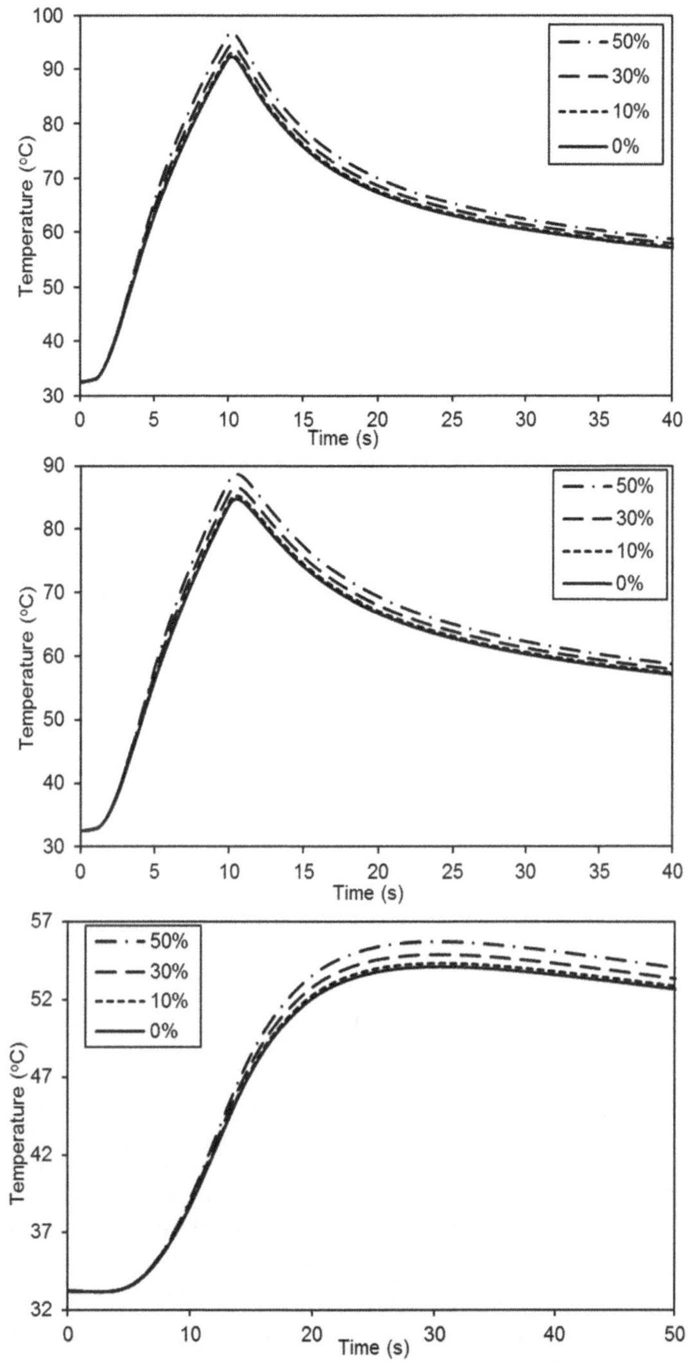

Figure 4. Effect of a reduction in the air gap width on the temperature distribution within the skin: (a) epidermis surface, (b) basal layer, (c) dermal base

The influence of a variation in the percentage reduction in the gap width on the temperature distributions within the skin is shown in Figure 4. The temperatures of the epidermis surface (Figure 4a), basal layer (Figure 4b) and dermal base (Figure 4c) increase with the increase in the percentage reduction. However, the increase in the skin layers temperatures is not linearly proportional to the increase in the percentage reduction. Moreover, the closer the fabric to the skin due to the shrinkage, the higher the influence of the variation in the percentage reduction on the temperature distributions within the skin. In addition, the divergences in the temperatures of the epidermis surface and the basal layer become evident by the end of the exposure period while the divergence in the dermal base temperature develops during the cool down period. Nevertheless, the divergence in the temperatures of the skin layers last during the cool down period.

The influence of a variation in the reduction in the gap width due to fabric shrinkage on times to skin burn injuries is shown in Figure 5. Times to first– and second–degree burns, shown in Figure 5a, decrease almost linearly with the increase in the percentage reduction from 0% (no shrinkage) to 50%. Any reduction in the air gap width beyond 50% does not affect times to first– and second–degree burns. On the other hand, times to third–degree burns, shown in Figure 5b, continuously decrease with the increase in the percentage reduction.

(a)

(b)

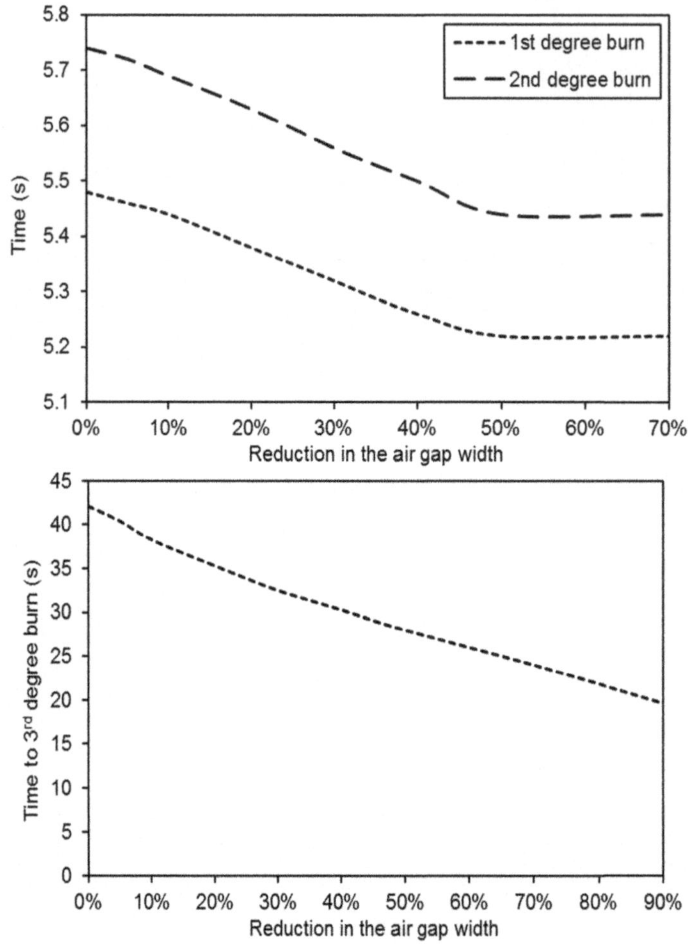

Figure 5. Effect of the reduction in the air gap width on skin burn predictions: (a) first– and second–degree burns, (b) third–degree burns

Figure 6 illustrates the effect of a variation in the reduction rate (fabric shrinkage rate) on the energy transfer within the gap. As shown in Figure 6a, any deviation in the reduction rate from the case of 0%/s (no shrinkage) causes a notable increase in the conduction heat flux within the gap. In addition, although the conduction heat flux increases as the reduction rate increases, the divergence in the conduction heat flux decreases with the increase in the reduction rate. This divergence drops off by the end of the exposure period and the deviation in the conduction heat flux from the case of no shrinkage vanishes gradually during the cool down period. On the other hand, the radiation heat flux within the gap decreases with the increase in the reduction rate, as shown in Figure 6b. However, like the distribution of conduction heat flux, the divergence in the radiation heat flux decreases with the increase in the reduction rate. Nevertheless, the divergence in the radiation heat flux with a variation in the reduction rate is less than that of the conduction heat flux.

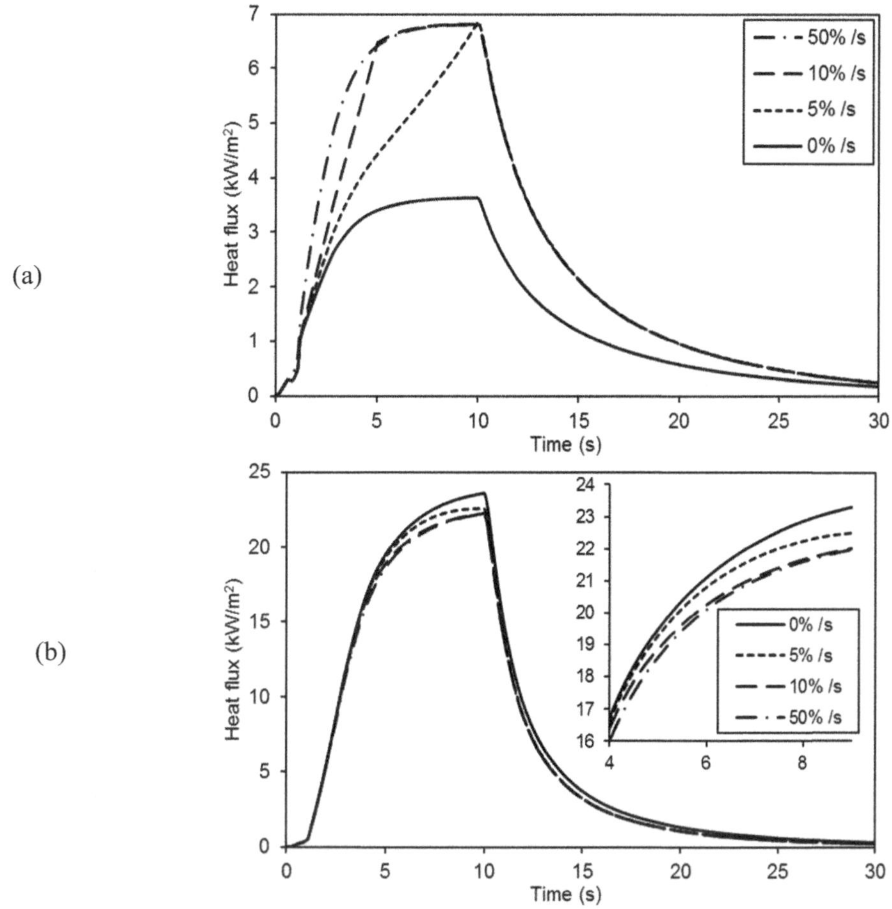

Figure 6. Effect of the reduction rate in the air gap width on energy transfer through the gap: (a) conduction heat flux at the fabric air interface, (b) emitted radiation from the fabric backside

The variation in the temperature distributions within the skin is illustrated in Figure 7. Expectedly, skin layers temperatures increase as the reduction rate increases. Nevertheless, the divergence in the skin layers temperatures decreases as the reduction rate increases; a rapid reduction in the air gap width does not make a big difference in the temperature distributions within the skin as compared to a slow one. For example, the increase in the skin layers temperatures due to a variation in the reduction rate from 5%/s to 50%/s is minimal compared to that caused by a variation from 0%/s (no shrinkage) to 5%/s. However, the divergence in the skin temperatures from the case of no shrinkage lasts during the cool down period.

(a)
(b)
(c)

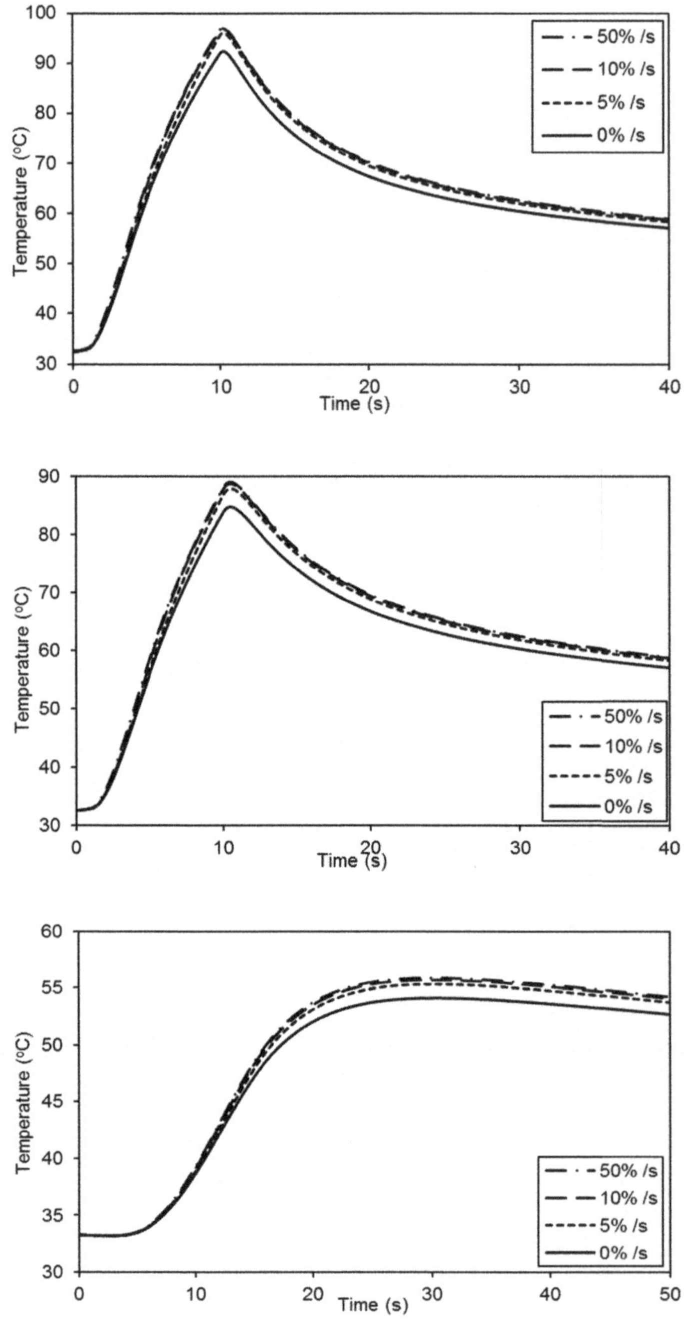

Figure 7. Effect of the reduction rate in the air gap width on the temperature distribution within the skin: (a) epidermis surface, (b) basal layer, (c) dermal base

The influence of a variation in the reduction rate on skin burn predictions is illustrated by Figure 8. Times to first– and second–degree burns, shown in Figure 8a, decrease with the increase in the reduction rate from 0%/s to 20%/s. There is almost no change in times to first–and second–degree burns for reduction rates that are greater than 20%/s. Times to third–degree burns, shown in Figure 8b, drop sharply as the reduction rate increases from 0%/s to 10%/s. However, for reduction rates that are greater than 20%/s, times to third–degree burns tend to be almost constant.

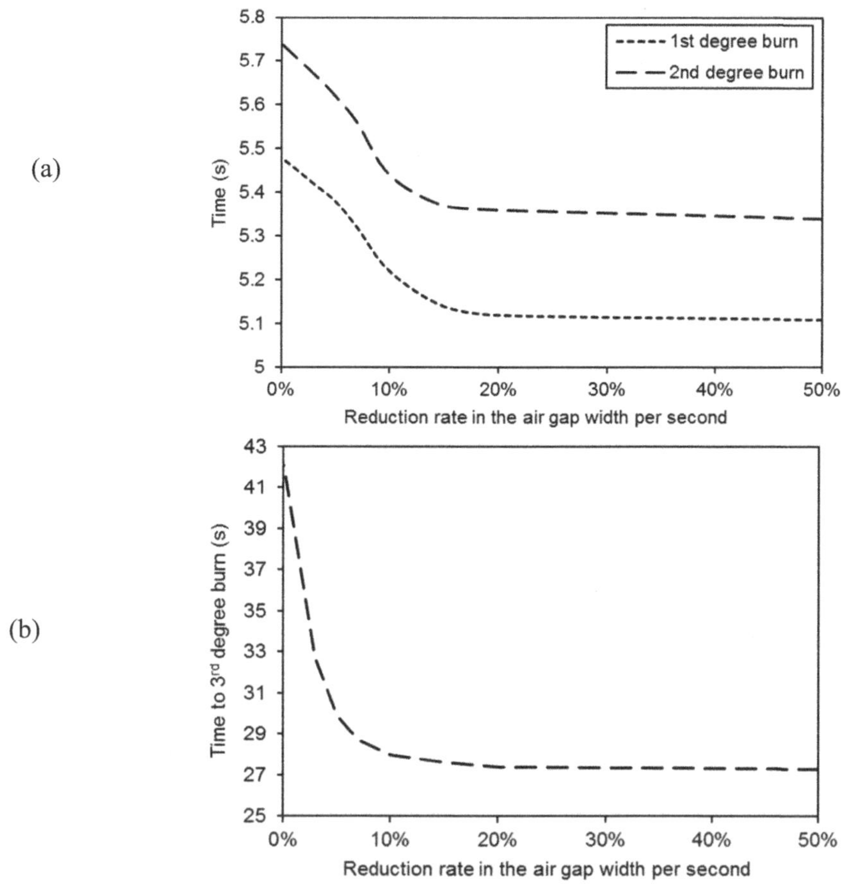

Figure 8. Effect of the reduction rate in the air gap width on skin burn predictions: (a) first–and second–degree burns, (b) third–degree burns

Figure 9 illustrates the influence of a variation in the reduction rate on skin burn predictions for different reduction percentages in the gap width. Similar to the trends shown in Figure 8, times to first (Figure 9a), second (Figure 9b) and third (Figure 9c) degree burns decrease with the increase in the reduction rate for low reduction rates then they become almost constant for higher reduction rates. The reduction rates at which times to skin burns become steady increase as the percentage reduction increases. In other words, the influence of the reduction rate on times to skin burn injuries increases with the increase in the percentage reduction in the air gap width. Lastly, Figure 9 shows the ultimate reduction in the gap width due to the fabric thermal shrinkage is more influential in determining the protective performance of the clothing than the rate at which this reduction takes place.

5. Conclusions

A finite volume model was developed to investigate the effect of the fabric thermal shrinkage on the performance of protective clothing during fire exposure. The model carefully accounted for the changes in the air gap between the fabric and the skin during the fabric thermal shrinkage. The influences of the reduction in the clothing perimeter around the body and the shrinkage rate of the fabric were numerically investigated. The following conclusions can be drawn from the study.

1) The protection provided by the clothing is considerably reduced by the thermal shrinkage of the fabric during fire exposure, particularly by the reduction in the clothing perimeter and the shrinkage rate of the fabric. Both parameters unequally affect the heat transfer from the fabric to the skin and hence the clothing protective performance.

2) A variation in the clothing perimeter or the fabric shrinkage rate nonlinearly alters the conduction and radiation heat fluxes from the clothing to the skin. However, the variation in the conduction heat flux from the fabric to the skin is bigger than that in the radiation heat flux.

3) The clothing protective performance linearly decreases with the reduction in the clothing perimeter around the body while its reduction with the increase in the fabric shrinkage rate is limited to small shrinkage rates.

4) The reduction in the clothing perimeter is more influential on the clothing protective performance than the shrinkage rate of the fabric. The closer the clothing is to the skin due to the fabric thermal shrinkage, the less the protective performance of the clothing.

5) The reduction in the clothing protective performance caused by an increase in the shrinkage rate of the fabric is worsened by the reduction of the clothing perimeter.

Finally, the paper, in all events, demonstrates the influential role that the air gap between protective clothing and the skin plays in determining the clothing performance during fire exposure. Furthermore, the paper, in general context, contributes to the knowledge of the combined conduction-radiation heat transfer in enclosures with dynamically changing widths filled with a radiation participating medium and bounded with high temperature and heat flux conditions.

(a)

(b)

(c)

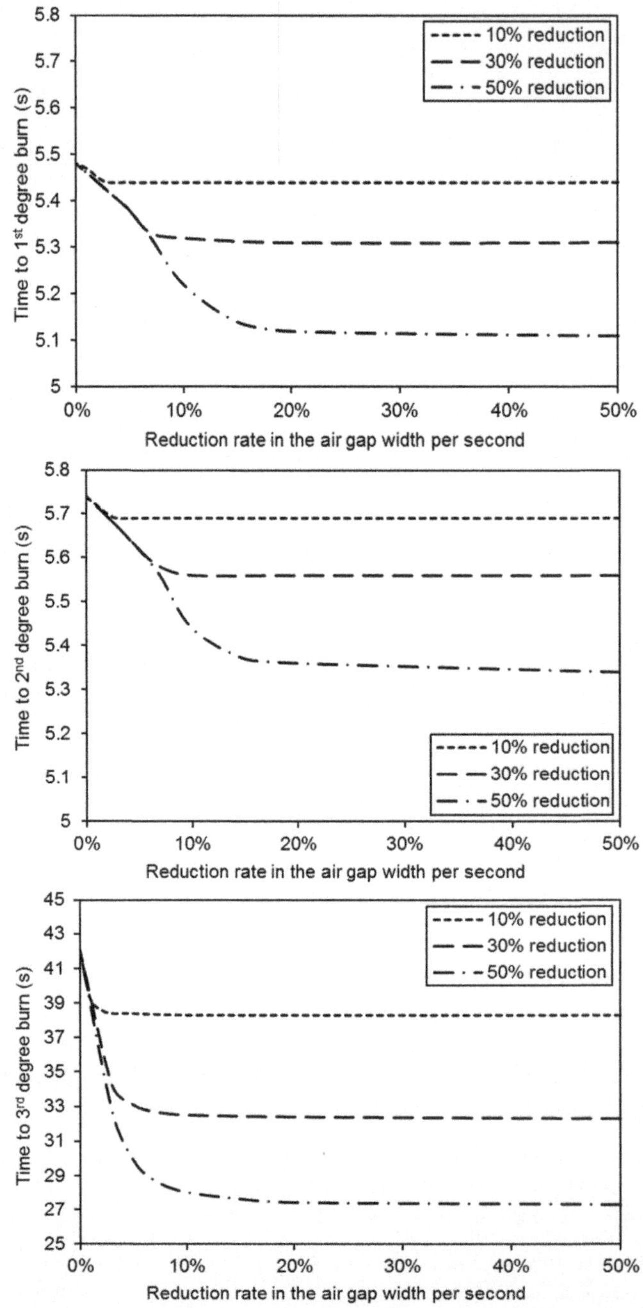

Figure 9. Skin burn predictions for different reduction rates and reduction percentages in the gap width: (a) first–degree burns, (b) second–degree burns, (c) third–degree burns

References

American Society for Testing Materials. (1987). *ASTM D 4108 Standard Test Method for Thermal Protective Performance of Materials and Clothing by open-Flame Method.* West Conshohocken, Pennsylvania.

American Society for Testing Materials. (1999). *ASTM F 1939 a Standard Test Method for Radiant Protective Performance of Flame Resistant Clothing Materials.* West Conshohocken, Pennsylvania.

American Society for Testing Materials. (2000). *ASTM F 1930 Standard Test Method for Evaluation of Flame Resistant Clothing for Protection Against Flash Fire Simulations Using an Instrumented Thermal Manikin.* West Conshohocken, Pennsylvania.

Chai, J. C., & Patankar, S. V. (2000). Finite-Volume Method for Radiation Heat Transfer. In J. P. Hartnett, & T. F. Ivrine (Ed.), *Advances in Numerical Heat Transfer* (Vol. 2, pp. 109-141). New York: Academic Press.

Chitrphiromsri, P., & Kuznetsov, A. V. (2005). Modeling Heat and Moisture Transport in Firefighter Protective Clothing during Flash Fire Exposure. *Heat and Mass Transfer, 41*, 206-215. http://dx.doi.org/10.1007/s00231-004-0504-x

Ghazy, A. (2011). *Air Gaps in Protective Clothing during Flash Fire Exposure* (Doctoral dissertation). University of Saskatchewan, Saskatoon, Saskatchewan.

Ghazy, A. (2013). *Numerical Study of the Air Gap between Fire-Protective Clothing and the Skin.* Industrial Textiles. http://dx.doi.org/ 10.1177/1528083713483784

Ghazy, A., & Bergstrom, D. J. (2010). Numerical Simulation of Transient Heat Transfer in a Protective Clothing System during a Flash Fire Exposure. *Numerical Heat Transfer A, 58*, 702-724. http://dx.doi.org/10.1080/10407782.2010.516691

Ghazy, A., & Bergstrom, D. J. (2011). Influence of the Air Gap between Protective Clothing and Skin on clothing Performance during Flash Fire Exposure. *Heat Mass Transfer, 47*, 1275-1288. http://dx.doi.org/10.1007/s00231-011-0791-y

Ghazy, A., & Bergstrom, D. J. (2012). Numerical Simulation of Heat Transfer in Firefighters' Protective Clothing with Multiple Air Gaps during Flash Fire Exposure. *Numerical Heat Transfer A, 61*, 569-593. http://dx.doi.org/10.1080/10407782.2012.666932

Henriques, F. C., Jr., & Moritz, A. R. (1947). Studies of thermal injuries I: The conduction of heat to and through skin and the temperatures attained therein. A theoretical and experimental investigation. *American Journal of Pathology, 23*, 531-549.

International Organization for Standardization. (1995). *ISO 9151 Protective Clothing against Heat and Flame Determination of Heat Transmission on Exposure to Flame.* Geneva, Switzerland.

Kim, I. Y., Lee, C., Li, P., Corner, B. D., & Paquette, S. (2002). Investigation of Air Gaps Entrapped in Protective Clothing Systems. *Fire and Materials, 26*, 121-126. http://dx.doi.org/10.1002/fam.790

Mah, T., & Song, G. (2010a). Investigation of the Contribution of Garment Design to Thermal Protection. Part 1: Characterizing Air Gaps using Three-dimensional body Scanning for Women's Protective Clothing. *Textile Research Journal, 80*, 1317-1329. http://dx.doi.org/10.1177/0040517509358795

Mah, T., & Song, G. (2010b). Investigation of the Contribution of Garment Design to Thermal Protection. Part 2: Instrumented Female Mannequin Flash-fire Evaluation System. *Textile Research Journal, 80*, 1473-1487. http://dx.doi.org/10.1177/0040517509358796

Mell, W. E., & Lawson, J. R. (2000). Heat Transfer Model for Fire Fighter's Protective Clothing. *Fire Technology, 36*, 39-68. http://dx.doi.org/10.1023/A:1015429820426

Mercer, G. N., & Sidhu, H. S. (2007). Mathematical modeling of the effect of fire exposure on a new type of protective clothing. *ANZIAM, 49*, C289-C305.

Mercer, G. N., & Sidhu, H. S. (2009). A Theoretical Investigation into Phase Change Clothing Benefits for Firefighters under Extreme Conditions. *Chemical Product and Process Modeling, 4*(3), http://dx.doi.org/10.2202/1934-2659.1349

Modest, M. F. (2003). *Radiative Heat Transfer* (2nd ed.), New York, NY: Academic Press.

National Fire Protection Association. (2007). *NFPA 1971 Standard on Protective Ensemble for Structural Fire Fighting, Quincy, Massachusetts.*

Patankar, S. V. (1980). *Numerical Heat Transfer and Fluid Flow*. Washington, DC: Taylor & Francis.

Pennes, H. H. (1948). Analysis of tissue and arterial blood temperatures in resting human forearm. *Applied Physiology, 1*, 93-122.

Sawcyn, C. M. J., & Torvi, D. A. (2009). Improving Heat Transfer Models of Air Gaps in Bench Top Tests of Thermal Protective Fabrics. *Textile Research Journal, 79*, 632-644. http://dx.doi.org/10.1177/0040517508093415

Song, G., Barker, R. L., Hamouda, H., Kuznetsov, A. V., Chitrphiromsri, P., & Grimes, R. V. (2004). Modeling the Thermal Protective Performance of Heat Resistant Garments in Flash Fire Exposures. *Textile Research Journal, 74*, 1033-1040. http://dx.doi.org/10.1177/004051750407401201

Song, G., Chitrphiromsri, P., & Ding, D. (2008). Numerical Simulation of Heat and Moisture Transport in Thermal Protective Clothing under Flash Fire Conditions. *International Journal of Occupational Safety and Ergonomics, 14*, 89-106. http://dx.doi.org/ciop/pl/25866

Takata, A. N., Rouse, J., & Stanley, T. (1973). Thermal analysis program. Illinois Institute of Technology, Chicago.

Talukdar, P., Torvi, D. A., Simonson, C. J., & Sawcyn, C. M. J. (2010). Coupled CFD and Radiation Simulation of Air Gaps in Bench Top Protective Fabric Tests. *International Journal of Heat and Mass Transfer, 53*, 526-539. http://dx.doi.org/10.1016/j.ijheatmasstransfer.2009.04.041

Tan, Y., Crown, E. M., & Capjack, L. (1998). Design and Evaluation of Thermal Protective Flightsuits. Part I: The Design Process and Prototype Development. *Clothing and Textiles Research Journal, 16*, 47-55. http://dx.doi.org/10.1177/0887302X9801600106

Torvi, D. A. (1997). *Heat Transfer in Thin Fibrous Materials under High Heat Flux Conditions* (Doctoral dissertation). University of Alberta, Edmonton, Alberta.

Torvi, D. A., & Dale, J. D. (1999). Heat Transfer in Thin Fibrous Materials under High Heat Flux. *Fire Technology, 35*, 210-231. http://dx.doi.org/10.1023/A:1015484426361

Torvi, D. A., & Threlfall, T. G. (2006). Heat Transfer Model of Flame Resistant Fabrics during Cooling After Exposure to Fire. *Fire Technology, 42*, 27-48. http://dx.doi.org/10.1007/s10694-005-3733-8

Torvi, D. A., Dale, J. D., & Faulkner, B. (1999). Influence of Air Gaps on Bench Top Test Results of Flame Resistant Fabrics. *Fire Protection Engineering, 10*, 1-12. http://dx.doi.org/10.1177/104239159901000101

Weaver, J. A., & Stoll, A. M. (1969). Mathematical Model of Skin Exposed to Thermal Radiation. *Aerospace Medicine, 40*, 24-30.

Zhu, F., & Zhang, W. (2009). Modeling Heat Transfer for Heat-resistant Fabrics Considering Pyrolysis Effect under an External Heat Flux. *Fire Sciences, 27*, 81-96. http://dx.doi.org/10.1177/0734904108094960

21

Incremental Inverse Kinematics of Wire-Suspended Parallel Mechanical System Taking into Account Many-Worlds Situation

Kazuyuki Hanahara[1]

[1] Graduate School of System Informatics, Kobe University, Kobe, Japan

Correspondence: Kazuyuki Hanahara, Graduate School of System Informatics, Kobe University, Kobe, Japan.
E-mail: hanahara@cs.kobe-u.ac.jp

Abstract

Wire-suspended platform is a parallel mechanical system consisting of a rigid platform and a number of length-adjustable wires. In order to deal with the kinematics of this type of mechanical system, we have to take into account the conspicuous characteristic of a wire that it can take only a tensile force. It has been demonstrated in the previous studies that this leads to plural possible kinematic states corresponding to a given set of wire lengths. In this study, we formulate an inverse kinematics problem in an incremental form based on the linearization of the kinematic relation of the system. Displacement of the mass center position of the platform is also taken into account for the articulation variables, in addition to the wire member lengths. A solution procedure that deals with the many-worlds situation is developed under the assumption of some adequate sensing devices. Numerical experiments based on computer simulation are carried out and the feasibility of the proposed approach is demonstrated with the attained kinematic motions of the simulated platform system.

Keywords: parallel mechanism, wire, inverse kinematics, many-worlds interpretation, Interval arithmetic

1. Introduction

Parallel robots (Merlet, 2006) or platforms (Stewart, 1965-66) have a number of different features compared with the conventional serial robotic mechanisms. One of such features is that they can have a number of non-rigid links of wire or cable. A platform supported by a number of wires or cables, which are actuated and length-adjustable, is a typical mechanical system of this type. Such mechanical systems are referred to as robocrane (Albus et al., 1992), wire-actuated parallel manipulator (Agahi & Notash, 2009), wire-driven parallel mechanism (Yamaguchi et al., 2004), cable robot (Ghasemi et al., 2009) (Borgstrom et al., 2009), cable-driven robot (Bosscher et al., 2006), cable-driven parallel robot (Tadokoro et al., 2003) (Carricato & Merlet 2013) (Carricato, 2013), cable-suspended robot (Roberts et al., 1998) (Oh & Agrawal, 2005), cable-suspended parallel robot (Pusey et al., 2004), cable-driven parallel mechanism (Perreault & Gosselin, 2008), tendon-based manipulator (Hiller et al., 2005), and so on. In this study, we refer this type of mechanical system to as a wire-suspended platform (Hanahara, 2015). A mechanical system of this type is well represented by a suspended platform of rigid body, a number of length-adjustable wires for the suspension, and the corresponding fixed connection points of the wires at the wall or at the ceiling. In order to control the six degrees of freedom (DOFs) in 3D space, or the three DOFs in 2D space, of the position and orientation of the platform, it is natural that the number of wires is six or three. However, such a mechanical system having less wires is possible (Carricato & Merlet, 2013) (Carricato, 2013); the equilibrium condition determines the position and orientation to be attained by the platform in this case. In contrast to the platform system consisting only of rigid links, a platform having a redundant number of wires (Oh & Agrawal, 2005) is also possible without difficulty.

There are several issues that have to be taken into consideration in the case of dealing with the kinematics of mechanical system of this type: the conspicuous characteristic of a wire that they cannot support any compressive force, the static equilibrium condition under gravity and the influence of displacement of mass center of the platform. The previous study (Hanahara, 2015) deals with these issues and shows a general formulation and the solution of the incremental forward kinematics problem of wire-suspended platform based on the interval arithmetic (Moore et al., 2009) as well as the many-worlds interpretation (Vaidman, 2008). It has been demonstrated that even in this linearized incremental case, the possible configuration of the platform is not unique in general; we have to deal with the plural possibilities simultaneously.

We deal with the inverse kinematics of mechanical system of this type. We adopt the approach developed in the previous study, which is based on the incremental formulation with the interval arithmetic and the many-worlds interpretation. We assume that some sort of sensing device is available in the current study, since the formulated incremental inverse kinematics problem has often no feasible solutions without confining the kinematical possibilities, which are to be obtained as the consequence of the many-worlds interpretation. It should be noted that the unavoidable error in the measurement results of the sensors is naturally taken into account in the form of the interval in our approach.

We conduct a number of numerical experiments with a simulated platform system assumed to have simple binary tension sensors. The feasibility of the proposed incremental inverse kinematics approach as well as the obtained characteristic behavior of the wire-suspended platform is demonstrated.

2. Incremental Forward Kinematics Based on Many-Worlds Interpretation

In order to make the article self-contained, we summarize the incremental forward kinematics problem of the wire-suspended platform as well as the management of the solution based on the many-worlds interpretation.

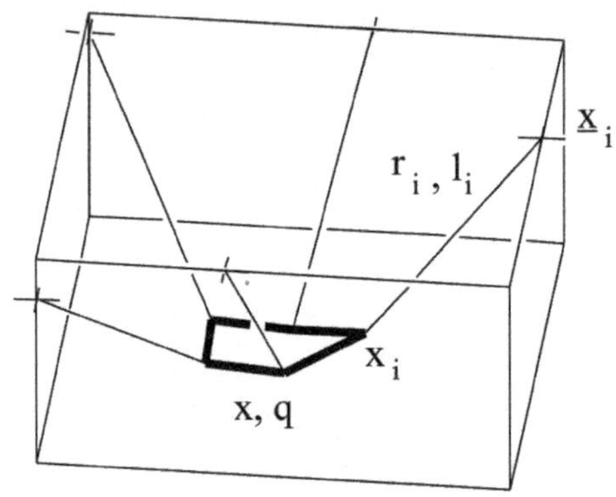

Figure 1. Conceptual illustration of wire-actuated platform

2.1 Basic Equation in General Form

Figure 1 shows the conceptual illustration of the wire-suspended platform system dealt with in the current study; a platform of rigid body is suspended by a number of length-adjustable wires from the fixed connection points. We deal with the kinematics of the system based on the following two basic equations in general form:

$$\boldsymbol{l} = \boldsymbol{l}(\boldsymbol{X}) \tag{1}$$

$$\boldsymbol{F} = \boldsymbol{F}(\boldsymbol{X}, \boldsymbol{\tau}, \hat{\boldsymbol{c}}) = 0 \tag{2}$$

Equation (1) is the geometry condition and Eq.(2) is the equilibrium condition under the gravity. In these equations, $\boldsymbol{l} = [l_1, \cdots, l_N]^T$ and l_i is the distance between the fixed connection point $\underline{\boldsymbol{x}}_i$ and the connection point $\boldsymbol{x}_i$ at the platform, which is referred to as the geometrical length of i-th wire, $\boldsymbol{X} = [\boldsymbol{x}^T, \boldsymbol{\theta}^T]^T$ denotes the position and orientation of the platform, $\boldsymbol{F} = [\boldsymbol{f}^T, \boldsymbol{\eta}^T]^T$ denotes the force and moment acting on the platform, $\boldsymbol{\tau} = [\tau_1, \cdots, \tau_N]^T$ and τ_i is the tension of the i-th wire and $\hat{\boldsymbol{c}}$ is the position of the mass center of the platform referring to the platform-fixed coordinate. Note that $\hat{\boldsymbol{c}}$ is not assumed constant in this study.

The following two conditions have to be satisfied for a wire-suspended system:

$$\boldsymbol{\rho} \geq \boldsymbol{l} \tag{3}$$

$$\boldsymbol{\tau} \geq 0 \tag{4}$$

where $\boldsymbol{\rho} = [\rho_1, \cdots, \rho_N]^T$ and ρ_i is the kinematical length of the i-th wire, that is, the wire length directly controlled by the actuation mechanism. Equation (3) is the geometrical constraint for the wire lengths and Eq.(4) denotes that the tension in a wire must have a positive value.

The following incremental relations of Eqs.(1) and (2) are used in this study:

$$\Delta \boldsymbol{l} = \frac{\partial \boldsymbol{l}}{\partial \boldsymbol{X}} \Delta \boldsymbol{X} \tag{5}$$

$$\Delta \boldsymbol{F} = \frac{\partial \boldsymbol{F}}{\partial \boldsymbol{X}} \Delta \boldsymbol{X} + \frac{\partial \boldsymbol{F}}{\partial \boldsymbol{\tau}} \Delta \boldsymbol{\tau} + \frac{\partial \boldsymbol{F}}{\partial \hat{\boldsymbol{c}}} \Delta \hat{\boldsymbol{c}} \tag{6}$$

The detail of the formulation of Eqs.(1) and (2) as well as their incremental relations (5) and (6) for the general case of the wire-suspended platform system is given in (Hanahara, 2015).

2.2 Incremental Forward Kinematics Solution

The forward kinematics based on the increment $\Delta \boldsymbol{\rho}$ in wire lengths and the displacement $\Delta \hat{\boldsymbol{c}}$ in platform mass center position is calculated based the following equilibrium condition in incremental form taking into account the unbalanced force $\boldsymbol{F}_\epsilon$ to be balanced:

$$\boldsymbol{F}_\epsilon + \frac{\partial \boldsymbol{F}}{\partial \boldsymbol{X}} \Delta \boldsymbol{X} + \frac{\partial \boldsymbol{F}}{\partial \boldsymbol{\tau}} \Delta \boldsymbol{\tau} + \frac{\partial \boldsymbol{F}}{\partial \hat{\boldsymbol{c}}} \Delta \hat{\boldsymbol{c}} = 0 \tag{7}$$

The geometrical and tension constraints (3) and (4) in the incremental form are expressed as

$$\boldsymbol{\rho} + \Delta \boldsymbol{\rho} \geq \boldsymbol{l} + \frac{\partial \boldsymbol{l}}{\partial \boldsymbol{X}} \Delta \boldsymbol{X} \tag{8}$$

$$\boldsymbol{\tau} + \Delta \boldsymbol{\tau} \geq 0 \tag{9}$$

The incremental forward kinematics problem is to find $\Delta \boldsymbol{X}$ and $\Delta \boldsymbol{\tau}$ corresponding to given $\Delta \boldsymbol{\rho}$ and $\Delta \hat{\boldsymbol{c}}$ that satisfy Eqs.(7), (8) and (9).

The tension state of the wires is denoted by the two sets: A for taut wires and S for slack wires. Assuming the tension state of wires A and S after the kinematic increment, the equilibrium condition (7) is rewritten as

$$\frac{\partial \boldsymbol{F}}{\partial \boldsymbol{X}} \Delta \boldsymbol{X} + \frac{\partial \boldsymbol{F}}{\partial \boldsymbol{\tau}_{\mathsf{A}}} \Delta \boldsymbol{\tau}_{\mathsf{A}} = \Delta \boldsymbol{F}_{\tau_{\mathsf{S}} \hat{c}} \tag{10}$$

where

$$\Delta \boldsymbol{F}_{\tau_{\mathsf{S}} \hat{c}} = -\boldsymbol{F}_\epsilon - \frac{\partial \boldsymbol{F}}{\partial \boldsymbol{\tau}_{\mathsf{S}}} \Delta \boldsymbol{\tau}_{\mathsf{S}} - \frac{\partial \boldsymbol{F}}{\partial \hat{\boldsymbol{c}}} \Delta \hat{\boldsymbol{c}} \tag{11}$$

The tension increment vector for wires to be slack is immediately determined as

$$\Delta \boldsymbol{\tau}_{\mathsf{S}} = -\boldsymbol{\tau}_{\mathsf{S}} \quad (\because \boldsymbol{\tau}_{\mathsf{S}} + \Delta \boldsymbol{\tau}_{\mathsf{S}} = 0) \tag{12}$$

For the wires to be taut, the following equation is then obtained as a part of the geometrical condition (8):

$$\frac{\partial \boldsymbol{l}_{\mathsf{A}}}{\partial \boldsymbol{X}} \Delta \boldsymbol{X} = \Delta \boldsymbol{l}_{\mathsf{A}} \tag{13}$$

where

$$\Delta \boldsymbol{l}_{\mathsf{A}} = \boldsymbol{\rho}_{\mathsf{A}} + \Delta \boldsymbol{\rho}_{\mathsf{A}} - \boldsymbol{l}_{\mathsf{A}} \tag{14}$$

On the basis of Eqs.(10) and (13), the following equation is obtained:

$$\boldsymbol{J} \Delta \boldsymbol{v} = \Delta \boldsymbol{z} \tag{15}$$

where

$$\boldsymbol{J} = \begin{bmatrix} \dfrac{\partial \boldsymbol{F}}{\partial \boldsymbol{X}} & \dfrac{\partial \boldsymbol{F}}{\partial \boldsymbol{\tau}_{\mathsf{A}}} \\ \dfrac{\partial \boldsymbol{l}_{\mathsf{A}}}{\partial \boldsymbol{X}} & 0 \end{bmatrix} \tag{16}$$

is the combined Jacobian matrix and

$$\Delta \boldsymbol{v} = \begin{bmatrix} \Delta \boldsymbol{X} \\ \Delta \boldsymbol{\tau}_{\mathsf{A}} \end{bmatrix}, \quad \Delta \boldsymbol{z} = \begin{bmatrix} \Delta \boldsymbol{F}_{\tau_{\mathsf{S}} \hat{c}} \\ \Delta \boldsymbol{l}_{\mathsf{A}} \end{bmatrix} \tag{17}$$

are the corresponding known and unknown vectors. Equation (15) can be solved as

$$\Delta \boldsymbol{v} = \boldsymbol{J}^{-1}\Delta \boldsymbol{z} \tag{18}$$

We denote the interior of the matrix $\boldsymbol{J}^{-1}$ in Eq.(18) as

$$\boldsymbol{J}^{-1} = \begin{bmatrix} \dfrac{\partial \boldsymbol{X}}{\partial \boldsymbol{F}} & \dfrac{\partial \boldsymbol{X}}{\partial \boldsymbol{l}_{\mathsf{A}}} \\ \dfrac{\partial \boldsymbol{\tau}_{\mathsf{A}}}{\partial \boldsymbol{F}} & \dfrac{\partial \boldsymbol{\tau}_{\mathsf{A}}}{\partial \boldsymbol{l}_{\mathsf{A}}} \end{bmatrix} \tag{19}$$

It should be noted that $\dfrac{\partial \boldsymbol{X}}{\partial \boldsymbol{F}} \neq \left(\dfrac{\partial \boldsymbol{F}}{\partial \boldsymbol{X}}\right)^{-1}$, $\dfrac{\partial \boldsymbol{X}}{\partial \boldsymbol{l}_{\mathsf{A}}} \neq \left(\dfrac{\partial \boldsymbol{l}_{\mathsf{A}}}{\partial \boldsymbol{X}}\right)^{-1}$ and $\dfrac{\partial \boldsymbol{\tau}_{\mathsf{A}}}{\partial \boldsymbol{F}} \neq \left(\dfrac{\partial \boldsymbol{F}}{\partial \boldsymbol{\tau}_{\mathsf{A}}}\right)^{-1}$ in general even in the case that $\mathrm{Dim}(\boldsymbol{l}_{\mathsf{A}}) = \mathrm{Dim}(\boldsymbol{X})$ since these are obtained as the partial matrices of $\boldsymbol{J}^{-1}$. These notations are consistently used as the same meanings in this paper. On the basis of the expression in Eq.(19), Eq.(18) is rewritten as

$$\Delta \boldsymbol{X} = \frac{\partial \boldsymbol{X}}{\partial \boldsymbol{F}}\Delta \boldsymbol{F}_{\tau_{\mathsf{S}}\hat{c}} + \frac{\partial \boldsymbol{X}}{\partial \boldsymbol{l}_{\mathsf{A}}}\Delta \boldsymbol{l}_{\mathsf{A}} \tag{20}$$

$$\Delta \boldsymbol{\tau}_{\mathsf{A}} = \frac{\partial \boldsymbol{\tau}_{\mathsf{A}}}{\partial \boldsymbol{F}}\Delta \boldsymbol{F}_{\tau_{\mathsf{S}}\hat{c}} + \frac{\partial \boldsymbol{\tau}_{\mathsf{A}}}{\partial \boldsymbol{l}_{\mathsf{A}}}\Delta \boldsymbol{l}_{\mathsf{A}} \tag{21}$$

The increment $\Delta \boldsymbol{X}$ has to satisfy the following geometrical condition for the slack wires, which is a part of Eq.(8):

$$\boldsymbol{\rho}_{\mathsf{S}} + \Delta \boldsymbol{\rho}_{\mathsf{S}} \geq \boldsymbol{l}_{\mathsf{S}} + \frac{\partial \boldsymbol{l}_{\mathsf{S}}}{\partial \boldsymbol{X}}\Delta \boldsymbol{X} \tag{22}$$

The increment $\Delta \boldsymbol{\tau}_{\mathsf{A}}$ has to satisfy the following condition for the taut wires, which is a part of Eq.(9):

$$\boldsymbol{\tau}_{\mathsf{A}} + \Delta \boldsymbol{\tau}_{\mathsf{A}} \geq 0 \tag{23}$$

In the case that the both conditions (22) and (23) are satisfied, the incremental forward kinematics problem based on the given increments $\Delta \boldsymbol{\rho}$ and $\Delta \hat{\boldsymbol{c}}$ is solved under the tension state assumption of A and S and the solution increments $\Delta \boldsymbol{X}$ and $\Delta \boldsymbol{\tau}$ are obtained, where the tension increment vector $\Delta \boldsymbol{\tau}$ is obtained partially as $\Delta \boldsymbol{\tau}_{\mathsf{A}}$ and partially as $\Delta \boldsymbol{\tau}_{\mathsf{S}}$. It is also possible to deal with the case that $\boldsymbol{J}$ is singular in a similar manner (Hanahara, 2015).

2.3 Interval Arithmetic Approach

It is practically impossible to eliminate measurement error in the wire lengths as well as in the platform mass center position. In the current study, we deal with these values in the form of interval (Moore et al., 2009) as

$$[\boldsymbol{\rho}] = \boldsymbol{\rho} + [\boldsymbol{\rho}_{\mathrm{err}}], \quad [\hat{\boldsymbol{c}}] = \hat{\boldsymbol{c}} + [\hat{\boldsymbol{c}}_{\mathrm{err}}] \tag{24}$$

where $[\boldsymbol{\rho}_{\mathrm{err}}]$ and $[\hat{\boldsymbol{c}}_{\mathrm{err}}]$ denote the uncertainty magnitude of $\boldsymbol{\rho}$ and $\hat{\boldsymbol{c}}$; the actual error values are bounded as follows:

$$\underline{[\boldsymbol{\rho}_{\mathrm{err}}]} \leq \boldsymbol{\rho}_{\mathrm{err}} \leq \overline{[\boldsymbol{\rho}_{\mathrm{err}}]}, \quad \underline{[\hat{\boldsymbol{c}}_{\mathrm{err}}]} \leq \hat{\boldsymbol{c}}_{\mathrm{err}} \leq \overline{[\hat{\boldsymbol{c}}_{\mathrm{err}}]}$$

Calculating equations in the previous section by taking into account these interval values, Eqs.(20) and (21) are rewritten in the following form as

$$[\Delta \boldsymbol{X}] = \frac{\partial \boldsymbol{X}}{\partial \boldsymbol{F}}[\Delta \boldsymbol{F}_{\tau_{\mathsf{S}}\hat{c}}] + \frac{\partial \boldsymbol{X}}{\partial \boldsymbol{l}_{\mathsf{A}}}[\Delta \boldsymbol{l}_{\mathsf{A}}] \tag{25}$$

$$[\Delta \boldsymbol{\tau}_{\mathsf{A}}] = \frac{\partial \boldsymbol{\tau}_{\mathsf{A}}}{\partial \boldsymbol{F}}[\Delta \boldsymbol{F}_{\tau_{\mathsf{S}}\hat{c}}] + \frac{\partial \boldsymbol{\tau}_{\mathsf{A}}}{\partial \boldsymbol{l}_{\mathsf{A}}}[\Delta \boldsymbol{l}_{\mathsf{A}}] \tag{26}$$

The interval of the geometrical wire lengths increment corresponding to the slack wires is obtained as

$$[\Delta \boldsymbol{l}_{\mathsf{S}}] = \frac{\partial \boldsymbol{l}_{\mathsf{S}}}{\partial \boldsymbol{X}}[\Delta \boldsymbol{X}] \tag{27}$$

Referring to the non-interval geometrical condition (22) for the slack wires, the following condition is obtained:

$$\overline{[\boldsymbol{\rho}_{\mathsf{S}}]} + \Delta\boldsymbol{\rho}_{\mathsf{S}} \geq \boldsymbol{l}_{\mathsf{S}} + \underline{[\Delta\boldsymbol{l}_{\mathsf{S}}]} \tag{28}$$

Similarly, the following condition for the taut wires is obtained based on the non-interval condition (23):

$$\boldsymbol{\tau}_{\mathsf{A}} + \overline{[\Delta\boldsymbol{\tau}_{\mathsf{A}}]} \geq 0 \tag{29}$$

In the case that the both conditions (28) and (29) are satisfied, the incremental forward kinematics has the possibility to be feasible under the tension state assumption of A and S.

2.4 Many-Worlds Interpretation

This interval-based incremental forward kinematics leads to a number of possible solutions each of which corresponds to different tension states. In order to cope with the situation, a kind of parallelism referred to as many-worlds interpretation is adopted to deal with the possible states simultaneously. The phrase 'many-worlds interpretation' is taken from the quantum mechanics theory (e.g. Vaidman, 2008); in brief, the interpretation says that each of the different possible states belongs to its corresponding different world and all of these worlds exist simultaneously.

In the incremental kinematics based on the many-worlds interpretation, basically, we deal with all the possible states as evenly actual and existent. It should be noted, however, that we also adopt the so-called unification process in order to suppress the number of possible states from the practicality point of view. The detail is also discussed in (Hanahara, 2015).

3. Formulation of Inverse Kinematics Problem

Taking into account the situation of the many-worlds interpretation, the inverse kinematics of the wire-suspended platform has to deal with the plural possible kinematic states.

3.1 How to Deal with the Problem

The inverse kinematics of the wire-suspended platform in incremental form is to find the increment $\Delta\boldsymbol{\rho}$ in kinematic wire lengths as well as the increment $\Delta\hat{\boldsymbol{c}}$ in platform mass center position, that is corresponding to the desired incremental motion of the platform. An important issue that has to be taken into consideration in this study is that the desired increment $\underline{\Delta\boldsymbol{X}}$ in platform position and orientation itself differs in accordance with the kinematic state of respective possibilities. It should be noted that the increments $\Delta\boldsymbol{\rho}$ and $\Delta\hat{\boldsymbol{c}}$ are common to all the possibilities under consideration but the desired increment $\underline{\Delta\boldsymbol{X}}$ is specific to each possibility; that is, it is generally impossible to find $\Delta\boldsymbol{\rho}$ and $\Delta\hat{\boldsymbol{c}}$ that meet various values of $\underline{\Delta\boldsymbol{X}}$ of all the possibilities. On the basis of the consideration, we deal with the inverse kinematics problem as a minimization problem. The difference between the desired platform position and orientation and the platform positions and orientations attained by the respective possibilities under consideration is adopted as the objective function to be minimized.

3.2 Formulation as Minimization Problem

Suppose we have P possibilities of kinematic states, which are obtained as the result of forward kinematics calculation. Sensor information such as the wire tension values as well as the platform position and orientation should be adopted. Accordingly, we confine the number of possible kinematic states to be taken into account for the inverse kinematics to $P' \leq P$ by means of the sensor information.

For the p-th possible kinematic state ($p = 1, \cdots, P'$) taken into account for the inverse kinematics in the current incremental step, the equilibrium condition (7) is expressed as follows:

$$\boldsymbol{F}_{\epsilon}^{(p)} + \frac{\partial \boldsymbol{F}^{(p)}}{\partial \boldsymbol{X}} \Delta\boldsymbol{X}^{(p)} + \frac{\partial \boldsymbol{F}^{(p)}}{\partial \boldsymbol{\tau}} \Delta\boldsymbol{\tau}^{(p)} + \frac{\partial \boldsymbol{F}^{(p)}}{\partial \hat{\boldsymbol{c}}} \Delta\hat{\boldsymbol{c}} = 0 \tag{30}$$

where superscript (p) denotes that the values are specific to the p-th possibility. It should be noted that the increment $\Delta\hat{\boldsymbol{c}}$ in platform mass center position referring to the platform-fixed coordinate is common to all the possibilities. Similarly, the geometrical condition (8) is rewritten as

$$\boldsymbol{\rho} + \Delta\boldsymbol{\rho} \geq \boldsymbol{l}^{(p)} + \frac{\partial \boldsymbol{l}^{(p)}}{\partial \boldsymbol{X}} \Delta\boldsymbol{X}^{(p)} \tag{31}$$

Note that the kinematical wire lengths $\boldsymbol{\rho}$ and its increment $\Delta\boldsymbol{\rho}$ are common to all the possibilities as well.

Assuming the tension state $\mathsf{A}^{(p)}$ and $\mathsf{S}^{(p)}$ for the p-th possible kinematic state after the increment, the equilibrium condition (30) is rewritten as

$$\frac{\partial \boldsymbol{F}^{(p)}}{\partial \hat{\boldsymbol{c}}}\Delta\hat{\boldsymbol{c}} + \frac{\partial \boldsymbol{F}^{(p)}}{\partial \boldsymbol{X}}\Delta\boldsymbol{X}^{(p)} + \frac{\partial \boldsymbol{F}^{(p)}}{\partial \boldsymbol{\tau}_{\mathsf{A}}}\Delta\boldsymbol{\tau}_{\mathsf{A}}^{(p)} = -\boldsymbol{F}_{\epsilon}^{(p)} - \frac{\partial \boldsymbol{F}^{(p)}}{\partial \boldsymbol{\tau}_{\mathsf{S}}}\Delta\boldsymbol{\tau}_{\mathsf{S}}^{(p)} \tag{32}$$

In order to be exact, the subscripts A and S in Eq.(32) should be denoted as $\mathsf{A}^{(p)}$ and $\mathsf{S}^{(p)}$; however, the superscript (p) for these subscripts are omitted for the sake of simplicity. Similarly, the geometrical condition (31) is rewritten separately for the taut and slack wires as

$$-\Delta\boldsymbol{\rho}_{\mathsf{A}^{(p)}} + \frac{\partial \boldsymbol{l}_{\mathsf{A}}^{(p)}}{\partial \boldsymbol{X}}\Delta\boldsymbol{X}^{(p)} = \boldsymbol{\rho}_{\mathsf{A}^{(p)}} - \boldsymbol{l}_{\mathsf{A}}^{(p)} \tag{33}$$

$$\frac{\partial \boldsymbol{l}_{\mathsf{S}}^{(p)}}{\partial \boldsymbol{X}}\Delta\boldsymbol{X}^{(p)} + \boldsymbol{l}_{\mathsf{S}}^{(p)} \leq \Delta\boldsymbol{\rho}_{\mathsf{S}^{(p)}} + \boldsymbol{\rho}_{\mathsf{S}^{(p)}} \tag{34}$$

It should be noted that the superscript (p) of the subscripts A and S for $\boldsymbol{\rho}$ and $\Delta\boldsymbol{\rho}$ should not be omitted in the above equations, since the *values* are common to all the possibilities but the *tension state* is specific to each of the possibilities. The tension constraint in incremental form (9) is similarly rewritten as

$$\boldsymbol{\tau}^{(p)} + \Delta\boldsymbol{\tau}^{(p)} \geq 0 \tag{35}$$

Taking into account all of the possible kinematic states under consideration, we introduce the following objective function to be minimized for platform position and orientation:

$$g_X = \frac{1}{P'}\sum_{p=1}^{P'}\frac{1}{2}\left(\Delta\boldsymbol{X}^{(p)} - \underline{\Delta\boldsymbol{X}^{(p)}}\right)^T\begin{bmatrix} K_x\boldsymbol{I} & 0 \\ 0 & K_\theta\boldsymbol{I}\end{bmatrix}\left(\Delta\boldsymbol{X}^{(p)} - \underline{\Delta\boldsymbol{X}^{(p)}}\right) \tag{36}$$

where

$$\underline{\Delta\boldsymbol{X}^{(p)}} = \underline{\boldsymbol{X}} - \boldsymbol{X}^{(p)} \tag{37}$$

is the target value expressed in terms of the specified platform position and orientation $\underline{\boldsymbol{X}}$ at the current incremental step, K_x and K_θ are the penalty coefficients respectively corresponding to position and orientation, and $\boldsymbol{I}$ is the three-dimensional identity matrix. The following auxiliary objective functions are also introduced:

$$g_{\rho\hat{c}} = \frac{1}{2}K_\rho\Delta\boldsymbol{\rho}^T\Delta\boldsymbol{\rho} + \frac{1}{2}K_{\hat{c}}\Delta\hat{\boldsymbol{c}}^T\Delta\hat{\boldsymbol{c}} \tag{38}$$

$$g_\tau = \frac{1}{P}\sum_{p=1}^{P}\frac{1}{2}K_\tau(\Delta\boldsymbol{\tau}_{\mathsf{A}}^{(p)} - \underline{\Delta\boldsymbol{\tau}_{\mathsf{A}}^{(p)}})^T(\Delta\boldsymbol{\tau}_{\mathsf{A}}^{(p)} - \underline{\Delta\boldsymbol{\tau}_{\mathsf{A}}^{(p)}}) \tag{39}$$

with the weight coefficients K_ρ, $K_{\hat{c}}$ and K_τ. Equation (38) is introduced intended to suppress the eccentricity of the increments in wire lengths as well as the platform mass center position. Equation (39) is introduced to avoid extreme values of wire tensions, where

$$\underline{\Delta\boldsymbol{\tau}_{\mathsf{A}}^{(p)}} = \underline{\boldsymbol{\tau}}_{\mathsf{A}^{(p)}} - \boldsymbol{\tau}_{\mathsf{A}}^{(p)} \tag{40}$$

and $\underline{\boldsymbol{\tau}}$ is an adequate reference tension value vector.

On the basis of these objective functions, the incremental inverse kinematics problem is formulated as

$$\begin{aligned} \text{Minimize} \quad g = g_X + g_{\rho\hat{c}} + g_\tau \quad & \text{with respect to} \quad \Delta\boldsymbol{\rho}, \Delta\hat{\boldsymbol{c}}, \Delta\boldsymbol{X}^{(p)}, \Delta\boldsymbol{\tau}_{\mathsf{A}}^{(p)} \ (p = 1, \cdots, P') \\ & \text{subject to} \quad (32),\ (33),\ (34),\ (35) \end{aligned} \tag{41}$$

The principal aim to solve this problem is to find $\Delta\boldsymbol{\rho}$ and $\Delta\hat{\boldsymbol{c}}$ that minimizes g_X. The values specific to each of the possibilities, $\Delta\boldsymbol{X}^{(p)}$ as well as $\Delta\boldsymbol{\tau}_{\mathsf{A}}^{(p)}$, are determined so as to satisfy the geometrical and equilibrium conditions for the p-th possible kinematic state.

Introducing $\Delta\boldsymbol{q} = [\Delta\boldsymbol{\rho}^T, \Delta\hat{\boldsymbol{c}}^T]^T$, Eqs.(32) and (33) are rewritten as

$$\boldsymbol{A}^{(p)}\Delta\boldsymbol{q} + \boldsymbol{J}^{(p)}\Delta\boldsymbol{v}^{(p)} = \boldsymbol{y}^{(p)} \tag{42}$$

where

$$A^{(p)} = \begin{bmatrix} 0 & \dfrac{\partial F^{(p)}}{\partial \hat{c}} \\ B_{\mathsf{A}^{(p)}} & 0 \end{bmatrix} \tag{43}$$

is the coefficient matrix of Δq specific to the p-th possibility, with the boolean matrix $B_{\mathsf{A}^{(p)}}$ to extract the taut wire elements corresponding to $\mathsf{A}^{(p)}$ and

$$\Delta y^{(p)} = \begin{bmatrix} -F_{\epsilon}^{(p)} - \dfrac{\partial F^{(p)}}{\partial \tau_{\mathsf{S}}} \Delta \tau_{\mathsf{S}}^{(p)} \\ \rho_{\mathsf{A}^{(p)}} - l_{\mathsf{A}}^{(p)} \end{bmatrix} \tag{44}$$

is consisting of the right-hand sides of Eqs.(32) and (33), which is also specific to the p-th possibility. Jacobian $J^{(p)}$ and vector $\Delta v^{(p)}$ are respectively corresponding to J and Δv in Eqs.(16) and (17) and specific to the p-th possibility as well. The objective function g of problem (41) is also rewritten as

$$g = \frac{1}{2}\Delta q^T C_q \Delta q + \frac{1}{P'} \sum_{p=1}^{P'} \frac{1}{2} \left(\Delta v^{(p)} - \Delta \underline{v}^{(p)}\right)^T C_v^{(p)} \left(\Delta v^{(p)} - \Delta \underline{v}^{(p)}\right) \tag{45}$$

where

$$C_q = \begin{bmatrix} K_\rho I & 0 \\ 0 & K_{\hat{c}} I \end{bmatrix} \tag{46}$$

$$C_v^{(p)} = \begin{bmatrix} K_x I & 0 & 0 \\ 0 & K_\theta I & 0 \\ 0 & 0 & K_\tau I^{(p)} \end{bmatrix} \tag{47}$$

The dimension of identity matrix $I^{(p)}$ in the above equation corresponds to the number of assumed taut wires $\mathrm{Dim}(\mathsf{A}^{(p)})$. On the basis of these representations, the inverse kinematics problem in minimization form (41) is rewritten as follows:

$$\text{Minimize} \quad g \quad \text{with respect to} \quad \Delta q, \Delta v^{(p)}\ (p = 1, \cdots, P') \quad \text{subject to} \quad (34),\ (35),\ (42) \tag{48}$$

It should be noted that the solution of this problem significantly depends on the assumed pattern of tension states $\mathsf{A}^{(p)}$ and $\mathsf{S}^{(p)}$ $(p = 1, \cdots, P')$ of the P' possibilities in consideration. How to choose the pattern of these tension states is quite important for this incremental inverse kinematics approach.

4. Approach to Inverse Kinematics Problem

Problem (48) consists of objective function (45) in quadratic form, linear constraint (42) and inequality constraints (34) and (35). We develop a solution of minimization problem of this type and apply it to problem (48). An approach to deal with the incremental inverse kinematics problem is also proposed on the basis of a procedure to generate patterns of tension states.

4.1 Solution Based on Assumed Pattern of Tension States

Problem (48) without inequality constraints (34) and (35) is expressed as

$$\text{Minimize} \quad g \quad \text{with respect to} \quad \Delta q, \Delta v^{(p)}\ (p = 1, \cdots, P') \quad \text{subject to} \quad (42) \tag{49}$$

This is a quadratic minimization problem characterized by the linear constraint (42), in which the rows have partially common variables; that is, Δq is common to all the rows but $\Delta v^{(p)}$ is specific to the p-th row. Assuming a pattern of tension states of $\mathsf{A}^{(p)}$ and $\mathsf{S}^{(p)}$ $(p = 1, \cdots, P')$ after the incremental step, problem (49) can be solved by means of the procedure dealt with in Appendix A. Obtained Δq and $\Delta v^{(p)}$ can also be a feasible solution of problem (48), in the case that $\Delta \rho$, $\Delta X^{(p)}$ and $\Delta \tau^{(p)}$ within Δq and $\Delta v^{(p)}$ satisfy inequality constraints (34) and (35) as well.

Accordingly, the incremental inverse kinematics problem is solved and the solution $\Delta q = [\Delta \rho^T, \Delta \hat{c}^T]^T$ is obtained. It should be noted that the optimality of the obtained solution in terms of the objective function g significantly depends on the corresponding assumed pattern of tension states.

4.2 Examining Patterns of Tension State

It is not practical to examine all the patterns of tension states $\mathsf{A}^{(p)}$ and $\mathsf{S}^{(p)}$ $(p = 1, \cdots, P')$, since even in a case of $P' = 3$ or $P' = 4$ which typically occurs for a six-wire system, we have to examine $2^{6\times3} = 262,144$ or $2^{6\times4} = 1.7 \times 10^7$ patterns of tension states in principle. It is clear, however, that the inverse kinematics solution should have more taut wires in order to maintain active kinematical DOFs, since the slack wires are practically non-existent from the viewpoint of kinematic motion. In this study, we propose the following procedure that examines patterns of tension states from that consisting of more taut wires, to that consisting of less taut wires.

1. Introduce tension state A^* and S^* to be adopted as the common starting sets for all of the possible kinematic states. The initial values are given as $\mathsf{A}^* = \{1, 2, \cdots, N\}$ and $\mathsf{S}^* = \phi$ in the case of $N \leq 6$, or adequate sets such that $\mathrm{Dim}(\mathsf{A}^*) = 6$ and $\mathrm{Dim}(\mathsf{S}^*) = N - 6$ in the case of $N > 6$.

2. Set $\mathsf{A}^{(p)} = \mathsf{A}^*$ and $\mathsf{S}^{(p)} = \mathsf{S}^*$ $(p = 1, \cdots, P')$.

3. Solve problem (49) based on the current assumed pattern of tension states $\mathsf{A}^{(p)}$ and $\mathsf{S}^{(p)}$ $(p = 1, \cdots, P')$.

4. In the case that the obtained solution satisfies constraints (34) and (35), adopt the obtained $\Delta \boldsymbol{q}$ as the solution of the incremental inverse kinematics problem.

5. If the obtained solution violates *only* the tension constraint (35), the wire n of maximum violation, $n \in \mathsf{A}^{(q)}$, is transferred to $\mathsf{S}^{(q)}$. Continue from step 3 with updated $\mathsf{A}^{(p)}$ and $\mathsf{S}^{(p)}$ $(p = 1, \cdots, P')$.

6. The obtained solution violates the geometrical constraint (34) for some of the possible kinematic states. The series starting from current A^* and S^* is terminated. Generate next common starting tension state A^* and S^*. Repeat from step 2.

7. If there is no other combination of A^* and S^* to be examined at step 6, the inverse kinematics problem is judged to have no solution.

The approach is to examine patterns of tension states based on an assumed tension state A^* and S^* common for all the P' possibilities $\mathsf{A}^{(p)}$ and $\mathsf{S}^{(p)}$ $(p = 1, \cdots, P')$, which is to be modified in one-by-one manner at step 5, according to the situation of tension constraint violation. The number of combinations of A^* and S^* is limited to 2^N. The modification of pattern of tension states at step 5 is unidirectional, that is, only from taut wire set to slack wire set. The procedure certainly stops.

It should be noted that there can be a possible solution for the incremental inverse kinematics problem even in the case that the procedure cannot find any feasible solution, because the procedure does not examine all of the combinations of pattern of tension states $\mathsf{A}^{(p)}$ and $\mathsf{S}^{(p)}$ $(p = 1, \cdots, P')$.

5. Numerical Experiments by Means of Simulated Platform System

In order to demonstrate the feasibility of the developed incremental inverse kinematics approach, we conduct numerical experiments based on a simulated wire-suspended platform system.

5.1 Assumed Condition

Figure 2(a) shows the initial configuration and the wire numbering of the six-wire system adopted in the following numerical experiments. The size and the mass of the platform are 1m and 100kg; the wires are assumed to be massless. The mass center of the platform coincides with its geometrical center in this initial configuration. The dimension of the room where the system is installed is 6(W) × 6(D) × 3(H)m. In the case that the uncertainty is taken into account in the form of an interval of ±0.01m for the wire lengths and the mass center position, the possible kinematic states shown in Figure 2(b) are obtained by means of the incremental forward kinematics calculation with $\Delta \boldsymbol{\rho} = 0$ and $\Delta \hat{\boldsymbol{c}} = 0$; the number of possible kinematic states in this configuration is 25. Slack wires are represented by broken lines.

Figure 2(c) shows the initial configuration of the simulated platform, which is corresponding to the configuration shown in Figure 2(a). This platform system has randomly generated initial error within a range of ±0.01m for the wire lengths and the mass center position. Stiffness of the wire is assumed as $EA = 100$MN; the value is not used in the forward and inverse kinematics calculations but adopted for the calculation of kinematic motion of this simulated practical platform. The calculation is performed based on the steepest descent method which continuously searches the local minima of the sum of the potential energy of gravity and the strain energy of the wires.

Among P possibilities in each incremental step obtained by means of the forward kinematics calculation, P' kinematic states having all the taut wires of the simulated platform are taken into consideration for the inverse kinematics calculation. The condition is expressed as $\mathsf{A}^{(p)} \supset \mathsf{A}^{Simul.}$ for any kinematic state p to be taken into account, where $\mathsf{A}^{Simul.}$ is the taut wire set of the simulated platform. This information is obtained by means of binary-valued wire tension sensors, which are assumed to be installed on the simulated platform in this study. The threshold tension value of the sensor is 150N; the sensing result is obtained in terms of the calculated wire tension of the simulated platform including a measurement noise of ±50N. The threshold of the adopted binary tension sensor is assumed to be of relatively high value. This is because in order to take into account the influence of elasticity of wires of the simulated practical platform, which is not taken into account in the forward and inverse kinematics calculations.

5.2 Simulation Results

Figures 3(a) and (b) show example simulated motions of the platform introduced as in Figure 2(c), based on the proposed inverse kinematics approach. The wire lengths as well as the platform mass center position are adopted as the articulation variables in these motions. The configuration shown in Figure 3(a) is attained based on a translational target motion of 1m right and 1m up from the initial configuration shown in Figure 2(c). The configuration shown in Figure 3(b) is attained based on an additional rotational target motion of 30deg clockwise. Figures 4(a) and (b) show the attained configurations corresponding to the same target motions; however, only the wire lengths are adopted as the articulation variables in this case.

The penalty coefficients K_x and K_θ and the weight coefficients K_ρ, $K_{\hat{c}}$ and K_τ adopted in the inverse kinematics calculations are $K_x = K_\theta = 1.0 \times 10^4$, $K_\rho = 1.0 \times 10^2$, $K_{\hat{c}} = 1.0 \times 10^4$ and $K_\tau = 1.0 \times 10^{-8}$, respectively. The large value of $K_{\hat{c}}$ compared to K_ρ is determined to suppress undesirable large motion of platform mass center position. The particularly small value of K_τ is because we are currently not imposing any limitation on the wire tension values. It should be noted, however, that $K_\tau = 0$ cannot be used even in this case in order to apply the solution procedure dealt with in Appendix A. The number of incremental steps for the forward and inverse kinematics calculations is 200 for both of the translational and rotational motions.

The obtained configurations of the simulated platform shown in Figures 3(a) and (b) are considered to be acceptable in view of the circumstances of the introduced initial error and the corresponding fluctuated initial configuration shown in Figure 2(c). In the case of the motion based only on the adjustment of wire lengths, the degree of attainment of the target motions shown in Figures 4(a) and (b) is lower than the case shown in Figures 3(a) and (b), especially concerning the platform orientation. This is because the right-side wires 1 and 6 cannot push down the right edge of the platform in order to attain the horizontal as well as the right-leaning configurations, since the wires cannot take any compressive force. These results well demonstrate the influence and the importance of the mass distribution of the platform in the case of the mechanical system of this type.

Figures 3(c) and 4(c) show the confined possible configurations to be taken into consideration for the inverse kinematics calculation respectively corresponding to the configurations of the simulated platform shown in Figures 3(b) and 4(b). In both cases, there is a single slack-judged wire as shown in Figures 3(b) and 4(b). The tension value of one of the rest five taut wires of the simulated platform shown in Figure 3(b) is, however, 83N and is judged as slack based on the assumed binary-valued tension sensor; that is, the number of taut wires used for the confinement process is four instead of five. This is the reason that more possibilities are to be taken into consideration for the inverse kinematics in the case of the configuration shown in Figure 3(c).

Figure 5 shows an example of terminated motion due to the failure of the inverse kinematics calculation. The intended motion of the platform is a translation of 3m right and 1m up from the initial configuration shown in Figure 2(c), in 400 incremental steps. The inverse kinematics fails in this case at the 368th step since the procedure cannot find any feasible solution of the minimization problem (48). The reason is as follows. The tension of rear central wire 2 of the simulated platform is only 16N and determined as slack by means of the assumed binary tension sensor. Accordingly, all of existing nine possibilities in the step obtained as the forward kinematics calculation have to be taken into account for the inverse kinematics calculation. It is as a matter of course that the attained configuration is nearly at the rightmost feasible position of the platform, since the wires can only take a tensile force. In other words, the crucial reason of this failure is not due to the inverse kinematics procedure itself, but due to the intrinsic limitation of wires that any of them cannot *push* the platform in order to attain a more rightward

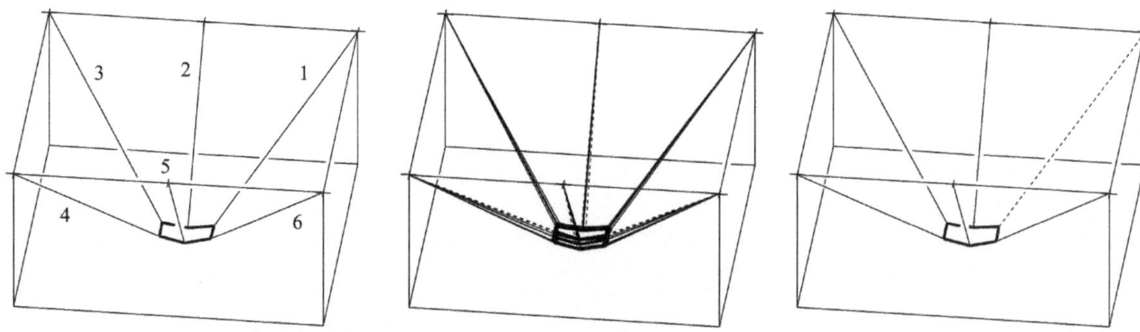

(a) Without uncertainty (wire numbering)
$\boldsymbol{X}$ = [0.000, 0.000, 0.000 (m), 0.0, 0.0, 0.0 (deg)]

(b) Uncertainty taken into account
(superimpose of 25 configurations)

(c) Simulated practical platform
$\boldsymbol{X}$ = [−0.000, −0.017, −0.003 (m), 4.2, −0.7, −1.0 (deg)]

Figure 2. Six-wire system and initial configuration

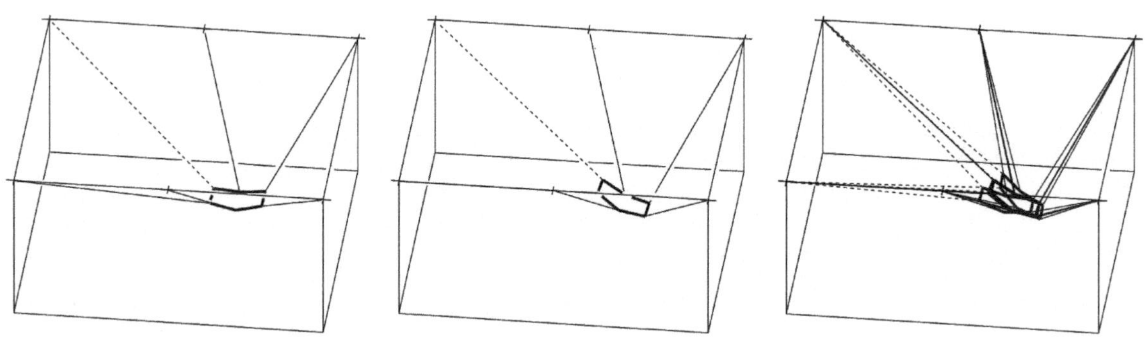

(a) Target: move 1m right, 1m up
$\boldsymbol{X}$ = [0.997, −0.005, 0.995 (m), 3.6, −1.1, −1.3 (deg)]

(b) Target: rotate 30deg clockwise
$\boldsymbol{X}$ = [0.976, 0.015, 0.986 (m), −0.5, 24.8, −3.7 (deg)]

(c) Example confined possibilities
(superimpose of 5 configurations)

Figure 3. Kinematic motion using mass center position as well

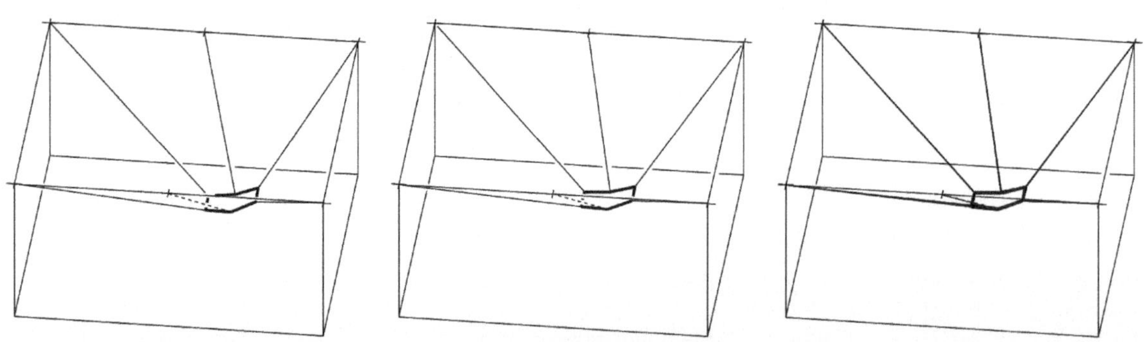

(a) Target: move 1m right, 1m up
$\boldsymbol{X}$ = [0.913, −0.008, 1.027 (m), 2.7, −15.4, −0.3 (deg)]

(b) Target: rotate 30deg clockwise
$\boldsymbol{X}$ = [0.747, −0.012, 1.087 (m), 3.0, −11.9, −0.3 (deg)]

(c) Example confined possibilities
(superimpose of 2 configurations)

Figure 4. Kinematic motion based only on wire lengths adjustment

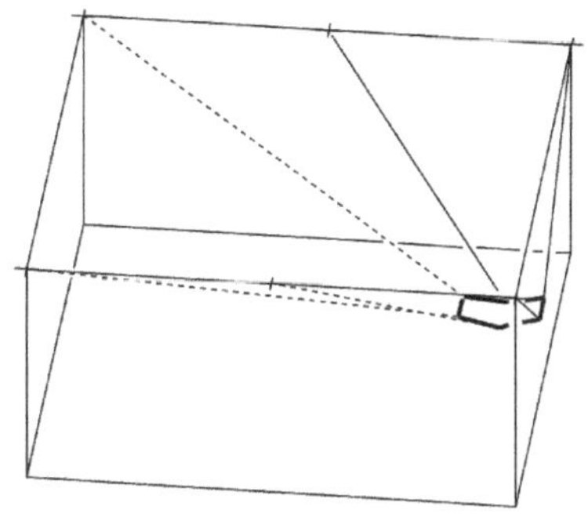

Figure 5. Terminated motion due to failure of inverse kinematics
$X = [\ 2.483, -0.004, 0.908\ \text{(m)}, -0.5, -2.6, -0.4\ \text{(deg)}\ ]$

position. We use the platform mass center position as the articulation variable as well in this kinematic motion; it is confirmed that the mass center is placed nearly at the right edge of the platform in the configuration shown in Figure 5.

6. Concluding Remarks

We have dealt with an incremental inverse kinematics approach for wire-suspended parallel mechanical systems. The incremental forward kinematics calculation of this type of mechanical system results in plural possibilities, due to the conspicuous mechanical characteristic of wire. In the current study, the formulation of the inverse kinematics problem in an incremental form was developed based on the many-worlds interpretation approach adopted in the previous study. The displacement of the mass center position of the platform is also taken into consideration in addition to the change in wire lengths. The problem has the form of a quadratic minimization problem, which is characterized by the partially common variables. The solution of the problem in general form is given in Appendix A. We also developed the procedure to solve the inverse kinematics problem by examining the possible patterns of tension states.

The developed inverse kinematics approach is quite general. It can be applied to wire-suspended platforms of various numbers of wires, various placements of connection points of wires and various patterns of mass distribution of platform. The feasibility of the proposed incremental inverse kinematics has been demonstrated based on the numerical experiments. The simulated wire-suspended platform is assumed to have error in wire lengths and mass center position and to be equipped with the binary-valued tension sensors. On the basis of this condition, the developed approach was able to generate the intended kinematic motion. The influence as well as the significance of the position of platform mass center was also demonstrated.

Since the developed inverse kinematics problem has a combinatorial nature, we introduce the common initial tension state and confine significantly the patterns of tension states to be examined. This can lead to a condition that the approach cannot find any solution within the examined patterns, but there can exist some solution within the non-examined patterns. Although this is an issue of trade-off, developing an approach that examines more patterns of tension states is considered to be one of the future works.

The wire-suspended platform system has the intrinsic kinematic limitation due to the mechanical condition, as shown in Figure 5. From this point of view, another future works is the design of the platform system in terms of the number of wires, the placement of the connection points and the platform mass distribution taking some specific applications into consideration.

Appendix A. Partially Common Quadratic Minimization

Problem (49) can be expressed in the following general form of partially common quadratic minimization problem:

$$\begin{aligned} \text{Minimize} \quad & g = \frac{1}{2}(\boldsymbol{x} - \underline{\boldsymbol{x}})^T \boldsymbol{N}_x (\boldsymbol{x} - \underline{\boldsymbol{x}}) + \sum_{i=1}^{n} \frac{1}{2}(\boldsymbol{y}_i - \underline{\boldsymbol{y}}_i)^T \boldsymbol{N}_i (\boldsymbol{y}_i - \underline{\boldsymbol{y}}_i) \\ & \text{with respect to} \quad \boldsymbol{x}, \boldsymbol{y}_i\ (i = 1, \cdots, n) \\ & \text{subject to} \quad \boldsymbol{A}_i \boldsymbol{x} + \boldsymbol{B}_i \boldsymbol{y}_i = z_i\ (i = 1, \cdots, n) \end{aligned} \tag{50}$$

Introducing Lagrange multipliers $\lambda_1, \cdots, \lambda_n$, this problem is rewritten as

$$\text{Minimize} \quad g' = g + \sum_{i=1}^{n} \boldsymbol{\lambda}_i^T (\boldsymbol{A}_i \boldsymbol{x} + \boldsymbol{B}_i \boldsymbol{y}_i - \boldsymbol{z}_i) \quad \text{with respect to} \quad \boldsymbol{x}, \boldsymbol{y}_i, \boldsymbol{\lambda}_i \ (i = 1, \cdots, n) \tag{51}$$

Differentiating g' with respect to $\boldsymbol{x}$ and $\boldsymbol{y}_i$ and solving for them, we obtain

$$\boldsymbol{x} = \underline{\boldsymbol{x}} - \boldsymbol{N}_x^{-1} \sum_{i=1}^{n} \boldsymbol{A}_i^T \boldsymbol{\lambda}_i \tag{52}$$

$$\boldsymbol{y}_i = \underline{\boldsymbol{y}}_i - \boldsymbol{N}_i^{-1} \boldsymbol{B}_i^T \boldsymbol{\lambda}_i \tag{53}$$

Differentiation of g' with respect to $\boldsymbol{\lambda}_i$ gives

$$\boldsymbol{A}_i \boldsymbol{x} + \boldsymbol{B}_i \boldsymbol{y}_i - \boldsymbol{z}_i = 0 \tag{54}$$

Substituting Eqs.(52) and (53) for Eq.(54), we obtain

$$\boldsymbol{A}_i \left(\underline{\boldsymbol{x}} - \boldsymbol{N}_x^{-1} \sum_{j=1}^{n} \boldsymbol{A}_j^T \boldsymbol{\lambda}_j \right) + \boldsymbol{B}_i (\underline{\boldsymbol{y}}_i - \boldsymbol{N}_i^{-1} \boldsymbol{B}_i^T \boldsymbol{\lambda}_i) - \boldsymbol{z}_i = 0 \tag{55}$$

This equation is modified as

$$\sum_{j=1}^{n} \boldsymbol{A}_i \boldsymbol{N}_x^{-1} \boldsymbol{A}_j^T \boldsymbol{\lambda}_j + \boldsymbol{B}_i \boldsymbol{N}_i^{-1} \boldsymbol{B}_i^T \boldsymbol{\lambda}_i = \boldsymbol{A}_i \underline{\boldsymbol{x}} + \boldsymbol{B}_i \underline{\boldsymbol{y}}_i - \boldsymbol{z}_i \tag{56}$$

Rearranging Eq.(56) for $i = 1, \cdots, n$, we obtain

$$\sum_{j=1}^{n} \boldsymbol{C}_{ij} \boldsymbol{\lambda}_j = \boldsymbol{A}_i \underline{\boldsymbol{x}} + \boldsymbol{B}_i \underline{\boldsymbol{y}}_i - \boldsymbol{z}_i \ (i = 1, \cdots, n) \tag{57}$$

where

$$\boldsymbol{C}_{ij} = \begin{cases} \boldsymbol{A}_i \boldsymbol{N}_x^{-1} \boldsymbol{A}_i^T + \boldsymbol{B}_i \boldsymbol{N}_i^{-1} \boldsymbol{B}_i^T & (i = j) \\ \boldsymbol{A}_i \boldsymbol{N}_x^{-1} \boldsymbol{A}_j^T & (i \neq j) \end{cases} \tag{58}$$

Solving linear equations (57) and substituting obtained $\lambda_1, \cdots, \lambda_n$ for Eqs.(52) and (53), we obtain the solution $\boldsymbol{x}$, $\boldsymbol{y}_1, \cdots, \boldsymbol{y}_n$ of problem (50).

It should be noted that norm matrices $\boldsymbol{N}_x$ and $\boldsymbol{N}_i$ $(i = 1, \cdots, n)$ must not be singular in this solution procedure.

References

Agahi, M., & Notash, L. (2009). Redundancy resolution of wire-actuated parallel manipulators. *Transactions of the Canadian Society for Mechanical Engineering, 33*(4), 561–573.

Albus, J., Bostelman, R., & Dagalakis, N. (1992). The NIST robocrane. *Journal of Research of the National Institute of Standards and Technology, 97*(3), 373–385. http://dx.doi.org/10.6028/jres.097.016

Borgstrom, P. H., Borgstrom, N. P., Stealey, M. J., Jordan, B., Sukhatme, G. S., Batalin, M. A., & Kaiser, W. J. (2009). Design and implementation of NIMS3D, a 3-D cabled robot for actuated sensing applications. *IEEE Transactions on Robotics, 25*(2), 325–339. http://dx.doi.org/10.1109/TRO.2009.2012339

Bosscher, P., Riechel, A. T., & Ebert-Uphoff, I. (2006). Wrench-feasible workspace generation for cable-driven robots. *IEEE Transactions on Robotics, 22*(5), 890–902. http://dx.doi.org/10.1109/TRO.2006.878967

Carricato, M. (2013). Direct geometorico-static problem of underconstrained cable-driven parallel robots with three cables. *Journal of Mechanisms and Robotics, 5*(3), Paper 031008. http://dx.doi.org/10.1115/1.4024293

Carricato, M., & Merlet, J.-P. (2013). Stability analysis of underconstrained cable-driven parallel robots. *IEEE Transactions on Robotics, 29*(1), 288–296. http://dx.doi.org/10.1109/TRO.2012.2217795

Ghasemi, A., Eghtesad, M. & Farid, M. (2009). Workspace analysis for planar and spatial redundant cable robots. *Journal of Mechanisms and Robotics, 1*(4), Paper 044502. http://dx.doi.org/10.1115/1.3211026

Hanahara, K. (2015). Incremental forward kinematics of wire-suspended parallel mechanical system (a many-worlds interpretation approach). *Journal of Mechanisms and Robotics, 7*(4), Paper 041021. http://dx.doi.org/10.1115/1.4030163

Hiller, M., Fang, S., Mielczarek, S., Verhoeven, R., & Franitza, D. (2005). Design, analysis and realization of tendon-based parallel manipulators. *Mechanism and Machine Theory, 40*, 429–445. http://dx.doi.org/10.1016/j.mechmachtheory.2004.08.002

Merlet, J.-P. (2006). *Parallel Robots* (2nd ed.). Springer.

Moore, R. E., Kearfott, R. B. & Cloud, J. J. (2009). *Introduction to Interval Analysis*. SIAM.

Oh, S.-R., & Agrawal, S. K. (2005). Cable suspended planar robots with redundant cables: controllers with positive tensions. *IEEE Transactions on Robotics, 21*(3), 457–465. http://dx.doi.org/10.1109/TRO.2004.838029

Perreault, S., & Gosselin, C. M. (2008). Cable-driven parallel mechanisms: application to a locomotion interface. *Journal of Mechanical Design, 130*, Paper 102301. http://dx.doi.org/10.1115/1.2965607

Pusey, J., Fattah, A., Agrawal, S., & Messina, E. (2004). Design and workspace analysis of a 6-6 cable-suspended parallel robot. *Mechanism and Machine Theory, 39*, 761–778. http://dx.doi.org/10.1016/j.mechmachtheory.2004.02.010

Roberts, R. G., Graham, T., & Lippitt, T. (1998). On the inverse kinematics, statics, and fault tolerance of cable-suspended robots. *Journal of Robotic Systems, 15*(10), 581–597. http://dx.doi.org/10.1002/(SICI)1097-4563(199810)15:10<581::AID-ROB4>3.0.CO;2-P

Stewart, D. (1965-66). A platform with six degrees of freedom. *Proceedings of Institution of Mechanical Engineers, 180*(15), Part 1, 371–386.

Tadokoro, S., Verhoeven, R., Zwiers, U., Hiller, M. & Takemura, F. (2003). Kinematic identification method for cable-driven rescue robots in unstructured environments. *Journal of Robotics and Mechatronics, 15*(5), 571–578.

Vaidman, L. (2008). Many-worlds interpretation of quantum mechanics. *The Stanford Encyclopedia of Philosophy* (Fall 2008 ed.). Retrieved from http://plato.stanford.edu/archives/fall2008/entries/qm-manyworlds/ .

Yamaguchi, D., Tagawa, Y., Hayatsu, M., & Yamada, M. (2004). Sequential identification technique of jacobian matrix for a power-assisted lifter using wire-driven parallel mechanism. *Journal of Robotics and Mechatronics, 16*(3), 228–236.

Permissions

All chapters in this book were first published in MER, by Canadian Center of Science and Education; hereby published with permission under the Creative Commons Attribution License or equivalent. Every chapter published in this book has been scrutinized by our experts. Their significance has been extensively debated. The topics covered herein carry significant findings which will fuel the growth of the discipline. They may even be implemented as practical applications or may be referred to as a beginning point for another development.

The contributors of this book come from diverse backgrounds, making this book a truly international effort. This book will bring forth new frontiers with its revolutionizing research information and detailed analysis of the nascent developments around the world.

We would like to thank all the contributing authors for lending their expertise to make the book truly unique. They have played a crucial role in the development of this book. Without their invaluable contributions this book wouldn't have been possible. They have made vital efforts to compile up to date information on the varied aspects of this subject to make this book a valuable addition to the collection of many professionals and students.

This book was conceptualized with the vision of imparting up-to-date information and advanced data in this field. To ensure the same, a matchless editorial board was set up. Every individual on the board went through rigorous rounds of assessment to prove their worth. After which they invested a large part of their time researching and compiling the most relevant data for our readers.

The editorial board has been involved in producing this book since its inception. They have spent rigorous hours researching and exploring the diverse topics which have resulted in the successful publishing of this book. They have passed on their knowledge of decades through this book. To expedite this challenging task, the publisher supported the team at every step. A small team of assistant editors was also appointed to further simplify the editing procedure and attain best results for the readers.

Apart from the editorial board, the designing team has also invested a significant amount of their time in understanding the subject and creating the most relevant covers. They scrutinized every image to scout for the most suitable representation of the subject and create an appropriate cover for the book.

The publishing team has been an ardent support to the editorial, designing and production team. Their endless efforts to recruit the best for this project, has resulted in the accomplishment of this book. They are a veteran in the field of academics and their pool of knowledge is as vast as their experience in printing. Their expertise and guidance has proved useful at every step. Their uncompromising quality standards have made this book an exceptional effort. Their encouragement from time to time has been an inspiration for everyone.

The publisher and the editorial board hope that this book will prove to be a valuable piece of knowledge for researchers, students, practitioners and scholars across the globe.

List of Contributors

Faysal Ayad Khaleel and Amir Sultan Dawood
Mechanical Engineering Department, Collage of Engineering, Mosul University, Mosul, Iraq

YinMun H'ng, Yusri Yusup, H. P. S. Abdul Khalil and TjoonTow Teng
School of Industrial Technology, Universiti Sains Malaysia, Penang, Malaysia

N. D. Agafonova and I. L. Paramonova
Saint-Petersburg State Polytechnic University, St.Petersburg, Russia

Yunfeng Li
School of Mechatronical Engineering, Henan University of Science and Technology, Luoyang, China

Kouroush Jenab
Faculty of College of Aeronautics, Embry-Riddle Aeronautical University, 600 S. Clyde Morris Blvd. Daytona Beach, FL 32114-3900, USA

Josh Herrin
Graduate Student, Embry-Riddle Aeronautical University, 600 S. Clyde Morris Blvd. Daytona Beach, FL 32114-3900, USA

Saeid Moslehpour
Faculty of College of Engineering, Hartford University, 200 Bloomfield Avenue, West Hartford, CT 06117, USA

Sam Khoury
Faculty of College of Business, Athens State University, 300 N. Beaty St., Athens, AL 35611, USA

A. N. Makarov
Tver' State Technical University, Russia

Ehsan Fadhil Abbas
Department of Refrigeration and Air Conditioning, Kirkuk Technical College, Foundation of Technical Education, Iraq

Chiu-Fan Hsieh and Hao-Yu Cheng
Department of Mechanical and Computer-Aided Engineering, National Formosa University, Taiwan

Tomohiro Izumikawa, Ryuichi Watanabe and Hiroyuki Yaguchi
Faculty of Engineering, Tohoku Gakuin University, Miyagi, Japan

Xiao-Song Wang
Institute of Mechanical and Power Engineering, Henan Polytechnic University, No. 2001, Century Avenue, Jiaozuo, Henan 454003, China

Magdalene Andrew-Munot and Ervina Junaidi
Department of Mechanical & Manufacturing Engineering, Faculty of Engineering, Universiti Malaysia Sarawak, Sarawak, Malaysia

Raafat N Ibrahim
Department of Mechanical and Aerospace Engineering, Faculty of Engineering, Monash University, Australia

Chiu-Fan Hsieh and You-Qing Zhu
Department of Mechanical and Computer-Aided Engineering, National Formosa University, Taiwan

Munzer S. Y. Ebaid
Department of Mechanical engineering, Philadelphia University, Amman, Jordan

Qusai Z. Al-hamdan
Perth College, University of the Highlands and Islands, Scotland, UK

Nikolay Goloshchapov
71 Grosvenor Road, London, N3 1EY, UK

Heikki Martikka
Himtech Oy, Engineering, Ollintie 4, FIN-54100 Joutseno, Finland

Ilkka Pöllänen
SAV Oy, Savonkatu 21, FIN-45100 Kouvola, Finland

Tatsuhiko Aizawa and Eijirou Masaki
Department of Engineering and Design, Shibaura Institute of Technology, Tokyo, Japan

Eiji Morimoto
Green Tool Co. Ltd., Okayama, Japan

Yoshio Sugita
YS-Electric Industry, Co. Ltd., Kofu, Japan

Adel T. Abbas
Department of Mechanical Engineering, King Saud University, Riyadh, Saudi Arabia

Karim Hamza
Department of Mechanical Engineering, University of Michigan, Ann Arbor, USA

Mohamed F. Aly
Department of Mechanical Design, American University in Cairo, New Cairo, Egypt

Pablo Ixtaina
Laboratorio de Acústica y Luminotecnia de la Comisión de Investigaciones, Científicas de la Provincia de Buenos Aires LAL CIC, Argentina

Matías Presso and Gustavo Marín
Instituto de Investigación en Tecnología Informática Avanzada INTIA - UNCPBA, Argentina

Pablo Ixtaina
Laboratorio de Acústica y Luminotecnia de la Comisión de Investigaciones, Científicas de la Provincia de Buenos Aires LAL CIC, Argentina

Matías Presso and Gustavo Marín
Instituto de Investigación en Tecnología Informática Avanzada INTIA - UNCPBA, Argentina

Ahmed Ghazy
Faculty of Engineering, Helwan University, Cairo, Egypt

Kazuyuki Hanahara
Graduate School of System Informatics, Kobe University, Kobe, Japan